BOLYAI SOCIETY
MATHEMATICAL STUDIES 23

BOLYAI SOCIETY MATHEMATICAL STUDIES

Editor-in-Chief:
Gábor Fejes Tóth

Series Editor:
Dezső Miklós

Publication Board:

Gyula O. H. Katona · László Lovász · Péter Pál Pálfy
András Recski · András Stipsicz · Domokos Szász

András Némethi
Ágnes Szilárd
(Eds.)

Deformations of
Surface Singularities

 Springer

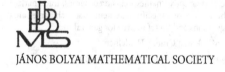
JÁNOS BOLYAI MATHEMATICAL SOCIETY

András Némethi
Alfréd Rényi Institute of Mathematics
Hungarian Academy of Sciences
Reáltanoda u. 13–15
Budapest 1053
Hungary
e-mail: nemethi.andras@renyi.mta.hu

Ágnes Szilárd
Alfréd Rényi Institute of Mathematics
Hungarian Academy of Sciences
Reáltanoda u. 13–15
Budapest 1053
Hungary
e-mail: szilard.agnes@renyi.mta.hu

Mathematics Subject Classification (2010):
14B05, 14B07, 14B12, 32B10, 32S05, 32S25, 32S30, 53D05, 53D10, 57M27

DOI 10.1007/978-3-642-39131-6
ISSN 1217-4696   ISBN 978-3-642-39131-6 (e-Book)
ISBN 978-3-662-52469-5 Springer Berlin Heidelberg New York
ISBN 978-963-9453-16-6 János Bolyai Mathematical Society, Budapest

Springer is a part of Springer Science+Business Media
springer.com

© 2013 János Bolyai Mathematical Society and Springer-Verlag
Softcover reprint of the hardcover 1st edition 2013

Cover design: WMXDesign GmbH, Heidelberg

Printed on acid-free paper   44/3142/db – 5 4 3 2 1 0

CONTENTS

PREFACE

The present publication contains a special collection of research and review articles on *deformations of surface singularities,* that put together serve as an introductory survey of results and methods of the theory, as well as open problems, important examples and connections to other areas of mathematics, such as the theory of Stein fillings and symplectic geometry.

We envision this volume as a guide for all those already doing or wishing to do research in this area, and thus it is intended to be especially useful for PhD students.

1. A SHORT INTRODUCTION IN DEFORMATION THEORY

Deformation theory appeared as the investigation of how complex structures may vary on a fixed compact, *smooth* manifold. In his famous paper "Theorie der abelschen Funktionen" [16], published in 1857, Riemann already mentioned the $3g - 3$ moduli determining the complex structure of an algebraic curve ('Riemann surface') of genus $g \geq 2$.

Looking for an analogous description in higher dimensions, Kodaira and Spencer started developing the machinery of what is called deformation theory today [12]. Because of the fact that (beginning in dimension two) a good moduli space does not always exist, they used a modified weaker concept: the versal (or semi-universal) deformation $f : X \to S$ of a manifold $X_0 = f^{-1}(0)$ $(0 \in S)$. This space parametrizes all possible deformations (but even the minimal one no longer provides a one-to-one correspondence between fibers and complex structures).

Kodaira and Spencer showed the existence of the mini-versal deformation space ('universal versal deformation space'), under a certain cohomological vanishing. Later Kuranishi [14] completed their work by allowing singular base spaces and eliminating the cohomological assumption.

In a similar manner, we may regard deformations of *germs of analytic spaces* (or equivalently, deformations of local, analytic $\mathbb{C}$-algebras) that are not necessarily smooth: that are 'singularities'. In this case one defines the possible deformation spaces, deformation functors, and the notion of mini-versal deformations as well. It is not hard to see that the mini-versal deformation (if it exists) is uniquely determined; moreover, in the case of complete intersections the base space is smooth. However, in general, the base space and the structure of the versal family might become extremely complicated.

The existence of the analytic mini-versal space was established (under some assumptions) by Schlessinger [18]. For normal surface singularities the existence-problem was completely solved by Tjurina, and the most general case by Grauert [6] in 1972.

Schlessinger's method leads to the construction of the versal deformations, and useful criterion to verify versality. The reader might consult for more information Artin's Lecture notes [4], Palamodov's large introduction [15], or J. Stevens Thesis [21] and his recent Springer LNM–book [23].

Deformation theory of *normal surface singularities* in the last decades witnessed an extraordinary development in spite of being one of the most difficult subjects of singularity theory, as it is based on hard machinery from algebraic geometry, sheaf-cohomology and algebraic topology.

As the singularity is encoded in the construction of the mini-versal deformation space (and/or its base-space), this space contains important information about the given germ and is a crucial source of numerical invariants. Its most important two ingredients are the *tangent space* and *obstruction space,* which became the subject of intense mathematical study in the last years. Let us list some key examples.

Riemenschneider [17] and Arndt [3] initiated the description of *cyclic quotient singularities,* particularly of their tangent space. Its construction, as well of the obstruction space, was completed by Christophersen [5] and Behnke. We emphasize that already in this particular case the deformation space is not smooth, and it contains many irreducible components. The fact that in the above construction one indeed obtains all the components of the deformation space was verified by J. Stevens [22] based on the article of Kollár and Shepherd–Barron [13].

The next steps were again seriously obstructed: it proved to be very hard to generalize the results valid for cyclic quotient singularities. Nevertheless, D. van Straten and T. de Jong using new ideas obtained positive results in the direction of (the still open) Kollár Conjecture (targeting the description of the base space of the mini-versal deformation) in case of *rational quadruple singularities* [7] and *minimal rational singularities* [8]. Moreover, in a series of articles they developed a whole 'deformation theory of non-isolated singularities' [9, 10]. Their theory was successfully used for many families of normal surface singularities, e.g. for 'sandwiched' and rational singularities, applied to their projection in $(\mathbb{C}^3, 0)$ [11]. Their article [11] describes the smoothings (i.e. those deformations which provide smooth deformation fibers) of sandwiched singularities,—an important family with testing characteristics for any new theory, introduced by M. Spivakovsky [19, 20].

Recently, smoothings of rational and sandwiched singularities became a focus of interest in Contact Geometry and Stein/symplectic fillings as well: they provide the most important models of the theory. Indeed, local Milnor fibers are particular Stein fillings of the corresponding singularity links.

Simultaneously, Teissier (and his school), Laufer and Wahl developed the theory of ('very weak', 'weak', 'strong') simultaneous resolutions, where deformation and resolution theory are combined (see e.g. [24]). This led to the development of *equisingularity theory* and its connections with commutative algebra (integral closures).

There is another class of singularities which is in the mainstream of the deformation research: those provided by toric geometry. Toric geometry is that part of algebraic geometry which identifies its object by combinatorial construction (etc. by integral polyhedrons, or rational fans), and targets the computation of all topological/algebraic/sheaf-theoretical invariants via combinatorics. Their deformation theory was developed by K. Altmann, (see e.g. [1], or [2]).

2. The 'Deformation Theory Conference' at Budapest

In the period 10–12 October, 2008, the Alfréd Rényi Institute of Mathematics (Hungarian Academy of Sciences in Budapest, Hungary), organized a meeting titled *Deformation of Surfaces* (organizers: A. Némethi and Á. Szilárd.)

At the meeting experts of this area—Klaus Altmann, Jan Christophersen, Helmut Hamm, Theo de Jong, Monique Lejeune-Jalabert, Mark Spivakovsky, Jan Stevens, Duco van Straten—gave lectures explaining the classical theory and constructions complemented with presentation of recent developments and open questions. A special emphasis was put on key classes of singularities such as *rational, sandwiched* and *minimal rational.*

Moreover, the local theory of surface singularities was related with the theory of affine surfaces and their deformations in the talk of M. Zaidenberg. (For a list of guiding open problems of Zaidenberg see [25].)

The talks given were:

D. van Straten: *Introduction (How should one think about deformation theory?).*

K. Altmann: *Introduction to the deformation theory of toric singularities.*

K. Altmann: *The smoothings of certain toric singularities described by quivers.*

J. Stevens: *Versal deformation of cyclic quotient singularities.*

J. Christophersen: *Deformations of Stanley-Reisner surfaces.*

M. Lejeune-Jalabert: *Integral closure of ideals and equisingularity.*

M. Spivakovsky: *Equisingular deformations of sandwiched surface singularities.*

T. de Jong & D. van Straten: *Deformation of minimal and sandwiched singularities I. and II.*

J. Stevens: *Open problems, interesting questions regarding deformations of surface singularities.*

H. A. Hamm: *Equisingularity of curves on surfaces.*

M. Zaidenberg: *Deformations of acyclic surfaces and of* $\mathbb{C}^*$-*actions.*

The meeting was an absolute success with 34 foreign participants, including a large number of PhD students. It was the great interest shown by the audience because of which the idea of a volume collecting *research and review articles* on deformations of surface singularities, that put together would serve as a comprehensive survey of the key results and methods in this area known today, as well as open problems, important examples and connections to other areas of mathematics was conceived.

Thus the aim of the present volume is to collect material that will help mathematicians already working or wishing to work in this area to deepen their insight and eliminate the technical barriers in this learning process. This also is supported by review articles providing some global picture and abundance of examples.

Additionally, we introduce some material which emphasizes the newly found relationship with the theory of Stein fillings and Symplectic geometry (work of Eliashberg, Ono, Ohta, Etnyre, Lisca, Stipsicz). This links two main theories of mathematics: low dimensional topology with algebraic geometry.

REFERENCES

[1] Altman, K., Deformations of Toric Singularities, *Habilitationsschrift,* Universität Berlin.

[2] Altman, K., The versal deformation of an isolated toric Gorenstein singularity, *Invent. Math.,* **128** (1997), no. 3, 443–479.

[3] Arndt, J., Verselle Deformationen zyklischer Quotientensingularitäten, *Dissertation* Universität Hamburg, 1988.

[4] Artin, M., Lectures on deformation of singularities, *Bombay, Tata Institute of Fundamental research,* 1976.

[5] Christophersen, J. A., Obstruction spaces for rational singularities and deformation of cyclic quotients, *Thesis,* University of Oslo, 1989/90.

[6] Grauert, H., Über die Deformationen isolierter Singularitäten analytischer Mengen, *Inv. Math.,* **15** (1972), 171–198.

[7] de Jong, T. and van Straten, D., On the base space of a semi-universal deformation of rational quadruple points, *Ann. of Math.,* **134** (1991), no. 3, 653–678.

[8] de Jong, T. and van Straten, D., On the deformation theory of rational surface singularities with reduced fundamental cycle, *J. Algebraic Geom.,* **3** (1994), no. 1, 117–172.

[9] de Jong, T. and van Straten, D., A deformation theory for nonisolated singularities *Abh. Math. Sem. Univ. Hamburg,* **60** (1990), 177–208.

[10] de Jong, T. and van Straten, D., Deformations of the normalization of hypersurfaces, *Math. Ann.,* **288** (1990), no. 3, 527–547.

[11] de Jong, T. and van Straten, D., Deformation theory of sandwiched singularities, *Duke Math. J.,* **95** (1998), no. 3, 451–522.

[12] Kodaira, K. and Spencer, D. C., On deformation of complex analytic structures, I–II, III, *Ann. of Math.,* **67** (1958), 328–466; **71** (1960), 43–76.

[13] Kollár, J. and Shepherd–Barron, N. I., Threefolds and deformations of surface singularities, *Invent. Math.,* **91** (1988), no. 2, 299–338.

[14] Kuranishi, K., On the locally complete families of complex analytic structures, *Ann. of Math.,* **75** (1962), 536–577.

[15] Palamodov, V. P., Deformations of complex spaces, *Russian Math. Surveys,* **31** (1976), 129–197.

[16]   Riemann, B., Gesammelte mathematische und wissenschaftliche Werke, *Nachlas Teubner,* Leipzig, 1892–1902, reprinted Dover, New York 1953.

[17]   Riemenschneider, O., Deformationen von Quotientensingularitäten (nach zyklischen Gruppen), *Math. Ann.,* **209** (1974), 211–248.

[18]   Schlessinger, M., Functors of Artin rings, *Trans. AMS,* **128** (1967), 41–70.

[19]   Spivakovsky, M., Sandwiched singularities and desingularization of surfaces by normalized Nash transformations, *Ann. of Math.* (2), **131** (1990), no. 3, 411–491.

[20]   Spivakovsky, M., Sandwiched surface singularities and the Nash resolution problem, *Complex analytic singularities,* 583–598, Adv. Stud. Pure Math., 8, North-Holland, Amsterdam, 1987.

[21]   Stevens, J., Deformations of Singularities, *Habilitationsschrift,* Universität Hamburg, 1995.

[22]   Stevens, J., On the versal deformation of cyclic quotient singularities, *Singularity theory and its applications, Part I* (Coventry, 1988/1989), 302–319, Lecture Notes in Math., 1462, Springer, Berlin, 1991.

[23]   Stevens, J., Deformations of singularities, *Lecture Notes in Math.,* **1811**, Springer-Verlag, Berlin, 2003.

[24]   Teissier, B., Introduction to equisingularity problems, *Algebraic geometry* (Proc. Sympos. Pure Math., Vol. **29**, Humboldt State Univ., Arcata, Calif., 1974), pp. 593–632. Amer. Math. Soc., Providence, R.I., 1975.

[25]   Zaidenberg, M., Selected problems, arXiv:math/0501457.

András Némethi
Ágnes Szilárd
Rényi Institute of Mathematics
Budapest, Hungary

BOLYAI SOCIETY
MATHEMATICAL STUDIES, 23

Deformations of
Surface Singularities
pp. 13–55.

NEGATIVE DEFORMATIONS OF TORIC SINGULARITIES
THAT ARE SMOOTH IN CODIMENSION TWO

KLAUS ALTMANN and LARS KASTNER

Given a cone $\sigma \subseteq N_{\mathbb{R}}$ with smooth two-dimensional faces and, moreover, an element $R \in \sigma^{\vee} \cap M$ of the dual lattice, we describe the part of the versal deformation of the associated toric variety $\mathbb{TV}(\sigma)$ that is built from the deformation parameters of multidegree R.

The base space is (the germ of) an affine scheme $\bar{\mathcal{M}}$ that reflects certain possibilities of splitting $Q := \sigma \cap [R = 1]$ into Minkowski summands.

1. INTRODUCTION

1.1. The entire deformation theory of an isolated singularity is encoded in its so-called versal deformation. For complete intersection singularities this is a family over a smooth base space obtained by certain perturbations of the defining equations.

As soon as we leave this class of singularities, the structure of the family, and sometimes even the base space, will be more complicated. It is well known that the base space may consist of several components or may be non-reduced.

1.2. Let M, N be two mutually dual, free abelian groups of finite rank. Then affine toric varieties are constructed from rational, polyhedral cones $\sigma \subseteq N_{\mathbb{R}} := N \otimes \mathbb{R}$: One takes the dual cone

$$\sigma^{\vee} := \{ r \in M_{\mathbb{R}} \mid \langle a, r \rangle \geq 0 \text{ for each } a \in \sigma \},$$

and $Y := \mathbb{TV}(\sigma)$ is defined as the spectrum of the semigroup algebra $\mathbb{C}[\sigma^\vee \cap M]$. In particular, the equations of Y are induced from linear relations between lattice points of $\sigma^\vee \subseteq M_\mathbb{R}$. As usual for all other toric objects or notions, the toric deformation theory also comes with an M-grading. In particular, for any $R \in M$, we might speak of infinitesimal or versal deformations of degree $-R$.

With the latter, we mean the following: The vector space T^1 of infinitesimal deformations serves as the ambient space of the germ of the versal base space. Hence it makes sense to intersect it with the linear space obtained as the annihilator of the T^1 coordinates of degree $\neq R$. Equivalently, the versal deformation of degree $-R$ can be understood as the maximal extension of the infinitesimal deformations in degree $-R$.

1.3. For investigating versal deformation spaces, Gorenstein singularities are the easiest examples beyond complete intersections. It is a helpful coincidence that the Gorenstein property has a very nice description in the toric context – the cone σ should just be spanned by a lattice polytope Q sitting in an affine hyperplane $[R^* = 1]$ of height one. Note that $R^* \in M$ equals the degree of the volume form. This leads to the investigation of the deformation theory of toric Gorenstein singularities in [1] – the interesting deformations were contained in degree $-R^*$.

The present paper is meant as a generalization of this approach. We discard the Gorenstein assumption. For Y we just assume smoothness in codimension two (as was already done in the Gorenstein case), and for R we restrict to the case of a primitive $R \in \sigma^\vee \cap M$. Otherwise, one would leave the toric framework, cf. [2].

While the main ideas work along the lines of [1], we try to keep the paper as self-contained as possible.

1.4. The main tool to describe our results is the notion of Minkowski sums.

Definition. For two polyhedra $P, P' \subseteq \mathbb{R}^n$ we define their Minkowski sum as the polytope $P + P' := \{p + p' \mid p \in P, \ p' \in P'\}$. Obviously, this notion also makes sense for translation classes of polytopes. For instance, each polyhedron Q is the Minkowski sum of a compact polytope and the so-called tail cone Q^∞.

Let us fix a primitive element R of $\sigma^\vee \cap M$ and intersect the cone σ with the hyperplane defined by $[R = 1]$. This intersection defines a polyhedron named $Q := Q(R)$. For our investigations, this Q plays a similar role as the Q in the Gorenstein case. However, in the present paper, it neither needs to be a lattice polyhedron, nor compact. If a^i is one of the primitive generators of σ, then it leads to a lattice/non-lattice vertex of Q or to a generating ray of its tail cone Q^∞ iff $\langle a^i, R \rangle = 1, \geq 2$, or $= 0$, respectively.

Following the Gorenstein case we will construct a "moduli space" $C(Q)$ of Minkowski summands of multiples of Q – but in the present paper, we have to take care of their possible tail cones as well as the non-lattice vertices of Q. Attaching each Minkowski summand at the point that represents it in $C(Q)$ yields the so-called tautological cone $\tilde{C}(Q)$ together with a projection onto $C(Q)$. It can be seen as the universal Minkowski summand of Q. Indeed, applying the functor that makes toric varieties from cones will provide the main step toward constructing the versal base space of $Y = \mathbb{TV}(\sigma)$ in degree $-R$.

1.5. For a given polyhedron $Q \subseteq \mathbb{R}^n$ we begin in Sect. 2 by presenting an affine scheme $\bar{\mathcal{M}}$. It is related to $C(Q)$ and describes the possibilities of splitting Q into Minkowski summands. In Sect. 3 we study the tautological cone $\tilde{C}(Q)$. Applied to $Q(R)$, this leads in Sect. 4 to the construction of a flat family over $\bar{\mathcal{M}}$ with Y as special fiber. Now we can state the main theorem (6.1) of this paper.

Theorem. *The family* $\bar{X} \times_{\bar{S}} \mathcal{M} \to \bar{\mathcal{M}}$ *(cf. 4.1) with base space* $\bar{\mathcal{M}}$ *is the versal deformation of* Y *of degree* $-R$,

i.e. the Kodaira–Spencer map is an isomorphism in degree $-R$ (Sect. 5) and the obstruction map is injective (Sect. 6). Based on this an interesting question arises, namely whether it is possible to construct the part of the versal deformation of Y with negative degrees by repeatedly applying the principles of this paper.

The last section starts with describing the situation for $\dim Y = 3$ (in Theorem 7.1) and then continues with an explicit example. It shows how to compute the family using SINGULAR (cf. [6]) and NORMALIZ (cf. [5]).

1.6. Acknowledgement. We would like to thank the anonymous referee for the careful reading, for checking the calculations, and for the valuable hints.

2. THE BASE SPACE

2.1. Let $\sigma = \langle a^1, \ldots, a^M \rangle \subseteq N_{\mathbb{R}}$ be a cone such that the two-dimensional faces $\langle a^j, a^k \rangle < \sigma$ are smooth (i.e. $a^1, \ldots, a^M \in N$ are primitive, and $\{a^j, a^k\}$ could be extended to a $\mathbb{Z}$-basis of N). Let $R \in \sigma^\vee \cap M$ be a primitive element. Then one can define:

Definition. Let $R \in \sigma^\vee \cap M$ be primitive. We define the affine space $\mathbb{A} := [R = 1] := \{a \in N_{\mathbb{R}} \mid \langle a, R \rangle = 1\} \subseteq N_{\mathbb{R}}$ with lattice $\mathbb{L} := \mathbb{A} \cap N$. It contains the polyhedron $Q := Q(R) := \sigma \cap [R = 1]$ with tail cone $Q^\infty = \sigma \cap [R = 0]$. Note that $Q^\infty = 0$ if and only if $R \in \text{int } \sigma^\vee$.

Note that we can recover σ as $\sigma = \overline{\mathbb{R}_{\geq 0} \cdot (Q, 1)} = \mathbb{R}_{\geq 0} \cdot (Q, 1) \cup (Q^\infty, 0)$. The vertices of Q are $v^i = a^i / \langle a^i, R \rangle$ for those fundamental generators $a^i \in \sigma$ with $\langle a^i, R \rangle \geq 1$; they belong to $\mathbb{L}$ iff $\langle a^i, R \rangle = 1$. We will see that Y is rigid in degree $-R$ unless Q has at least one such $\mathbb{L}$-vertex. Assuming this, we fix one of the $\mathbb{L}$-vertices of Q to be the origin.

2.2. Denote by $d^1, \ldots, d^N \in N_{\mathbb{Q}}$ the compact edges of Q after choosing some orientation of each of them. Calling edges that meet in a common *non-lattice* vertex of Q "*connected*" implies that the set $\{d^1, \ldots, d^N\}$ may be uniquely decomposed into components according to this notion.

Definition. For every compact 2-face $\varepsilon < Q$ we can define the sign vector $\underline{\varepsilon} = (\varepsilon_1, \ldots, \varepsilon_N) \in \{0, \pm 1\}^N$ by

$$\varepsilon_i := \begin{cases} \pm 1 & \text{if } d^i \text{ is an edge of } \varepsilon \\ 0 & \text{otherwise} \end{cases}$$

such that the oriented edges $\varepsilon_i \cdot d^i$ fit into a cycle along the boundary of ε. This determines $\underline{\varepsilon}$ up to sign and we choose one of both possibilities. In particular, we have $\sum_i \varepsilon_i d^i = 0$ if $\varepsilon < Q$ is a compact 2-face.

Now we define the vector space $V(Q) \subseteq \mathbb{R}^N$ by

$$V(Q) := \left\{ (t_1, \ldots, t_N) \,\middle|\, \sum_i t_i \varepsilon_i d^i = 0 \text{ for every compact 2-face } \varepsilon < Q, \text{ and} \right.$$

$$\left. t_i = t_j \text{ if } d^i, d^j \text{ contain a common non-lattice vertex of } Q \right\}.$$

To simplify notation we are going to use $V := V(Q)$. For each component of edges there is a well defined associated coordinate function $V \to \mathbb{R}$. Now, $C(Q) := V \cap \mathbb{R}_{\geq 0}^N$ is a rational, polyhedral cone in V, and its points correspond to certain Minkowski summands of positive multiples of Q:

2.3. Lemma. *Each point $\underline{t} \in C(Q)$ define a Minkowski summand of a positive multiple of Q; its i-th compact edge equals $t_i d^i$. This yields a bijection between $C(Q)$ and the set of all Minkowski summands (of positive multiples of Q) that change components of edges just by a scalar.*

Proof. For an Element $\underline{t} \in C(Q)$ the corresponding summand $Q_{\underline{t}}$ is built by the edges $t_i \cdot d^i$ as follows: Each vertex v of Q can be reached from $0 \in Q$ by some walk along the compact edges d^i of Q. We obtain

$$v = \sum_{i=1}^{N} \lambda_i d^i \text{ for some } \underline{\lambda} = (\lambda_1, \dots, \lambda_N), \ \lambda_i \in \mathbb{Z}.$$

Now given an element $\underline{t} \in C(Q)$, we may define the corresponding vertex $v_{\underline{t}}$ by

$$v_{\underline{t}} := \sum_{i=1}^{N} t_i \lambda_i d^i,$$

and the linear equations defining V ensure that this definition does not depend on the particular path from v to 0 through the compact part of the 1-skeleton of Q. We define the Minkowski summand by $Q_{\underline{t}} := \text{conv}\{v_{\underline{t}}\} + Q^{\infty}$.
∎

2.4. Now, we define a higher degree analogous to the linear equations defining V:

Definition. For each compact 2-face $\varepsilon < Q$, and for each integer $k \geq 1$ we define the vector valued polynomial

$$g_{\varepsilon, k}(\underline{t}) := \sum_{i=1}^{N} t_i^k \varepsilon_i d^i.$$

Using coordinates of $\mathbb{A}$, the d^i turn into scalars, thus the $g_{\varepsilon, k}(\underline{t})$ turn into regular polynomials; for each pair (ε, k) we will get two linearly independent ones. Since

$$V^{\perp} = \text{span}\left\{ \left[\langle \varepsilon_1 d^1, c \rangle, \dots, \langle \varepsilon_N d^N, c \rangle \right] \mid \varepsilon < Q \text{ is a compact 2-face, } c \in \mathbb{A}^*; \right.$$

$$\left. [0, \dots, 1_i, \dots, -1_j, \dots, 0] \mid d^i, d^j \text{ have a common non-lattice } Q\text{-vertex} \right\},$$

they (together with $t_i - t_j$ for d^i, d^j sharing a common non-lattice vertex)
can be written as

$$g_{\underline{d},k}(\underline{t}) := \sum_{i=1}^{N} d_i t_i^k$$

with $\underline{d} \in V^\perp \cap \mathbb{Z}^N$ and $k \in \mathbb{N}$. We thus may define the ideal

$$\mathcal{J} := (g_{\varepsilon,k})_{\varepsilon,k \geq 1} + (t_i - t_j \mid d^i, d^j \text{ share a common non-lattice vertex})$$
$$= \left(g_{\underline{d},k}(\underline{t}) \mid \underline{d} \in V^\perp \cap \mathbb{Z}^N \right) \subseteq \mathbb{C}[\underline{t}]$$

which defines an affine closed subscheme

$$\mathcal{M} := \mathrm{Spec} \, \mathbb{C}[\underline{t}] \Big/ \mathcal{J} \subseteq V_{\mathbb{C}} \subseteq \mathbb{C}^N.$$

Denote by ℓ the canonical projection

$$\ell : \mathbb{C}^N \twoheadrightarrow \mathbb{C}^N \Big/ \mathbb{C} \cdot (1, \dots, 1) \, .$$

On the level of regular functions this corresponds to the inclusion $\mathbb{C}[t_i - t_j \mid 1 \leq i, j \leq N] \subseteq \mathbb{C}[\underline{t}]$. Note that the vector $\underline{1} = (1, \dots, 1) \in C(Q) \subseteq V$ encodes Q as a Minkowski summand of itself.

2.5. Theorem. (1) $\mathcal{J}$ is generated by polynomials from $\mathbb{C}[t_i - t_j]$, i.e. $\mathcal{M} = \ell^{-1}(\bar{\mathcal{M}})$ for the affine closed subscheme $\bar{\mathcal{M}} \subseteq V_{\mathbb{C}} \Big/ \mathbb{C} \cdot \underline{1} \subseteq \mathbb{C}^N \Big/ \mathbb{C} \cdot \underline{1}$ defined by $\mathcal{J} \cap \mathbb{C}[t_i - t_j]$.

(2) $\mathcal{J} \subseteq \mathbb{C}[t_1, \dots, t_N]$ is the smallest ideal that meets property (1) and, on the other hand, contains the "toric equations"

$$\prod_{i=1}^{N} t_i^{d_i^+} - \prod_{i=1}^{N} t_i^{d_i^-} \quad \text{with } \underline{d} \in V^\perp \cap \mathbb{Z}^N.$$

(For an integer h we denote

$$h^+ := \begin{cases} h & \text{if } h \geq 0 \\ 0 & \text{otherwise;} \end{cases} \qquad h^- := \begin{cases} 0 & \text{if } h \geq 0 \\ -h & \text{otherwise.} \end{cases})$$

The proof is similar to the one of [1, Theorem (2.4)].

3. THE TAUTOLOGICAL CONE

3.1. While $C(Q) \subseteq V(Q)$ were built to describe the base space, we turn now to the cone that will eventually lead to the total space of our deformation.

Definition. The tautological cone $\tilde{C}(Q) \subseteq \mathbb{A} \times V$ is defined as

$$\tilde{C}(Q) := \{(v, \underline{t}) \mid \underline{t} \in C(Q); \; v \in Q_{\underline{t}}\};$$

it is generated by the pairs $\left(v_{\underline{t}^l}^j, \underline{t}^l\right)$ and $(v^k, 0)$ where $\underline{t}^l$, v^j, and v^k run through the generators of $C(Q)$, vertices of Q, and generators of Q^∞, respectively.

Since $\sigma = \overline{\text{Cone}(Q)} \subseteq \mathbb{A} \times \mathbb{R} = N_{\mathbb{R}}$, we obtain a fiber product diagram of rational polyhedral cones:

$$
\begin{array}{ccc}
[\sigma \subseteq \mathbb{A} \times \mathbb{R}] & \overset{i}{\hookrightarrow} & [\tilde{C}(Q) \subseteq \mathbb{A} \times V] \\
\downarrow{\scriptstyle \text{pr}_{\mathbb{R}}} & & \downarrow{\scriptstyle \text{pr}_V} \\
\mathbb{R}_{\geq 0} & \overset{\cdot 1}{\hookrightarrow} & [C(Q) \subseteq V]
\end{array}
$$

The three cones $\sigma \subseteq \mathbb{A} \times \mathbb{R}$, $\tilde{C}(Q) \subseteq \mathbb{A} \times V$ and $C(Q) \subseteq V$ define affine toric varieties called Y, X and S, respectively. The corresponding rings of regular functions are

$$A(Y) = \mathbb{C}\left[\sigma^\vee \cap (\mathbb{L}^* \times \mathbb{Z})\right],$$

$$A(X) = \mathbb{C}\left[\tilde{C}(Q)^\vee \cap \tilde{M}\right], \quad \tilde{M} := \mathbb{L}^* \times V_{\mathbb{Z}}^*$$

$$A(S) = \mathbb{C}\left[C(Q)^\vee \cap V_{\mathbb{Z}}^*\right],$$

and we obtain the following diagram:

$$
\begin{array}{ccc}
Y & \overset{i}{\hookrightarrow} & X \\
\downarrow & & \downarrow{\scriptstyle \pi} \\
\mathbb{C} & \hookrightarrow & S.
\end{array}
$$

Unfortunately, this diagram does not need to be a fiber product diagram as we will explain in (3.6).

3.2. To each non-trivial $c \in (Q^\infty)^\vee$ we associate a vertex $v(c) \in Q$ and a number $\eta_0(c) \in \mathbb{R}$ meeting the properties

$$\langle Q, c \rangle + \eta_0(c) \geq 0 \qquad \text{and}$$

$$\langle v(c), c \rangle + \eta_0(c) = 0.$$

For $c = 0$ we define $v(0) := 0 \in \mathbb{L}$ and $\eta_0(0) := 0 \in \mathbb{R}$.

Remark. (1) With respect to Q, $c \neq 0$ is the inner normal vector of the affine supporting hyperplane $\left[\langle \bullet, c \rangle + \eta_0(c) = 0 \right]$ through $v(c)$. In particular, $\eta_0(c)$ is uniquely determined, while $v(c)$ is not.

(2) Since $0 \in Q$, the $\eta_0(c)$ are non-negative.

Moreover, if $c \in (Q^\infty)^\vee \cap \mathbb{L}^*$, we denote by $\eta_0^*(c)$ the smallest integer greater than or equal to $\eta_0(c)$, i.e. $\eta_0^*(c) = \lceil \eta_0(c) \rceil$. Then

$$\sigma^\vee = \left\{ \left[c, \eta_0(c) \right] \mid c \in (Q^\infty)^\vee \right\} + \mathbb{R}_{\geq 0} \cdot [\underline{0}, 1]$$

and

$$\sigma^\vee \cap M = \left\{ \left[c, \eta_0^*(c) \right] \mid c \in \mathbb{L}^* \cap (Q^\infty)^\vee \right\} + \mathbb{N} \cdot [\underline{0}, 1].$$

Note that $[\underline{0}, 1]$ equals the element $R \in M$ fixed in the beginning. In particular we can choose a generating set $E \subseteq \sigma^\vee \cap M$ as some

$$E = \left\{ [\underline{0}, 1], \left[c^1, \eta_0^*(c^1) \right], \ldots, \left[c^w, \eta_0^*(c^w) \right] \right\}.$$

3.3. Thinking of $C(Q)$ as a cone in $\mathbb{R}^N$ instead of V allows dualizing the equation $C(Q) = \mathbb{R}_{\geq 0}^N \cap V$ to get $C(Q)^\vee = \mathbb{R}_{\geq 0}^N + V^\perp$. Hence, for $C(Q)$ as a cone in V we obtain

$$C(Q)^\vee = \mathbb{R}_{\geq 0}^N + V^\perp \Big/ V^\perp = \mathrm{im}\, [\mathbb{R}_{\geq 0}^N \longrightarrow V^*].$$

The surjection $\mathbb{R}_{\geq 0}^N \twoheadrightarrow C(Q)^\vee$ induces a map $\mathbb{N}^N \longrightarrow C(Q)^\vee \cap V_\mathbb{Z}^*$ which does not need to be surjective at all. This leads to the following definition:

Definition. *On $V_\mathbb{Z}^*$ we introduce a partial ordering "$\geq$" by*

$$\underline{\eta} \geq \underline{\eta}' \quad \Longleftrightarrow \quad \underline{\eta} - \underline{\eta}' \in \mathrm{im}\, [\mathbb{N}^N \to V_\mathbb{Z}^*] \subseteq C(Q)^\vee \cap V_\mathbb{Z}^*.$$

On the geometric level, the non-saturated semigroup $\mathrm{im}\, [\mathbb{N}^N \to V_\mathbb{Z}^*] \subseteq C(Q)^\vee \cap V_\mathbb{Z}^*$ corresponds to the scheme theoretical image $\bar{S}$ of $p : S \to \mathbb{C}^N$, and $S \to \bar{S}$ is its normalization, cf. (4.2). The equations of $\bar{S} \subseteq \mathbb{C}^N$ are collected in the kernel of

$$\mathbb{C}[t_1, \ldots, t_N] = \mathbb{C}[\mathbb{N}^N] \xrightarrow{\varphi} \mathbb{C}[C(Q)^\vee \cap V_{\mathbb{Z}}^*] \subseteq \mathbb{C}[V_{\mathbb{Z}}^*],$$

and it is easy to see that

$$\ker \varphi = \left(\prod_{i=1}^N t_i^{d_i^+} - \prod_{i=1}^N t_i^{d_i^-} \,\Big|\, \underline{d} \in \mathbb{Z}^N \cap V^\perp \right)$$

is generated by the toric equations from (2.5).

3.4. To deal with the dual space V^* the following point of view will be useful: In the Gorenstein case we described its elements by using the surjection $\mathbb{R}^N \to V^*$. In particular, an element $\eta \in V^*$ was given by coordinates η_i corresponding to the edges d^i of Q. Now, in the general case, the set of edges of Q splits into several components, cf. (2.2). For each such component, not the single coordinates but only their sum along the entire component is well defined. However, this does not affect that the total summation map $\mathbb{R}^N \to \mathbb{R}$ factors through $V^* \to \mathbb{R}$. It will still be denoted as $\underline{\eta} \mapsto \sum_i \eta_i$.

Definition. (1) For $c \in (Q^\infty)^\vee$ choose some path from $0 \in Q$ to $v(c) \in Q$ through the 1-skeleton of Q and let $\underline{\lambda}^c := (\lambda_1^c, \ldots, \lambda_N^c) \in \mathbb{Z}^N$ be the vector counting how often (and in which direction) we went through each particular edge. Then

$$\underline{\eta}(c) := \left[-\lambda_1^c \langle d^1, c \rangle, \ldots, -\lambda_N^c \langle d^N, c \rangle \right] \in \mathbb{Q}^N$$

defines an element $\underline{\eta}(c) \in V^*$ not depending on the special choice of the path $\underline{\lambda}^c$.

(2) Let $v \in Q$ be a vertex not contained in the lattice $\mathbb{L}$. Then we denote by $e[v] \in V_{\mathbb{Z}}^*$ the element represented by $[0, \ldots, 0, 1_i, 0, \ldots, 0] \in \mathbb{Z}^N$ for some compact edge d^i containing v, i.e. $e[v]$ yields the entry t_i of $\underline{t} \in V$. (Note that $e[v]$ does not depend on the choice of d^i.)

(3) For $c \in (Q^\infty)^\vee \cap \mathbb{L}^*$ denote $\underline{\eta}^*(c) := \underline{\eta}(c) + \big(\eta_0^*(c) - \eta_0(c)\big) \cdot e[v(c)] \in V^*$. (If $v(c) \in \mathbb{L}$, then $\eta_0^*(c) = \eta_0(c)$ implies that we do not need $e[v(c)]$ in that case.)

Here are the essential properties of $\underline{\eta}^*(c)$:

3.5. Lemma. *Let* $c \in (Q^\infty)^\vee \cap \mathbb{L}^*$. *Then*

(i) $\underline{\eta}^*(c) \in \mathrm{im}[\mathbb{N}^N \to V_{\mathbb{Z}}^*] \subseteq C(Q)^\vee \cap V_{\mathbb{Z}}^*$, *and this element equals* $\underline{\eta}(c)$ *if and only if* $\langle v(c), c \rangle \in \mathbb{Z}$ (*in particular, if* $v(c) \in \mathbb{L}$).

(ii) *For* $c^\nu \in \mathbb{L}^* \cap (Q^\infty)^\vee$ *and* $g_\nu \in \mathbb{N}$ *we have* $\sum_\nu g_\nu \underline{\eta}^*(c^\nu) \geq \underline{\eta}^* \left(\sum_\nu g_\nu c^\nu \right)$ *in the sense of (3.3).*

(iii) $\sum_{i=1}^N \eta_i(c) = \eta_0(c)$ *and* $\sum_{i=1}^N \eta_i^*(c) = \eta_0^*(c)$.

Proof. (iii) By definition of $\underline{\lambda}^c$ we have $\sum_{i=1}^N \lambda_i^c d^i = v(c)$. In particular:

$$\sum_{i=1}^N \eta_i^*(c) = \sum_{i=1}^N \eta_i(c) + \sum_{i=1}^N \left(\eta_0^*(c) - \eta_0(c) \right) \cdot e_i \big[v(c) \big]$$

$$= \left(-\sum_{i=1}^N \langle \lambda_i^c d^i, c \rangle \right) + \eta_0^*(c) - \eta_0(c)$$

$$= -\langle v(c), c \rangle + \eta_0^*(c) - \eta_0(c) = \eta_0(c) + \eta_0^*(c) - \eta_0(c) = \eta_0^*(c).$$

The equality $\sum_{i=1}^N \eta_i(c) = \eta_0(c)$ follows from the previous argument by leaving out the $e\big[v(c) \big]$-terms.

(i) Now, for $c \in \mathbb{L}^* \cap (Q^\infty)^\vee$, we will show that $\underline{\eta}^*(c) \in V^*$ can be represented by an integral vector of $\mathbb{R}^N$ having only non-negative coordinates: We choose some path along the edges of Q passing $v^0 = 0, \ldots, v^p = v(c)$ and decreasing the value of c at each step. This provides some vector $\underline{\lambda}^c \in \mathbb{Z}^N$ yielding $\underline{\eta}(c)$ with $\eta_i(c) = -\lambda_i^c \langle d^i, c \rangle \geq 0$.

Denote by $v^{j_0}, \ldots, v^{j_q}$ ($\{j_0, \ldots, j_q\} \subseteq \{0, \ldots, p\}$) the $\mathbb{L}$-vertices on the path. Then, for $s = 1, \ldots, q$, the edges between $v^{j_{s-1}}$ and v^{j_s} (say $d^{i_1}, \ldots, d^{i_k}$) belong to the same "component". In particular, not the single $\eta_{i_1}^*(c), \ldots, \eta_{i_k}^*(c)$ but only their sums have to be considered:

$$\sum_{\mu=1}^k \eta_{i_\mu}^*(c) = \sum_{\mu=1}^k \eta_{i_\mu}(c) = \left\langle -\sum_{\mu=1}^k \lambda_{i_\mu}^c d^{i_\mu}, c \right\rangle = \langle v^{j_{s-1}} - v^{j_s}, c \rangle \in \mathbb{N}.$$

If $v(c)$ belongs to the lattice $\mathbb{L}$, then we are done. Otherwise, there might be at most one non-integer coordinate (assigned to $v(c) \notin \mathbb{L}$) in $\underline{\eta}^*(c)$. However, this cannot be the case, since the sum taken over all coordinates of $\underline{\eta}^*(c)$ yields the integer $\eta_0^*(c)$.

(ii) We define the following paths through the 1-skeleton of Q:

- $\underline{\lambda} :=$ path from $0 \in Q$ to $v\left(\sum_\nu g_\nu c^\nu\right) \in Q$,

- $\underline{\mu}^\nu :=$ path from $v\left(\sum_\nu g_\nu c^\nu\right) \in Q$ to $v(c^\nu) \in Q$ such that $\mu_i^\nu \langle d^i, c^\nu\rangle \le 0$ for each $i = 1, \ldots, N$.

Then $\underline{\lambda}^\nu := \underline{\lambda} + \underline{\mu}^\nu$ is a path from $0 \in Q$ to $v(c^\nu)$, and for $i = 1, \ldots, N$ we obtain

$$\sum_\nu g_\nu \eta_i(c^\nu) - \eta_i\left(\sum_\nu g_\nu c^\nu\right) = -\sum_\nu g_\nu(\lambda_i + \mu_i^\nu)\langle d^i, c^\nu\rangle + \lambda_i\left\langle d^i, \sum_\nu g_\nu c^\nu\right\rangle$$

$$= -\sum_\nu g_\nu \mu_i^\nu \langle d^i, c^\nu\rangle \ge 0.$$

This yields the (componentwise) inequality

$$\sum_\nu g_\nu \underline{\eta}^*(c^\nu) \ge \sum_\nu g_\nu \underline{\eta}(c^\nu) \ge \underline{\eta}\left(\sum_\nu g_\nu c^\nu\right).$$

On the other hand, $\underline{\eta}\left(\sum_\nu g_\nu c^\nu\right)$ and $\underline{\eta}^*\left(\sum_\nu g_\nu c^\nu\right)$ might differ in at most one coordinate (assigned to $a\left(\sum_\nu g_\nu c^\nu\right)$). If so, then by definition of $\underline{\eta}^*$ the latter one equals the smallest integer not smaller than the first one. Hence, we are done, since the left hand side of our inequality involves integers only. ∎

We obtain the following description of $\tilde{C}(Q)^\vee$:

3.6. Proposition. (1) $\tilde{C}(Q)^\vee = \left\{ [c, \underline{\eta}] \in (Q^\infty)^\vee \times V^* \subseteq \mathbb{A}^* \times V^* \mid \underline{\eta} - \underline{\eta}(c) \in C(Q)^\vee \right\}$.

(2) In particular, $[c, \underline{\eta}(c)] \in \tilde{C}(Q)^\vee$; it is the only preimage of $[c, \eta_0(c)] \in \sigma^\vee$ via the surjection $i^\vee : \tilde{C}(Q)^\vee \twoheadrightarrow \sigma^\vee$. Moreover, for $c \in \mathbb{L}^* \cap (Q^\infty)^\vee$, it holds $[c, \underline{\eta}^*(c)] \in \tilde{C}(Q)^\vee \cap \tilde{M}$. These elements are liftings of $[c, \eta_0^*(c)] \in \sigma^\vee \cap M$ – but, in general, they are not the only ones.

(3) $[c^1, \underline{\eta}^*(c^1)], \ldots, [c^w, \underline{\eta}^*(c^w)]$ and $C(Q)^\vee \cap V_{\mathbb{Z}}^*$, embedded as $[0, C(Q)^\vee]$, generate the semigroup $\Gamma := \left\{ [c, \underline{\eta}] \in (\mathbb{L}^* \cap (Q^\infty)^\vee) \times V_{\mathbb{Z}}^* \mid \underline{\eta} - \underline{\eta}^*(c) \in C(Q)^\vee \right\} \subseteq \tilde{C}(Q)^\vee \cap \tilde{M}$. Moreover, $\tilde{C}(Q)^\vee \cap \tilde{M}$ is the saturation of that subsemigroup.

Proof. The proof of (1) and (2) is similar to the proof of [1, Prop. (4.6)].

(3) First, the condition $\underline{\eta} - \underline{\eta}^*(c) \in C(Q)^\vee$ indeed describes a semigroup; this is a consequence of (ii) of Lemma 3.5. On the other hand, let $[c, \underline{\eta}^*(c)]$ be given. Using some representation $[c, \eta_0^*(c)] = \sum_{t=1}^w p_\nu [c^\nu, \eta_0^*(c^\nu)]$ $(p_\nu \in \mathbb{N})$, we obtain by the same lemma

$$\sum_\nu p_\nu \underline{\eta}^*(c^\nu) - \underline{\eta}^*(c) = \sum_\nu p_\nu \underline{\eta}^*(c^\nu) - \underline{\eta}^*\left(\sum_\nu p_\nu c^\nu\right) \in C(Q)^\vee \text{ (or even } \geq 0).$$

Since, at the same time, the sum taken over all coordinates of that difference vanishes, the whole difference has to be zero. Now we obtain

$$[c, \underline{\eta}] = [c, \underline{\eta}^*(c)] + [0, \underline{\eta} - \underline{\eta}^*(c)]$$

$$= \sum_\nu p_\nu [c^\nu, \underline{\eta}^*(c^\nu)] + [0, \underline{\eta} - \underline{\eta}^*(c)].$$

Finally, for every $[c, \underline{\eta}] \in \tilde{C}(Q)^\vee$ with $\underline{\eta} - \underline{\eta}^*(c) \notin C(Q)^\vee$ there exists a $k \in \mathbb{N}_{\geq 1}$ such that $\underline{\eta}^*(k \cdot c) = \underline{\eta}(k \cdot c)$, since $v(c) = a(k \cdot c)$ yields $\underline{\eta}(k \cdot c) = k \cdot \underline{\eta}(c)$ and $\underline{\eta}(c) \in \mathbb{Q}^N$. Then we obtain

$$k \cdot \underline{\eta} - \underline{\eta}^*(k \cdot c) = k \cdot \underline{\eta} - \underline{\eta}(k \cdot c) = k \cdot (\underline{\eta} - \underline{\eta}(c)) \in C(Q)^\vee$$

by part (i) of this proposition. ∎

3.7. Now we will provide an example for the case $\Gamma \neq \tilde{C}(Q)^\vee \cap \tilde{M}$:

Example. Let $N = \mathbb{Z}^3$ be a lattice. Define the cone σ by

$$\sigma := \big\langle (0, 0, 1), (6, -1, 2), (5, 0, 1), (5, 1, 1), (24, 7, 5), (6, 5, 2), (2, 3, 1) \big\rangle$$
$$\subseteq \mathbb{Q}^3 = N_\mathbb{Q}.$$

We choose $R := [0, 0, 1] \in M = \mathbb{Z}^3$ and obtain the following polygon Q:

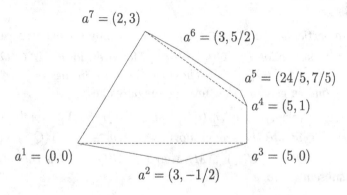

$a^7 = (2, 3)$

$a^6 = (3, 5/2)$

$a^5 = (24/5, 7/5)$

$a^4 = (5, 1)$

$a^1 = (0, 0)$

$a^3 = (5, 0)$

$a^2 = (3, -1/2)$

The paths along the edges of Q are denoted as follows:

$$d^1 = \begin{pmatrix} 3 \\ -\frac{1}{2} \end{pmatrix}, \quad d^2 = \begin{pmatrix} 2 \\ \frac{1}{2} \end{pmatrix}, \quad d^3 = \begin{pmatrix} 0 \\ 1 \end{pmatrix}, \quad d^4 = \begin{pmatrix} -\frac{1}{5} \\ \frac{2}{5} \end{pmatrix},$$

$$d^5 = \begin{pmatrix} -\frac{9}{5} \\ \frac{11}{10} \end{pmatrix}, \quad d^6 = \begin{pmatrix} -1 \\ \frac{1}{2} \end{pmatrix}, \quad d^7 = \begin{pmatrix} -2 \\ -3 \end{pmatrix}.$$

Let us consider $V(Q)$. We identify t_i and t_j if the corresponding edges have a common non-lattice vertex. Then $V(Q)$ as a subspace of $\mathbb{R}^4$ is the kernel of the following matrix obtained by the 2-face equation of Q:

$$\begin{pmatrix} 5 & 0 & -3 & -2 \\ 0 & 1 & 2 & -3 \end{pmatrix}.$$

It is generated by $\underline{t}^1 := (13, 0, 15, 10)$ and $\underline{t}^2 := (2, 15, 0, 5)$, and this leads to $C(Q) = \mathbb{R}_{\geq 0} \cdot \underline{t}^1 \oplus \mathbb{R}_{\geq 0} \cdot \underline{t}^2$.

Let $c := [-1, -1] \in M$, then $v(c) = a^5$ and $\underline{\lambda}^c = (1, 1, 1, 1, 0, 0, 0)$. Now we compute $\underline{\eta}(c)$ as described in (3.4):

$$\underline{\eta}(c) = [5/2, 5/2, 1, 1/5, 0, 0, 0].$$

Since we only described $V(Q)$ as a subspace of $\mathbb{R}^4$, we can also denote $\underline{\eta}(c)$ by $\underline{\eta}(c) = [5, 1, 1/5, 0]$, which corresponds to taking the sum on components of Q. Let $\underline{\eta} := [5, 1, 3/5, 2/5] \in V^*$, so that it is also contained in $V_{\mathbb{Z}}^*$: Since the first row of the matrix defining $V(Q)$ yields $5t_1 = 3t_3 + 2t_4$, the sum on the right hand side has 5 as a divisor if we only consider integral solutions. We could also regard $\underline{\eta}$ as $[6, 1, 0, 0]$ as element of $V_{\mathbb{Z}}^*$. Let us consider $\underline{\eta} - \underline{\eta}(c)$:

$$\underline{\eta} - \underline{\eta}(c) = [5 - 5, 1 - 1, 3/5 - 1/5, 2/5 - 0] = [0, 0, 2/5, 2/5].$$

Obviously $\underline{\eta} - \underline{\eta}(c)$ is contained in $C(Q)^{\vee}$, as it has positive entries only. Hence, $[c, \underline{\eta}] \in \tilde{C}(Q)^{\vee} \cap \tilde{M}$.

We build up $\underline{\eta}^*(c)$ as described in (3.4): $\underline{\eta}^*(c) = [5, 1, 1, 0]$. This yields

$$\underline{\eta} - \underline{\eta}^*(c) = [5 - 5, 1 - 1, 3/5 - 1, 2/5 - 0] = [0, 0, -2/5, 2/5].$$

Now we apply this to $\underline{t}^1$:

$$\langle t_1, \underline{\eta} - \underline{\eta}^*(c) \rangle = \langle (13, 0, 15, 10), [0, 0, -2/5, 2/5] \rangle$$

$$= -2/5 \cdot 15 + 2/5 \cdot 10$$

$$= -6 + 4 = -2.$$

Hence, we gain $\underline{\eta} - \underline{\eta}^*(c) \notin C(Q)^{\vee}$ and $[c, \underline{\eta}] \notin \Gamma$.

Remark. If one replaces the semigroup $\tilde{C}(Q)^{\vee} \cap \tilde{M}$ by its non-saturated subgroup Γ and X by $X' := \operatorname{Spec} \mathbb{C}[\Gamma]$, respectively, then the diagram of (3.1) becomes a fiber product diagram. Moreover, X' equals the scheme theoretical image of the map $X \to \mathbb{C}^w \times S$ induced by the elements $[c^1, \underline{\eta}^*(c^1)], \dots, [c^w, \underline{\eta}^*(c^w)] \in \Gamma$. And, in return, X is the normalization of X'.

4. A Flat Family over $\bar{\mathcal{M}}$

We use the previous constructions to provide a deformation of Y over $\bar{\mathcal{M}}$:

4.1. Theorem. *Denote by $\bar{X}$ and $\bar{S}$ the scheme theoretical images of X and S in $\mathbb{C}^w \times \mathbb{C}^N$ and $\mathbb{C}^N$, respectively. Then*

(1) *$X \to \bar{X}$ and $S \to \bar{S}$ are the normalization maps,*

(2) *$\pi : X' \to S$ induces a map $\bar{\pi} : \bar{X} \to \bar{S}$ such that π can be recovered from $\bar{\pi}$ via base change $S \to \bar{S}$, and*

(3) *restricting to $\mathcal{M} \subseteq \bar{S}$ and composing with ℓ turns $\bar{\pi}$ into a family*

$$\bar{X} \times_{\bar{S}} \mathcal{M} \xrightarrow{\bar{\pi}} \mathcal{M} \xrightarrow{\ell} \bar{\mathcal{M}}.$$

It is flat in $0 \in \bar{\mathcal{M}} \subseteq \mathbb{C}^{N-1}$, and the special fiber equals Y.

The proof of this theorem will fill Section 4.

4.2. The ring of regular functions $A(\bar{S})$ is given as the image of $\mathbb{C}[t_1, \dots, t_N] \to A(S)$. Since $\mathbb{Z}^N \twoheadrightarrow V_{\mathbb{Z}}^*$ is surjective, the rings $A(\bar{S}) \subseteq A(S) \subseteq \mathbb{C}[V_{\mathbb{Z}}^*]$ have the same field of fractions.

On the other hand, while t-monomials with negative exponents might be involved in $A(S)$, the surjectivity of $\mathbb{R}_{\geq 0}^N \twoheadrightarrow C(Q)^{\vee}$ tells us that sufficiently high powers of those monomials always come from $A(\bar{S})$. In particular, $A(S)$ is normal over $A(\bar{S})$.

$A(\bar{X})$ is given as the image $A(\bar{X}) = \operatorname{im}\left(\mathbb{C}[Z_1, \dots, Z_w, t_1, \dots, t_N] \to A(X')\right)$ with $Z_i \mapsto$ [monomial associated to $[c^i, \eta_0^*(c^i)]$]. Since $A(X')$ is generated by these monomials over its subring $A(S)$, cf. Proposition 3.6(3), the same arguments as for S and $\bar{S}$ apply. Hence, Part (1) of the previous theorem is proved.

4.3. Denoting by $z_1, \ldots, z_w, t$ the variables mapping to the $A(Y)$-monomials with exponents $[c^1, \eta_0^*(c^1)], \ldots, [c^w, \eta_0^*(c^w)], [0,1] \in \sigma^\vee \cap M$, respectively, we obtain the following equations for $Y \subseteq \mathbb{C}^{w+1}$:

$$f_{(a,b,\alpha,\beta)}(\underline{z}, t) := t^\alpha \prod_{t=1}^w z_\nu^{a_\nu} - t^\beta \prod_{t=1}^w z_\nu^{b_\nu}$$

$$\text{with} \quad a, b \in \mathbb{N}^w : \sum_\nu a_\nu c^\nu = \sum_\nu b_\nu c^\nu \quad \text{and}$$

$$\alpha, \beta \in \mathbb{N} : \sum_\nu a_\nu \eta_0^*(c^\nu) + \alpha = \sum_\nu b_\nu \eta_0^*(c^\nu) + \beta.$$

Defining $c := \sum_\nu a_\nu c^\nu = \sum_\nu b_\nu c^\nu$, we can lift them to the following elements of $A(\bar{S})[Z_1, \ldots, Z_w]$ (described by using liftings to $\mathbb{C}[Z_1, \ldots, Z_w, t_1, \ldots, t_N]$):

$$F_{(a,b,\alpha,\beta)}(\underline{Z}, t) := f_{(a,b,\alpha,\beta)}(\underline{Z}, t_1)$$
$$- \underline{Z}^{[c,\underline{\eta}^*(c)]} \cdot \left(\underline{t}^{\alpha e_1 + \sum_\nu a_\nu \underline{\eta}^*(c^\nu)} - \underline{t}^{\beta e_1 + \sum_\nu b_\nu \underline{\eta}^*(c^\nu)} \right) \cdot \underline{t}^{-\underline{\eta}^*(c)}.$$

Remark. (1) The symbol $\underline{Z}^{[c,\underline{\eta}^*(c)]}$ means $\prod_{v=1}^w Z_\nu^{p_\nu}$ with natural numbers $p_\nu \in \mathbb{N}$ such that $[c, \underline{\eta}^*(c)] = \sum_\nu p_\nu [c^\nu, \underline{\eta}^*(c^\nu)]$, or equivalently $[c, \eta_0^*(c)] = \sum_\nu p_\nu [c^\nu, \eta_0^*(c^\nu)]$. This condition does not determine the coefficients p_ν uniquely – choose one of the possibilities. Choosing other coefficients q_ν with the same property yields

$$Z_1^{p_1} \cdot \ldots \cdot Z_w^{p_w} - Z_1^{q_1} \cdot \ldots \cdot Z_w^{q_w} = F_{(p,q,0,0)}(\underline{Z}, t) = f_{(p,q,0,0)}(\underline{Z}, t).$$

(2) By part (iii) of Lemma 3.5, we have $\sum_\nu a_\nu \underline{\eta}^*(c^\nu), \sum_\nu b_\nu \underline{\eta}^*(c^\nu) \geq \underline{\eta}^*(c)$ in the sense of (3.3). In particular, representatives of the $\underline{\eta}^*$'s can be chosen such that all t-exponents occurring in monomials of F are non-negative integers, i.e. F indeed defines an element of $A(\bar{S})[Z_1, \ldots, Z_w]$.

4.4. Lemma. *The polynomials $F_{(a,b,\alpha,\beta)}$ generate* $\mathrm{Ker}\left(A(\bar{S})[\underline{Z}] \to A(X') \right)$, *i.e. they can be used as equations for $X \subseteq \mathbb{C}^w \times \bar{S}$.*

Proof. Mapping F into $A(X') = \oplus_{[c,\eta]} \mathbb{C} x^{[c,\eta]}$ ($[c,\eta]$ runs through all elements of $\Gamma \cap (\mathbb{L}^* \times V_{\mathbb{Z}}^*)$; $Z_\nu \mapsto x^{[c^\nu, \underline{\eta}^*(c^\nu)]}$, $t_i \mapsto x^{[0, e_i]}$) yields

$$F_{(a,b,\alpha,\beta)} = \left(t_1^\alpha \prod_\nu Z_\nu^{a_\nu} - \underline{Z}^{[c,\underline{\eta}^*(c)]} t^{\alpha e_1 + \sum_\nu a_\nu \underline{\eta}^*(c^\nu) - \underline{\eta}^*(c)} \right)$$
$$- \left(t_1^\beta \prod_\nu Z_\nu^{b_\nu} - \underline{Z}^{[c,\underline{\eta}^*(c)]} t^{\beta e_1 + \sum_\nu b_\nu \underline{\eta}^*(c^\nu) - \underline{\eta}^*(c)} \right)$$

$$\mapsto \left(x^{\alpha[0,e_1]+\sum_\nu a_\nu[c^\nu,\underline{\eta}^*(c^\nu)]} - x^{[c,\underline{\eta}^*(c)]+\alpha[0,e_1]+\sum_\nu a_\nu[0,\underline{\eta}^*(c^\nu)]-[0,\underline{\eta}^*(c)]}\right)$$

$$- \left(x^{\beta[0,e_1]+\sum_\nu b_\nu[c^\nu,\underline{\eta}^*(c^\nu)]} - x^{[c,\underline{\eta}^*(c)]+\beta[0,e_1]+\sum_\nu b_\nu[0,\underline{\eta}^*(c^\nu)]-[0,\underline{\eta}^*(c)]}\right)$$

$$= 0 - 0 = 0.$$

On the other hand, $\mathrm{Ker}\left(A(\bar S)[\underline Z] \to A(X')\right)$ is obviously generated by the binomials $\underline t^{\underline\eta} Z_1^{a_1} \cdots Z_w^{a_w} - \underline t^{\underline\mu} Z_1^{b_1} \cdots Z_w^{b_w}$ such that

$$\sum_\nu a_\nu\left[c^\nu, \underline\eta^*(c^\nu)\right] + [0, \underline\eta] = \sum_\nu b_\nu\left[c^\nu, \underline\eta^*(c^\nu)\right] + [0, \underline\mu],$$

$$\text{i.e. } c := \sum_\nu a_\nu c^\nu = \sum_\nu b_\nu c^\nu \text{ and}$$

$$\sum_\nu a_\nu \underline\eta^*(c^\nu) + \underline\eta = \sum_\nu b_\nu \underline\eta^*(c^\nu) + \underline\mu.$$

However,

$$\underline t^{\underline\eta} \underline Z^a - \underline t^{\underline\mu} \underline Z^b = \underline t^{\underline\eta} \cdot \left(\prod_\nu Z_\nu^{a_\nu} - \underline Z^{[c,\underline\eta^*(c)]} \underline t^{\sum_\nu a_\nu \underline\eta^*(c^\nu) - \underline\eta^*(c)}\right)$$

$$- \underline t^{\underline\mu} \cdot \left(\prod_\nu Z_\nu^{b_\nu} - \underline Z^{[c,\underline\eta^*(c)]} \underline t^{\sum_\nu b_\nu \underline\eta^*(c^\nu) - \underline\eta^*(c)}\right)$$

$$= \underline t^{\underline\eta} \cdot F_{(a,p,0,\alpha)} - \underline t^{\underline\mu} \cdot F_{(b,p,0,\beta)}$$

with $p \in \mathbb{N}^w$ such that $\sum_\nu p_\nu\left[c^\nu, \underline\eta^*(c^\nu)\right] = \left[c, \underline\eta^*(c)\right]$, $\alpha = \sum_\nu a_\nu \eta_0^*(c^\nu) - \eta_0^*(c)$, and $\beta = \sum_\nu b_\nu \eta_0^*(c^\nu) - \eta_0^*(c)$. $\blacksquare$

Using exponents $\eta, \mu \in \mathbb{Z}^N$ (instead of $\mathbb{N}^N$), the binomials $\underline t^{\underline\eta} \underline Z^a - \underline t^{\underline\mu} \underline Z^b$ generate the kernel of the map

$$A(S)[\underline Z] = A(\bar S)[\underline Z] \otimes_{A(\bar S)} A(S) \twoheadrightarrow A(\bar X) \otimes_{A(\bar S)} A(S) \twoheadrightarrow A(X').$$

Since

$$\underline Z^a \otimes \underline t^{\underline\eta} - \underline Z^b \otimes \underline t^{\underline\mu} = \underline Z^{[c,\underline\eta^*(c)]} \otimes \left(\underline t^{\sum_\nu a_\nu \underline\eta^*(c^\nu) - \underline\eta^*(c) + \underline\eta} - \underline t^{\sum_\nu b_\nu \underline\eta^*(c^\nu) - \underline\eta^*(c) + \underline\mu}\right) = 0$$

in $A(\bar X) \otimes_{A(\bar S)} A(S)$, this implies that the surjection $A(\bar X) \otimes_{A(\bar S)} A(S) \twoheadrightarrow A(X')$ is injective, too. In particular, part (2) of our theorem is proved.

We are going to use the following well known criterion of flatness:

4.5. Theorem. *Let $\tilde{\pi} : \tilde{X} \hookrightarrow \mathbb{C}^{w+1} \times \bar{\mathcal{M}} \twoheadrightarrow \bar{\mathcal{M}}$ be a map with special fiber $Y = \tilde{\pi}^{-1}(0)$; in particular, $Y \subseteq \mathbb{C}^{w+1}$ is defined by the restrictions to $0 \in \bar{\mathcal{M}}$ of the equations defining $\tilde{X} \subseteq \mathbb{C}^{w+1} \times \bar{\mathcal{M}}$. Then $\tilde{\pi}$ is flat, if and only if each linear relation between the (restricted) equations for Y lifts to some linear relation between the original equations for $\tilde{X}$.*

Proof. According to [7, (20.C), Theorem 49], flatness of $\tilde{\pi}$ in $0 \in \bar{\mathcal{M}}$ is equivalent to the vanishing of $\mathrm{Tor}_1^{\mathcal{O}_{\bar{\mathcal{M}},0}} \big((\tilde{\pi}_* \mathcal{O}_{\tilde{X}})_0, \mathbb{C} \big)$ where $\mathbb{C}$ becomes an $\mathcal{O}_{\bar{\mathcal{M}},0}$-module via evaluating in $0 \in \bar{\mathcal{M}}$.

Using the embedding $\tilde{X} \hookrightarrow \mathbb{C}^{w+1} \times \bar{\mathcal{M}}$ (together with the defining equations and linear relations between them) we obtain an $\mathcal{O}_{\bar{\mathcal{M}},0}[Z_0, \dots, Z_w]$-free (hence $\mathcal{O}_{\bar{\mathcal{M}},0}$-free) resolution of $(\tilde{\pi}_* \mathcal{O}_{\tilde{X}})_0$ up to the second term. Now, the condition that relations between Y-equations lift to those between $\tilde{X}$-equations is equivalent to the fact that our (partial) resolution remains exact under $\otimes_{\mathcal{O}_{\bar{\mathcal{M}},0}} \mathbb{C}$. ∎

For our special situation take $\tilde{X} := \bar{X} \times_{\bar{S}} \mathcal{M}$ (and $\bar{\mathcal{M}} := \mathcal{M}$, $Y := Y$); in (4.3) we have seen how the equations defining $Y \hookrightarrow \mathbb{C}^w \times \mathbb{C}$ can be lifted to those defining $\bar{X} \hookrightarrow \mathbb{C}^w \times \bar{S}$, hence $\bar{X} \times_{\bar{S}} \mathcal{M} \hookrightarrow \mathbb{C}^w \times \mathcal{M} \xrightarrow{\sim} \mathbb{C}^w \times \mathbb{C} \times \bar{\mathcal{M}}$.

In particular, to show (3) of Theorem 4.1, we only have to take the linear relations between the $f_{(a,b,\alpha,\beta)}$'s and lift them to relations between the $F_{(a,b,\alpha,\beta)}$'s.

4.6. According to the special shape of our generator set E, there are three types of relations between the $f_{(a,b,\alpha,\beta)}$'s:

(i) $f_{(a,r,\alpha,\gamma)} + f_{(r,b,\gamma,\beta)} = f_{(a,b,\alpha,\beta)}$ with

$$\sum_\nu a_\nu c^\nu = \sum_\nu r_\nu c^\nu = \sum_\nu b_\nu c^\nu \quad \text{and}$$

$$\sum_\nu a_\nu \eta_0^*(c^\nu) + \alpha = \sum_\nu r_\nu \eta_0^*(c^\nu) + \gamma = \sum_\nu b_\nu \eta_0^*(c^\nu) + \beta.$$

For this relation, the same equation between the F's is true.

(ii) $t \cdot f_{(a,b,\alpha,\beta)} = f_{(a,b,\alpha+1,\beta+1)}$ lifts to $t_1 \cdot F_{(a,b,\alpha,\beta)} = F_{(a,b,\alpha+1,\beta+1)}$.

(iii) $z^r \cdot f_{(a,b,\alpha,\beta)} = f_{(a+r,b+r,\alpha,\beta)}$.

With $c := \sum_\nu a_\nu c^\nu = \sum_\nu b_\nu c^\nu$, $\tilde{c} := c + \sum_\nu r_\nu c^\nu$ we obtain for (iii)

$$\underline{Z}^r \cdot F_{(a,b,\alpha,\beta)} - F_{(a+r,b+r,\alpha,\beta)}$$

$$= \underline{Z}^{[\tilde{c},\underline{\eta}^*(\tilde{c})]} \cdot \left(\underline{t}^{\alpha e_1 + \sum_\nu a_\nu \underline{\eta}^*(c^\nu) + \sum_\nu r_\nu \underline{\eta}^*(c^\nu)} \right.$$

$$\left. - \underline{t}^{\beta e_1 + \sum_\nu b_\nu \underline{\eta}^*(c^\nu) + \sum_\nu r_\nu \underline{\eta}^*(c^\nu)} \right) \cdot \underline{t}^{-\underline{\eta}^*(\tilde{c})}$$

$$- \underline{Z}^{[c,\underline{\eta}^*(c)]} \underline{Z}^r \cdot \left(\underline{t}^{\alpha e_1 + \sum_\nu a_\nu \underline{\eta}^*(c^\nu)} - \underline{t}^{\beta e_1 + \sum_\nu b_\nu \underline{\eta}^*(c^\nu)} \right) \cdot \underline{t}^{-\underline{\eta}^*(c)}$$

$$= \left(\underline{t}^{\alpha e_1 + \sum_\nu a_\nu \underline{\eta}^*(c^\nu) - \underline{\eta}^*(c)} - \underline{t}^{\beta e_1 + \sum_\nu b_\nu \underline{\eta}^*(c^\nu) - \underline{\eta}^*(c)} \right)$$

$$\cdot \left(\underline{t}^{\underline{\eta}^*(c) + \sum_\nu r_\nu \underline{\eta}^*(c^\nu) - \underline{\eta}^*(\tilde{c})} \underline{Z}^{[\tilde{c},\underline{\eta}^*(\tilde{c})]} - \underline{Z}^{[c,\underline{\eta}^*(c)]} \underline{Z}^r \right).$$

Now, the inequalities

$$\sum_\nu a_\nu \underline{\eta}^*(c^\nu), \sum_\nu b_\nu \underline{\eta}^*(c^\nu) \geq \underline{\eta}^*(c) \quad \text{and} \quad \underline{\eta}^*(c) + \sum_\nu r_\nu \underline{\eta}^*(c^\nu) - \underline{\eta}^*(\tilde{c}) \geq 0$$

imply that the first factor is contained in the ideal defining $0 \in \bar{\mathcal{M}}$ and that the second factor is an equation of $\bar{X} \subseteq \mathbb{C}^w \times \bar{S}$ (called $F_{(q,p+r,\xi,0)}$ in (6.7)). In particular, we have found a lift for the third relation, too. The proof of Theorem 4.1 is complete.

5. THE KODAIRA–SPENCER MAP

5.1. To each vertex $v^j \in Q$ we associate the subset

$$E_j := E_{v^j} := \left\{ [c, \eta_0^*(c)] \in E \mid \langle v^j, c \rangle + \eta_0^*(c) < 1 \right\}.$$

Additionally define the sets

$$E_0 := \bigcup_j E_j, \quad E_{ij} := E_i \cap E_j.$$

Let $r = [c, \eta_0^*(c)] \in E$ be given. Then we have

$$\langle v^j, c \rangle + \eta_0^*(c) = \left\langle (v^j, 1) [c, \eta_0^*(c)] \right\rangle = \frac{\langle a^j, r \rangle}{\langle a^j, R \rangle}$$

and we obtain the following alternative description of E_j:

$$E_j := \left\{ r \in E \mid \langle a^j, r \rangle < \langle a^j, R \rangle \right\}.$$

In other words: The primitive generators a^j of σ define the facets of the dual cone $\sigma^\vee$, i.e. they define hyperplanes such that $\sigma^\vee$ is the intersection of the halfspaces above these hyperplanes:

$$\sigma^\vee = \bigcup_j \left\{ m \in M_{\mathbb{R}} \mid \langle a^j, m \rangle \geq 0 \right\} \subseteq M_{\mathbb{R}}.$$

Now E^j contains those elements of E that are closer to the facet of σ defined by a^j than R.

We also get the following alternative description of $\underline{\eta}^*(c)$ compared with its definition in (3.4):

5.2. Lemma. *Assume that $\left[c, \eta_0^*(c) \right]$ is contained in E_j. Then*

$$\underline{\eta}^*(c) = \langle v_\bullet^j, -c \rangle + \left(\langle v^j, c \rangle + \eta_0^*(c) \right) \cdot e[v^j]$$

where $v_\bullet^j$ denotes the map assigning $\underline{t} \in V(Q)$ the vertex $v_{\underline{t}}^j$ of the (generalized) Minkowski summand $Q_{\underline{t}}$.

Proof. If $v^j \in \mathbb{L}$, then the condition $\langle v^j, c \rangle + \eta_0^*(c) < 1$ is equivalent to $\langle v^j, -c \rangle = \eta_0(c) = \eta_0^*(c)$. Hence, the second summand in our formula vanishes, and we are done.

On the other hand, if $v^j \notin \mathbb{L}$, then there is not any lattice point contained in the strip $\langle v^j, c \rangle \geq \langle \bullet, c \rangle > \langle v(c), c \rangle$. In particular, every edge on the path from v^j to $v(c)$ (decreasing the c-value at each step) belongs to the "component" induced by v^j, cf. (2.2). Now, our formula follows from the definition of $\underline{\eta}^*(c)$. ∎

5.3. Denoting by $L(\bullet)$ the abelian group of $\mathbb{Z}$-linear relations of the argument, we consider the bilinear map

$$\Phi : {}^{V_{\mathbb{Z}}}\!/_{\mathbb{Z} \cdot \underline{1}} \times L(\cup_j E_j) \longrightarrow \mathbb{Z}$$

$$\underline{t} \quad , \quad q \quad \mapsto \quad \sum_{v,i} t_i q_v \eta_i^*(c^v).$$

It is correctly defined, and we obtain $\Phi(\underline{t}, q) = 0$ for $q \in L(E_j)$. Indeed,

$$\Phi(\underline{t}, q) = \sum_\nu q_\nu \langle \underline{t}, \underline{\eta}^*(c^\nu) \rangle$$

$$= \sum_\nu q_\nu \cdot \left(\langle v_{\underline{t}}^j, -c^\nu \rangle + \left(\langle v^j, c^\nu \rangle + \eta_0^*(c^\nu) \right) \cdot t_{vj} \right)$$

$$= \left\langle v_{\underline{t}}^j, -\sum_\nu q_\nu c^\nu \right\rangle + \left(\sum_\nu q_\nu \eta_0^*(c^\nu) - \left\langle v^j, \sum_\nu q_\nu c^\nu \right\rangle \right) \cdot t_{vj} = 0.$$

5.4. Theorem. *The Kodaira–Spencer map of the family* $\bar{X} \times_{\bar{S}} \mathcal{M} \to \bar{\mathcal{M}}$ *of (4.1) equals the map*

$$T_0 \bar{\mathcal{M}} = {}^{V_{\mathbb{C}}}\!/_{\mathbb{C} \cdot \underline{1}} \longrightarrow \left(L_{\mathbb{C}}(E \cap \partial\sigma^\vee) \Big/ \sum_j L_{\mathbb{C}}(E_j) \right)^* = T_Y^1(-R)$$

induced by the previous pairing. Moreover, this map is an isomorphism.

Proof. Using the same symbol $\mathcal{J}$ for the ideal $\mathcal{J} \subseteq \mathbb{C}[t_1, \ldots, t_N]$ as well as for the intersection $\mathcal{J} \cap \mathbb{C}[t_i - t_j \mid 1 \leq i, j \leq N]$, cf. (2.4), our family corresponds to the flat $\mathbb{C}[t_i - t_j]/\mathcal{J}$-module $\mathbb{C}[\underline{Z}, \underline{t}]/(\mathcal{J}, F_\bullet(\underline{Z}, \underline{t}))$. Now, we fix a non-trivial tangent vector $\underline{t}^0 \in V_{\mathbb{C}}$. Via $t_i \mapsto t + t_i^0 \varepsilon$, it induces the infinitesimal family given by the flat $\mathbb{C}[\varepsilon]/_{\varepsilon^2}$-module

$$A_{\underline{t}^0} := {}^{\mathbb{C}[\underline{z}, t, \varepsilon]}\!\Big/_{\left(\varepsilon^2, F_\bullet(\underline{z}, t + \underline{t}^0 \varepsilon) \right)}.$$

To obtain the associated $A(Y)$-linear map $I/I^2 \to A(Y)$ with $I := \left(f_\bullet(\underline{z}, t) \right)$ denoting the ideal of Y in $\mathbb{C}^{w+1}$, we have to compute the images of $f_\bullet(\underline{z}, t)$ in $\varepsilon A(Y) \subseteq A_{\underline{t}^0}$ and divide them by ε: Using the notation of (4.3), in $A_{\underline{t}^0}$ it holds true that

$$0 = F_{(a,b,\alpha,\beta)}(\underline{z}, t + \underline{t}^0 \varepsilon)$$

$$= f_{(a,b,\alpha,\beta)}(\underline{z}, t + t_1^0 \varepsilon)$$

$$- \underline{z}^{[c, \underline{\eta}^*(c)]} \cdot \left((t + \underline{t}^0 \varepsilon)^{\alpha e_1 + \sum_\nu a_\nu \underline{\eta}^*(c^\nu) - \underline{\eta}^*(c)} \right.$$

$$\left. - (t + \underline{t}^0 \varepsilon)^{\beta e_1 + \sum_\nu b_\nu \underline{\eta}^*(c^\nu) - \underline{\eta}^*(c)} \right).$$

The relation $\varepsilon^2 = 0$ yields

$$f_{(a,b,\alpha,\beta)}(\underline{z}, t + t_1^0\varepsilon) = f_{(a,b,\alpha,\beta)}(\underline{z}, t) + \varepsilon \cdot (\alpha t^{\alpha-1} t_1^0 \underline{z}^a - \beta t^{\beta-1} t_1^0 \underline{z}^b),$$

and similarly we can expand the other terms. Eventually, we obtain

$$f_{(a,b,\alpha,\beta)}(\underline{z}, t) = -\varepsilon t_1^0 \big(\alpha t^{\alpha-1} \underline{z}^a - \beta t^{\beta-1} \underline{z}^b \big) + \varepsilon \underline{z}^{[c, \underline{\eta}^*(c)]} t^{\alpha + \sum_\nu a_\nu \eta_0^*(c^\nu) - \eta_0^*(c) - 1}$$

$$\cdot \left[t_1^0(\alpha - \beta) + \sum_i t_i^0 \left(\sum_\nu (a_\nu - b_\nu) \eta_i^*(c^\nu) \right) \right]$$

$$= \varepsilon \cdot x^{\sum_\nu a_\nu [c^\nu, \eta_0^*(c^\nu)] + [0, \alpha - 1]} \cdot \left(\sum_i t_i^0 \left(\sum_\nu (a_\nu - b_\nu) \eta_i^*(c^\nu) \right) \right).$$

Note that, in $\varepsilon A(Y)$, we were able to replace the variables t and z_ν by $x^{[0,1]}$ and $x^{[c^\nu, \eta_0^*(c^\nu)]}$, respectively.

On the other hand, the explicit description of $T_Y^1(-R)$ as $L_{\mathbb{C}}(E \cap \partial\sigma^\vee)/\sum_j L_{\mathbb{C}}(E_j)$ was given in [3, Theorem (3.4)]. It even says that the map $I/I^2 \to A(Y)$ with $\big(t^\alpha \underline{z}^a - t^\beta \underline{z}^b \big) \mapsto \big(\sum_{i,v} t_i^0 (a_\nu - b_\nu) \eta_i^*(c^\nu) \big) \cdot x^{\sum_\nu a_\nu [c^\nu, \eta_0^*(c^\nu)] + [0, \alpha - 1]}$ corresponds to $q \mapsto \sum_{i,v} t_i^0 q_\nu \eta_i^*(c^\nu) = \Phi(\underline{t}^0, q)$.

In [2] it was already proven that there is an isomorphism $\Psi : T_Y^1(-R) \to V/\underline{1}$ if Y is smooth in codimension two. Now we want to show that the composition $\Psi \circ \Phi$ yields the identity on $V/\underline{1}$. Thus we have to take a closer look at the construction of Ψ.

Let us switch from the notion of $L(E_j)$ to the notion of span (E_j). The advantage lies in that span (E_j) is much easier to describe than $L(E_j)$:

Remark.

$$\text{span}_{\mathbb{R}} E_j = \begin{cases} 0 & \langle a^j, R \rangle = 0 \\ (a^j)^\perp & \langle a^j, R \rangle = 1 \\ M_{\mathbb{R}} & \langle a^j, R \rangle \geq 2. \end{cases}$$

To change between the two notions, let $q \in L(E_0)$ be given. Then decompose $q = \sum_j q^j$ with $q^j \in \mathbb{Z}^{E_j}$. Define $w^j := \sum_\nu q_\nu^j r^\nu$, $r^\nu \in E$ so that the vector $(w^1, \ldots, w^M)$ is contained in $\ker(\oplus_j \text{span} E_j \to M)$.

Let $\tau < \sigma$ be a face. We define the following set

$$E_\tau := \bigcap_{a^j \in \tau} E_j$$

and obtain a complex span $(E)_\bullet$ with

$$\text{span}_{-k}(E) = \bigoplus_{\substack{\tau \text{ a face of } \sigma \\ \dim \tau = k}} E_\tau$$

and the obvious differentials. Now the dual complex yields the following description of $T_Y^1(-R)$:

5.5. Theorem ([3] (6.1)). *The homogenous piece of T_Y^1 in degree $-R$ is given by*

$$T_Y^1(-R) = H^1\big(\text{span } (E)_\bullet^* \otimes_{\mathbb{Z}} \mathbb{C}\big),$$

i.e. $T_Y^1(-R)$ equals the complexification of the cohomology of the subsequence

$$N_{\mathbb{R}} \to \bigoplus_j (\text{span}_{\mathbb{R}} E_j)^* \to \bigoplus_{\langle a^j, a^j \rangle < \sigma} (\text{span}_{\mathbb{R}} E_{ij})^*$$

of the dual complex to span $(E)_\bullet$.

Given an element $b \in T_Y^1(-R)$, we can build an element $\underline{t} \in V/\underline{1}$. First we will show how to build $\underline{t} \in V$ from a given $b \in \bigoplus_j (\text{span}_{\mathbb{R}} E_j)^*$. Then we will show that the action of $N_{\mathbb{R}}$ equals the action of $\mathbb{R} \cdot \underline{1}$ on V.

Step 1: By the above remark, we can represent $b \in \bigoplus_j (\text{span}_{\mathbb{R}} E_j)^*$ by a family of

- $b^j \in N_{\mathbb{R}}$ if $\langle a^j, R \rangle \geq 2$ and

- $b^j \in N_{\mathbb{R}}/\mathbb{R} \cdot a^j$ if $\langle a^j, R \rangle = 1$.

We will only consider the b^j for $\langle a^j, R \rangle \geq 1$, otherwise span E_j will be zero. This corresponds to the fact that $v^j = a^j/\langle a^j, R \rangle$ is not a vertex of Q.

Dividing by the image of $N_{\mathbb{R}}$ means shifting the family by a common vector $c \in N_{\mathbb{R}}$. The condition of our family $\{b^j\}$ mapping onto 0 means that b^j and b^k have to be equal on $\text{span}_{\mathbb{R}} E_{ik}$ for each compact edge $\overline{v^j, v^k} < Q$. Since

$$(a^j, a^k)^{\perp} \subseteq \text{span}_{\mathbb{R}} E_{jk} \subseteq \text{span}_{\mathbb{R}} E_j \cap \text{span}_{\mathbb{R}} E_k$$

we obtain $b^j - b^k \in \mathbb{R}a^j + \mathbb{R}a^k$.

Step 2: Let us introduce new coordinates

$$\bar{b}^j := b^j - \langle b^j, R \rangle v^j \in R^{\perp}.$$

The condition $b^j - b^k \in \mathbb{R}a^j + \mathbb{R}a^j$ changes into the condition $\overline{b}^j - \overline{b}^k \in \mathbb{R}v^j + \mathbb{R}v^k$. We assume $\langle a^j, R \rangle, \langle a^k, R \rangle \neq 0$, i.e. $a^j, a^k \notin R^\perp$. On the other hand, we know $\overline{b}^j, \overline{b}^k \in R^\perp$, hence $\overline{b}^j - \overline{b}^k \in R^\perp$. This yields

$$\overline{b}^j - \overline{b}^k \in (\mathbb{R}v^j + \mathbb{R}v^k) \cap R^\perp = \mathbb{R}(v^j - v^k).$$

Thus we obtain

$$\overline{b}^j - \overline{b}^k = t_{jk} \cdot (v^j - v^k).$$

Now collect these t_{ij} for each compact edge $\overline{v^j, v^k} < Q$. Together they yield an element $\underline{t}_b \in \mathbb{R}^N$.

Step 3: Consider shifting the family by a common vector $c \in N_\mathbb{R}$, i.e. $b^{j'} := b^j + c$. We obtain

$$t'_{jk}(v^j - v^k) = \overline{b}^{j'} - \overline{b}^{k'}$$

$$= \left(b^j + c - \langle b^j, R \rangle v^j - \langle c, R \rangle v^j \right) - \left(b^k + c - \langle b^k, R \rangle v^k - \langle c, R \rangle v^k \right)$$

$$= \overline{b}^j - \overline{b}^k - \langle c, R \rangle \cdot (v^j - v^k) = \left(t_{jk} - \langle c, R \rangle \right) \cdot (v^j - v^k).$$

Hence, the action of $c \in N_\mathbb{R}$ comes down to an action of $\langle c, R \rangle$ only, and we obtain $\underline{t}_b \in \mathbb{R}^N / \underline{1}$.

Step 4: It is rather easy to see that $\underline{t}_b$ satisfies the 2-face equations of V. In [2] (2.7) it is proven that $\underline{t}_b$ also satisfies the equations given by non lattice vertices of Q since Y is smooth in codimension two. We obtain the following Corollary:

Corollary ([2] (2.6)). _If Y is smooth in codimension two_

$$\Psi : T_Y^1(-R) \longrightarrow V_\mathbb{C}/\underline{1}$$

$$b \longmapsto \underline{t}_b$$

is an isomorphism.

Step 5: Let us now combine Φ with this isomorphism. Denoting by t_j the coordinate of $\underline{t}$ corresponding to the component arising from a non-lattice vertex v^j and defining $g^j := -\sum_\nu q_\nu^j [c^\nu, \eta_0^*(c^\nu)]$, we obtain

$$\Phi(\underline{t}, q) = \sum_j \Phi(\underline{t}, q^j) = \sum_{j,\nu} q_\nu^j \langle \underline{t}, \underline{\eta}^*(c^\nu) \rangle$$

$$= \sum_{j,\nu} q^j_\nu \langle v^j_{\underline{t}}, -c^\nu \rangle + \sum_{j,\nu} q^j_\nu \left[\eta^*_0(c^\nu) + \langle v^j, -c^\nu \rangle \right] \cdot t_j$$

$$= \sum_{j,\nu} \left\langle (v^j_{\underline{t}}, 0), g^j \right\rangle - \left\langle (v^j, 1), g^j \right\rangle \cdot t_j$$

i.e. Φ assigns to $\underline{t}$ exactly the vertices of the corresponding Minkowski summand $Q_{\underline{t}}$. Thus, applying Ψ to $\Phi(\underline{t})$ yields the identity. ∎

6. THE OBSTRUCTION MAP

Now we can approach the main goal of this paper:

6.1. Theorem. *The family of Theorem 4.1 with base space $\bar{\mathcal{M}}$ is the versal deformation of Y of degree $-R$.*

By [4] we know that a deformation is versal if the Kodaira–Spencer-map is an isomorphism and the Obstruction map is injective. In Section 5 we proved the first condition for the degree-R-part of T^1_Y. The following section will prove the second condition, i.e. the injectivity of the obstruction map.

6.2. Dealing with obstructions in the deformation theory of Y involves the $A(Y)$-module T^2_Y. Usually it is defined in the following way: Let

$$m := \left\{ ([a,\alpha], [b,\beta]) \in \mathbb{N}^{w+1} \times \mathbb{N}^{w+1} \mid \sum_\nu a_\nu c^\nu = \sum_\nu b_\nu c^\nu \text{ and} \right.$$

$$\left. \sum_\nu a_\nu \eta^*_0(c^\nu) + \alpha = \sum_\nu b_\nu \eta^*_0(c^\nu) + \beta \right\}$$

denote the set parametrizing the equations $f_{(a,b,\alpha,\beta)}$ generating the ideal $I \subseteq \mathbb{C}[\underline{z}, t]$. Then

$$\mathcal{R} := \ker \left(\varphi : \mathbb{C}[\underline{z}, t]^m \twoheadrightarrow I \right)$$

is the module of linear relations between these equations; it contains the submodule $\mathcal{R}_0$ of the so-called Koszul relations, i.e. those of the form $f_j \cdot e_i - f_i \cdot e_j$ where f_i, f_j are generators of I and e_i, e_j are their corresponding preimages under φ.

Definition. $T_Y^2 := \dfrac{\operatorname{Hom}\left(\left.\mathcal{R}\middle/\mathcal{R}_0\right., A\right)}{\operatorname{Hom}\left(\mathbb{C}[\underline{z},t]^m, A\right)}.$

Recall that $R = [\underline{0}, 1]$. To obtain information about T^2 not only in degree $-R$ but also in its multiples $k \cdot R$, $k \geq 2$, we define, analogously to E_j, the following sets:

$$E_j^k := \left\{ \left[c^\nu, \eta_0^*(c^\nu)\right] \mid \langle a^j, c^\nu \rangle + \eta_0^*(c^\nu) < k \right\} \cup \{R\} \subseteq \sigma^\vee \cap M.$$

For the following theorem it is very important that σ has smooth two-dimensional faces, i.e. that Y is smooth in codimension two:

6.3. Theorem [3]. *The vector space T_Y^2 is M-graded, and in degree $-kR$ it equals*

$$T_Y^2(-kR) = \left(\frac{\ker\left(\oplus_j L_{\mathbb{C}}(E_j^k) \to L_{\mathbb{C}}(E) \right)}{\operatorname{im}\left(\oplus_{\langle v^i, v^j \rangle < Q} L_{\mathbb{C}}(E_i^k \cap E_j^k) \to \oplus_i L_{\mathbb{C}}(E_i^k) \right)} \right)^*.$$

6.4. In this section we build up the so-called obstruction map. It detects all infinitesimal extensions of our family over $\bar{\mathcal{M}}$ to a flat family over some larger base space. By $\mathcal{J}$ let us denote

$$\mathcal{J} := (g_{\underline{d},k}(\underline{t} - t_1 \mid \underline{d} \in V^\perp \cap \mathbb{Z}^N, \ k \geq 1) \subseteq \mathbb{C}[t_i - t_j]$$

the homogenous ideal of the base space $\bar{\mathcal{M}}$. Let $\mathcal{J}_1$ denote the degree 1 part of $\mathcal{J}$. We define the subideal $\tilde{\mathcal{J}} \subseteq \mathcal{J}$ by:

$$\tilde{\mathcal{J}} = (t_i - t_j)_{i,j} \cdot \mathcal{J} + \mathcal{J}_1 \cdot \mathbb{C}[t_i - t_j] \subseteq \mathbb{C}[t_i - t_j \mid 1 \leq i, j \leq N].$$

Then $W := \mathcal{J}/\tilde{\mathcal{J}}$ is a finite-dimensional, $\mathbb{Z}$-graded vector space. It comes as the kernel in the exact sequence

$$0 \to W \to \left.\mathbb{C}[t_i - t_j]\middle/\tilde{\mathcal{J}}\right. \to \left.\mathbb{C}[t_i - t_j]\middle/\mathcal{J}\right. \to 0.$$

Identifying t with t_1 and $\underline{z}$ with $\underline{Z}$, the tensor product with $\mathbb{C}[\underline{z},t]$ over $\mathbb{C}$ yields the important exact sequence

$$0 \to W \otimes \mathbb{C}[\underline{z},t] \to \left.\mathbb{C}[\underline{Z},\underline{t}]\middle/\tilde{\mathcal{J}} \cdot \mathbb{C}[\underline{Z},\underline{t}]\right. \to \left.\mathbb{C}[\underline{Z},\underline{t}]\middle/\mathcal{J} \cdot \mathbb{C}[\underline{Z},\underline{t}]\right. \to 0.$$

Now, let s be any relation with coefficients in $\mathbb{C}[\underline{z},t]$ between the equations $f_{(a,b,\alpha,\beta)}$, i.e.

$$\sum s_{(a,b,\alpha,\beta)} f_{(a,b,\alpha,\beta)} = 0 \quad \text{in} \quad \mathbb{C}[\underline{z},t].$$

By flatness of our family, cf. (4.6), the components of s can be lifted to $\mathbb{C}[\underline{Z}, \underline{t}]$ obtaining an $\tilde{s}$, such that

$$\sum \tilde{s}_{(a,b,\alpha,\beta)} F_{(a,b,\alpha,\beta)} = 0 \text{ in } \mathbb{C}[\underline{Z}, \underline{t}] \Big/ \mathcal{J} \cdot \mathbb{C}[\underline{Z}, \underline{t}].$$

In particular, each relation $s \in \mathcal{R}$ induces some element

$$\lambda(s) := \sum \tilde{s} F \in W \otimes \mathbb{C}[\underline{z}, \underline{t}] \subseteq \mathbb{C}[\underline{Z}, \underline{t}] \Big/ \tilde{\mathcal{J}} \cdot \mathbb{C}[\underline{Z}, \underline{t}].$$

which does not depend on choices after the additional projection to $W \otimes_{\mathbb{C}} A(Y)$. This procedure describes a certain element $\lambda \in T_Y^2 \otimes_{\mathbb{C}} W = \text{Hom}\,(W^*, T_Y^2)$ called the obstruction map.

The remaining part of Sect. 6 contains the proof of the following theorem:

6.5. Theorem. *The obstruction map* $\lambda : W^* \to T_Y^2$ *is injective.*

6.6. We have to improve our notation of Sects. 3 and 4. Since $\mathcal{M} \subseteq \bar{S} \subseteq \mathbb{C}^N$, we were able to use the toric equations, cf. (2.5) during computations modulo $\mathcal{J}$. In particular, the exponents $\underline{\eta} \in V^*$ of $\underline{t}$ needed only to be known modulo $V^\perp$; it was enough to define $\underline{\eta}^*(\bullet)$ as elements of $V_{\mathbb{Z}}^*$.

However, to compute the obstruction map, we have to deal with the smaller ideal $\tilde{\mathcal{J}} \subseteq \mathcal{J}$. Let us start with refining the definitions of (3.4):

(i) For each vertex $v \in Q$, we choose certain paths through the 1-skeleton of Q:

- $\underline{\lambda}(v) :=$ path from $0 \in Q$ to $v \in Q$.

- $\underline{\mu}^\nu(v) :=$ path from $v \in Q$ to $v(c^\nu) \in Q$ such that $\mu_i^\nu(v)\langle d^i, c^\nu \rangle \leq 0$ for each $i = 1, \ldots, N$.

- $\underline{\lambda}^\nu(v) := \underline{\lambda}(v) + \underline{\mu}^\nu(v)$ is then a path from $0 \in Q$ to $v(c^\nu)$ depending on v.

(ii) For each $c \in (Q^\infty)^\vee$, we use the vertex $v(c)$ to define

$$\underline{\eta}^c(c) := \big[-\lambda_1\big(v(c)\big)\langle d^1, c \rangle, \ldots, -\lambda_N\big(v(c)\big)\langle d^N, c \rangle \big] \in \mathbb{Q}^N$$

and

$$\underline{\eta}^c(c^\nu) := \big[-\lambda_1^\nu\big(v(c)\big)\langle d^1, c^\nu \rangle, \ldots, -\lambda_N^\nu\big(v(c)\big)\langle d^N, c^\nu \rangle \big] \in \mathbb{Q}^N.$$

Additionally, if $v(c) \notin \mathbb{L}$, we need to define

$$\underline{\eta}^{*c}(c) := \underline{\eta}^{c}(c) + \left[\eta_0^*(c) - \eta_0(c)\right] \cdot e[v(c)]$$

and

$$1\underline{\eta}^{*c}(c^\nu) := \underline{\eta}^{c}(c^\nu) + \left[\eta_0^*(c^\nu) - \eta_0(c^\nu)\right] \cdot e[v(c^\nu)].$$

(iii) For each $c \in (Q^\infty)^\vee \cap \mathbb{L}^*$ we fix a representation $c = \sum_\nu p_\nu^c c^\nu$ ($p_\nu^c \in \mathbb{N}$) such that $\left[c, \eta_0^*(c)\right] = \sum_\nu p_\nu^c \left[c^\nu, \eta_0^*(c^\nu)\right]$. (That means, c is represented only by those generators c^ν that define faces of Q containing the face defined by c itself.) Now, we improve the definition of the polynomials $F_\bullet(\underline{Z}, \underline{t})$ given in (4.3). Let $a, b \in \mathbb{N}^w$, $\alpha, \beta \in \mathbb{N}$ such that $\left([a, \alpha], [b, \beta]\right) \in m \subseteq \mathbb{N}^{w+1} \times \mathbb{N}^{w+1}$, i.e.

$$c := \sum_\nu a_\nu c^\nu = \sum_\nu b_\nu c^\nu \quad \text{and} \quad \sum_\nu a_\nu \eta_0^*(c^\nu) + \alpha = \sum_\nu b_\nu \eta_0^*(c^\nu) + \beta.$$

Then

$$F_{(a,b,\alpha,\beta)}(\underline{Z}, \underline{t}) := f_{(a,b,\alpha,\beta)}(\underline{Z}, t_1)$$
$$- \underline{Z}^{p^c} \cdot \left(t^{\alpha e_1 + \sum_\nu a_\nu \underline{\eta}^{*c}(c^\nu) - \underline{\eta}^{*c}(c)} - t^{\beta e_1 + \sum_\nu b_\nu \underline{\eta}^{*c}(c^\nu) - \underline{\eta}^{*c}(c)}\right).$$

6.7. We need to discuss the same three types of relations as we did in (4.6). Since there is only one single element $c \in \mathbb{L}$ involved in the relations (i) and (ii), computing modulo $\tilde{\mathcal{J}}$ instead of $\mathcal{J}$ makes no difference in these cases – we always obtain $\lambda(s) = 0$. Let us consider the third relation $s := \left[z^r \cdot f_{(a,b,\alpha,\beta)} - f_{(a+r,b+r,\alpha,\beta)} = 0\right]$ ($r \in \mathbb{N}^w$). We will use the following notation:

- $c := \sum_\nu a_\nu c^\nu = \sum_\nu b_\nu c^\nu$; $\quad \underline{p} := \underline{p}^c$; $\quad \underline{\eta}^* := \underline{\eta}^{*c}$;

- $\tilde{c} := \sum_\nu (a_\nu + r_\nu) c^\nu = \sum_\nu (b_\nu + r_\nu) c^\nu = \sum_\nu (p_\nu + r_\nu) c^\nu$;
  $\underline{q} := \underline{p}^{\tilde{c}}$; $\quad \tilde{\underline{\eta}}^* := \underline{\eta}^{*\tilde{c}}$;

- $\xi := \sum_i \left(\left(\sum_\nu (p_\nu + r_\nu) \tilde{\eta}_i^*(c^\nu)\right) - \tilde{\eta}_i^*(\tilde{c})\right) = \sum_\nu (p_\nu + r_\nu) \eta_0^*(c^\nu) - \eta_0^*(\tilde{c}).$

Using the same lifting of s to $\tilde{s}$ as in (4.6) yields

$$
\lambda(s) = \underline{Z}^r \cdot F_{(a,b,\alpha,\beta)} - F_{(a+r,b+r,\alpha,\beta)}
$$
$$
- \left(\underline{t}^{\alpha e_1 + \sum_\nu a_\nu \underline{\eta}^*(c^\nu) - \underline{\eta}^*(c)} - \underline{t}^{\beta e_1 + \sum_\nu b_\nu \underline{\eta}^*(c^\nu) - \underline{\eta}^*(c)} \right) \cdot F_{(q,p+r,\xi,0)}
$$
$$
= -\underline{Z}^{p+r} \cdot \left(\underline{t}^{\alpha e_1 + \sum_\nu (a_\nu - p_\nu)\underline{\eta}^*(c^\nu)} - \underline{t}^{\beta e_1 + \sum_\nu (b_\nu - p_\nu)\underline{\eta}^*(c^\nu)} \right)
$$
$$
+ \underline{Z}^q \cdot \left(\underline{t}^{\alpha e_1 + \sum_\nu (a_\nu + r_\nu - q_\nu)\tilde{\underline{\eta}}^*(c^\nu)} - \underline{t}^{\beta e_1 + \sum_\nu (b_\nu + r_\nu - q_\nu)\tilde{\underline{\eta}}^*(c^\nu)} \right)
$$
$$
- \left(\underline{t}^{\alpha e_1 + \sum_\nu (a_\nu - p_\nu)\underline{\eta}^*(c^\nu)} - \underline{t}^{\beta e_1 + \sum_\nu (b_\nu - p_\nu)\underline{\eta}^*(c^\nu)} \right)
$$
$$
\cdot \left(\underline{Z}^q \underline{t}^{\sum_\nu (p_\nu + r_\nu - q_\nu)\tilde{\underline{\eta}}^*(c^\nu)} - \underline{Z}^{p+r} \right)
$$
$$
= \underline{Z}^q \cdot \left(\underline{t}^{\alpha e_1 + \sum_\nu (a_\nu + r_\nu - q_\nu)\tilde{\underline{\eta}}^*(c^\nu)} \right.
$$
$$
\left. - \underline{t}^{\alpha e_1 + \sum_\nu (p_\nu + r_\nu - q_\nu)\tilde{\underline{\eta}}^*(c^\nu) + \sum_\nu (a_\nu - p_\nu)\underline{\eta}^*(c^\nu)} \right)
$$
$$
- \underline{Z}^q \cdot \left(\underline{t}^{\beta e_1 + \sum_\nu (b_\nu + r_\nu - q_\nu)\tilde{\underline{\eta}}^*(c^\nu)} \right.
$$
$$
\left. - \underline{t}^{\beta e_1 + \sum_\nu (p_\nu + r_\nu - q_\nu)\tilde{\underline{\eta}}^*(c^\nu) + \sum_\nu (b_\nu - p_\nu)\underline{\eta}^*(c^\nu)} \right).
$$

As in (4.6)(iii), we can see that $\lambda(s)$ vanishes modulo $\mathcal{J}$ (or even in $A(\bar{S})$)
– just identify $\underline{\eta}^*$ and $\tilde{\underline{\eta}}^*$.

6.8. In (6.2) we already mentioned the isomorphism

$$
W \otimes_{\mathbb{C}} \mathbb{C}[\underline{z}, t] \xrightarrow{\sim} \mathcal{J} \cdot \mathbb{C}[\underline{Z}, \underline{t}] \Big/ \tilde{\mathcal{J}} \cdot \mathbb{C}[\underline{Z}, \underline{t}]
$$

obtained by identifying t with t_1 and $\underline{z}$ with $\underline{Z}$. Now, with $\lambda(s)$, we have
obtained an element of the right hand side, which has to be interpreted as
an element of $W \otimes_{\mathbb{C}} \mathbb{C}[\underline{z}, t]$. For this, we quote from [1, Lemma (7.5)]:

6.9. Lemma. *Let $A, B \in \mathbb{N}^N$ such that $\underline{d} := A - B \in V^\perp$, i.e. $\underline{t}^A - \underline{t}^B \in$
$\mathcal{J} \cdot \mathbb{C}[\underline{Z}, \underline{t}]$. Then, via the previously mentioned isomorphism, $\underline{t}^A - \underline{t}^B$
corresponds to the element*

$$
\sum_{k \geq 1} c_k \cdot g_{\underline{d}, k}(\underline{t} - t_1) \cdot t^{k_0 - k} \in W \otimes_{\mathbb{C}} \mathbb{C}[\underline{z}, t],
$$

*where $k_0 := \sum_i A_i$, and c_k are some constants occurring in the context of
symmetric polynomials, cf. [1, (3.4)]. In particular, the coefficients from W_k
vanish for $k > k_0$.*

Corollary. Transferred to $W \otimes_{\mathbb{C}} \mathbb{C}[\underline{z}, t]$, the element $\lambda(s)$ equals

$$\sum_{k \geq 1} c_k \cdot g_{\underline{d}, k}(\underline{t} - t_1) \cdot \underline{z}^q \cdot t^{k_0 - k} \quad \text{with} \quad \underline{d} := \sum_\nu (a_\nu - b_\nu) \cdot \left(\underline{\tilde{\eta}}(c^\nu) - \underline{\eta}(c^\nu) \right),$$

$$k_0 := \alpha + \sum_\nu (a_\nu + r_\nu) \eta_0^*(c^\nu) - \eta_0^*(\tilde{c}).$$

The coefficients vanish for $k > k_0$.

Proof. Since the $e[v(c)]$-terms kill each other, one can easily see, that

$$\underline{d} = \sum_\nu (a_\nu - b_\nu) \cdot \left(\underline{\tilde{\eta}}(c^\nu) - \underline{\eta}(c^\nu) \right) = \sum_\nu (a_\nu - b_\nu) \cdot \left(\underline{\tilde{\eta}}^*(c^\nu) - \underline{\eta}^*(c^\nu) \right).$$

We apply the previous lemma to both the a- and the b-summand of the $\lambda(s)$-formula of (6.7). For the first one we obtain

$$\underline{d}^{(a)} = \left[\alpha e_1 + \sum_\nu (a_\nu + r_\nu - q_\nu) \underline{\eta}^*(c^\nu) \right]$$

$$- \left[\alpha e_1 + \sum_\nu (p_\nu + r_\nu - q_\nu) \underline{\eta}^*(c^\nu) + \sum_\nu (a_\nu - p_\nu) \underline{\eta}^*(c^\nu) \right]$$

$$= \sum_\nu (a_\nu - p_\nu) \cdot \left(\underline{\tilde{\eta}}^*(c^\nu) - \underline{\eta}^*(c^\nu) \right) \quad \text{and}$$

$$k_0 = \sum_i \left(\alpha e_1 + \sum_\nu (a_\nu + r_\nu - q_\nu) \underline{\tilde{\eta}}^*(c^\nu) \right)_i$$

$$= \alpha + \sum_\nu (a_\nu + r_\nu - q_\nu) \eta_0^*(c^\nu) = \alpha + \sum_\nu (a_\nu + r_\nu) \eta_0^*(c^\nu) - \eta_0^*(\tilde{c}).$$

k_0 has the same value for both the a- and b-summand, and

$$\underline{d} = \underline{d}^{(a)} - \underline{d}^{(b)}$$

$$= \sum_\nu (a_\nu - p_\nu) \cdot \left(\underline{\tilde{\eta}}^*(c^\nu) - \underline{\eta}^*(c^\nu) \right) - \sum_\nu (b_\nu - p_\nu) \cdot \left(\underline{\tilde{\eta}}^*(c^\nu) - \underline{\eta}^*(c^\nu) \right)$$

$$= \sum_\nu (a_\nu - b_\nu) \cdot \left(\underline{\tilde{\eta}}^*(c^\nu) - \underline{\eta}^*(c^\nu) \right). \quad \blacksquare$$

6.10. Now, we try to approach the obstruction map λ from the opposite direction. Using the description of T_Y^2 given in (6.2) we construct an element of $T_Y^2 \otimes_{\mathbb{C}} W$ that, afterwards, will turn out to equal λ.

For $\rho \in \mathbb{Z}^N$ induced from some path along the edges of Q, we will denote

$$\underline{d}(\rho, c) := \left[\langle \rho_1 d^1, c \rangle, \ldots, \langle \rho_N d^N, c \rangle \right] \in \mathbb{R}^N$$

the vector showing the behavior of $c \in L^*$ passing each particular edge. If ρ governs the walk between two lattice vertices and is regarded modulo $t_i - t_j$ (if d^i, d^j contain a common non-lattice vertex), then $\underline{d}(\rho, c)$ is contained in $\mathbb{Z}^N$. In particular, this property holds for closed paths. In this case $\underline{d}(\rho, c)$ will be contained in $V^\perp$.

On the other hand, for each $k \geq 1$, we can use the $\underline{d}$'s from $V^\perp$ to get elements $g_{\underline{d},k}(\underline{t} - t_1) \in W_k$ generating this vector space. Composing both procedures we obtain, for each closed path $\rho \in \mathbb{Z}^N$, a map

$$g^{(k)}(\rho, \bullet) : \mathbb{A}^* \longrightarrow V^\perp \longrightarrow W_k$$

$$c \quad\quad \mapsto \quad\quad g_{\underline{d}(\rho,c),k}(\underline{t} - t_1).$$

6.11. Lemma. (1) *Taking the sum over all compact 2-faces we get a surjective map*

$$\sum_{\varepsilon < Q} g^{(k)}(\underline{\varepsilon}, \bullet) : \oplus_{\varepsilon < Q} \mathbb{A}^* \otimes_{\mathbb{R}} \mathbb{C} \twoheadrightarrow W_k.$$

(2) *Let $c \in L^*$ be integral. If $\rho^1, \rho^2 \in \mathbb{Z}^N$ are two paths each connecting vertices $v, w \in Q$ such that*

- $\left| \langle v, c \rangle - \langle w, c \rangle \right| \leq k - 1$ *and*

- *c is monotone along both paths, i.e. $\left\langle \rho_i^{1/2} d^i, c \right\rangle \geq 0$ for $i = 1, \ldots, N$,*

then $\rho^1 - \rho^2 \in \mathbb{Z}^N$ will be a closed path yielding $g^{(k)}(\rho^1 - \rho^2, c) = 0$ in W_k.

Proof. The reason for (1) is the fact that the elements $\underline{d}(\varepsilon, c)$ ($\varepsilon < Q$ compact 2-face; $c \in L^*$) and $e_i - e_j$ (for d^i, d^j containing a common non-lattice vertex) generate $V^\perp$ as a vector space; since $t_i - t_j \in \mathcal{J}_1$ the latter type yields zero in W_k.

For the proof of (2), we consider $\underline{d} := \underline{d}(\rho^1 - \rho^2, c)$. Since $d_i = \left\langle \rho_i^1 d^i, c \right\rangle - \left\langle \rho_i^2 d^i, c \right\rangle$ is the difference of two non-negative integers, we obtain $d_i^+ \leq \left\langle \rho_i^1 d^i, c \right\rangle$. Hence,

$$\sum_i d_i^+ \leq \sum_i \left\langle \rho_i^1 d^i, c \right\rangle = \langle w, c \rangle - \langle v, c \rangle \leq k - 1,$$

and we obtain $g_{\underline{d},k}(\underline{t} - t_1) \in \tilde{\mathcal{J}}$ by the following corollary. ∎

Corollary. *Let* $k_0 := \sum \underline{d}(\rho_1 - \rho_2, c)^+$. *Then* $g_{\underline{d}(\rho_1 - \rho_2, c), k}(\underline{t} - t_1) \in \tilde{\mathcal{J}}$ *for* $k > k_0$.

Proof. Consider $\underline{d} \in V^\perp \cap \mathbb{Z}^N$. From [3] proposition (2.3) we know that $g_{\underline{d},k}(\underline{t} - t_1)$ can be written as a $\mathbb{C}[t_i - t_j]$-linear combinations of $g_{\underline{d},1}(\underline{t} - t_1)$, $\dots, g_{\underline{d},k_0}(\underline{t} - t_1)$ for $k > k_0$, where $k_0 = \sum_i d_i^+$.

Now $\underline{d} := \underline{d}(\rho_1 - \rho_2, c) \in V^\perp$ does not have to be contained in $\mathbb{Z}^N$. But since the path $\rho_1 - \rho_2$ is closed, $\underline{d}$ yields an integer as sum on every component. Since $\underline{d}_{ij} := [0, \dots, 0, 1_i, 0, \dots, 0, -1_j, 0, \dots, 0] \in V^\perp$ for d^i, d^j containing a common non-lattice vertex we are able to find some $\underline{\tilde{d}} \in V^\perp \cap \mathbb{Z}^N$ with $\sum_i \tilde{d}_i^+ = \sum_i d_i^+$ such that $g_{\underline{\tilde{d}},k}(\underline{t} - t_1) = g_{\underline{d},k}(\underline{t} - t_1) + \sum_{ij} q_{ij} \cdot g_{\underline{d}_{ij},k}(\underline{t} - t_1)$, with the usual assumptions for i, j. In particular, the q_{ij} do not depend on k. For the $g_{\underline{\tilde{d}},k}(\underline{t} - t_1)$ the first assumptions apply, and we obtain for $k > k_0$:

$$g_{\underline{\tilde{d}},k}(\underline{t} - t_1) = \sum_{n=1}^{k_0} a_n(\underline{t} - t_1) \cdot g_{\underline{\tilde{d}},n}(\underline{t} - t_1)$$

and

$$g_{\underline{d},k}(\underline{t} - t_1) = \sum_{n=1}^{k_0} a_n(\underline{t} - t_1) \cdot g_{\underline{\tilde{d}},n}(\underline{t} - t_1) - \sum_{ij} q_{ij} \cdot g_{\underline{d}_{ij},k}(\underline{t} - t_1).$$

Now we assume w.l.o.g. that $a_n(\underline{t} - t_1)$ is homogenous and has degree $k - n$. Hence, the first sum on the right hand side is contained in $(t_i - t_j)_{ij} \cdot \mathcal{J} \subseteq \tilde{\mathcal{J}}$. Now consider the second sum. We know $g_{\underline{d}_{ij},1}(\underline{t} - t_1) \in \mathcal{J}_1$ and $\sum_r (d_{ij}^+)_r = 1$. Thus $g_{\underline{d}_{ij},k}(\underline{t} - t_1) = f(\underline{t} - t_1) \cdot g_{\underline{d}_{ij},1}(\underline{t} - t_1)$ and this is contained in $\mathcal{J}_1 \cdot \mathbb{C}[t_i - t_j] \subseteq \tilde{\mathcal{J}}$. ∎

6.12. Recalling the sets E_j^k from (6.2), we can define the following linear maps:

$$\psi_j^{(k)} : L(E_j^k) \longrightarrow W_k$$
$$q \mapsto \sum_\nu q_\nu \cdot g^{(k)}\big(\underline{\lambda}(v^j) + \underline{\mu}^\nu(v^j) - \underline{\lambda}(v(c^\nu)), c^\nu\big).$$

(The q-coordinate corresponding to $R \in E_j^k$ is not used in the definition of $\psi_j^{(k)}$.)

6.13. Lemma. *Let $\langle v^j, v^l \rangle < Q$ be an edge of the polyhedron Q. Then, on $L(E_j^k \cap E_l^k) = L(E_j^k) \cap L(E_l^k)$, the maps $\psi_j^{(k)}$ and $\psi_l^{(k)}$ coincide. In particular (cf. Theorem 6.2), the $\psi_j^{(k)}$'s induce a linear map $\psi^{(k)} : T_Y^2(-kR)^* \to W_k$.*

Proof. The proof is similar to the proof of Lemma (7.6) in [1]. ∎

Now, both ends will meet and we obtain an explicit description of the obstruction map:

6.14. Proposition. $\sum_{k \geq 1} c_k \psi^{(k)}$ *equals λ^*, the adjoint of the obstruction map.*

Proof. Using Theorem 3.5 of [3], we can find an element of $\mathrm{Hom}\left(\mathcal{R}/\mathcal{R}_0, W_k \otimes A(Y)\right)$ representing $\psi^{(k)} \in T_Y^2 \otimes W_k$ – it sends relations of type (i), cf. (4.6), to 0 and deals with relations of type (ii) and (iii) in the following way:

$$\left[\underline{z}^r t^\gamma \cdot f_{(a,b,\alpha,\beta)} - f_{(a+r,b+r,\alpha+\gamma,\beta+\gamma)} = 0 \right]$$

$$\mapsto \psi_j^{(k)}(a-b) \cdot x^{\sum_\nu (a_\nu + r_\nu)} \left[c^\nu, \eta_0^*(c^\nu) \right] + (\alpha + \gamma - k)R,$$

if

$$\left\langle (Q,1), \sum_\nu (a_\nu + r_\nu) \left[c^\nu, \eta_0^*(c^\nu) \right] + (\alpha + \gamma - k)R \right\rangle \geq 0,$$

and j is such that

$$\left\langle (v^j, 1), \sum_\nu a_\nu \left[c^\nu, \eta_0^*(c^\nu) \right] + (\alpha - k)R \right\rangle < 0;$$

otherwise the relation is sent to 0 (in particular, if there is not any j meeting the desired condition).

On Q, the linear forms $c := \sum_\nu a_\nu c^\nu$ and $\tilde{c} = \sum_\nu (a_\nu + r_\nu) c^\nu$ admit their minimal values at the vertices $v(c)$ and $v(\tilde{c})$, respectively. Hence, we can transform the previous formula into

$$\left[\underline{z}^r t^\gamma \cdot f_{(a,b,\alpha,\beta)} - f_{(a+r,b+r,\alpha+\gamma,\beta+\gamma)} = 0 \right]$$

$$\mapsto \psi_{v(c)}^{(k)}(a-b) \cdot x^{\sum_\nu (a_\nu + r_\nu)} \left[c^\nu, \eta_0^*(c^\nu) \right] + (\alpha + \gamma - k)R$$

if $\sum_\nu (a_\nu + r_\nu)\eta_0^*(c^\nu) - \eta_0^*(\tilde{c}) + (\alpha + \gamma - k)$

$$= \left\langle (v(\tilde{c}),1), \sum_\nu (a_\nu + r_\nu)[c^\nu, \eta_0^*(c^\nu)] + (\alpha + \gamma - k)R \right\rangle \geq 0,$$

$\sum_\nu a_\nu \eta_0^*(c^\nu) - \eta_0^*(c) + (\alpha - k) =$

$$= \left\langle (v(c),1), \sum_\nu a_\nu [c^\nu, \eta_0^*(c^\nu)] + (\alpha - k)R \right\rangle < 0$$

and mapping onto 0 otherwise.

Adding the coboundary $h \in \mathrm{Hom}\left(\mathbb{C}[\underline{z},t]^m, W_k \otimes A(Y)\right)$

$$h_{(a,\alpha),(b,\beta)} := \begin{cases} \psi_{v(c)}^{(k)}(a-b) \cdot x^{\sum_\nu a_\nu [c^\nu, \eta_0^*(c^\nu)] + (\alpha - k)R} \\ \qquad \text{for } \sum_\nu a_\nu \eta_0^*(c^\nu) - \eta_0^*(c) + \alpha \geq k, \\ 0 \qquad \text{otherwise} \end{cases}$$

does not change the class in $T_Y^2(-kR) \otimes W_k$ (still representing $\psi^{(k)}$), but improves the representative from $\mathrm{Hom}\left(^\mathcal{R}/_{\mathcal{R}_0}, W_k \otimes A(Y)\right)$. It still maps type-(i)-relations to 0, and moreover

$$\left[\underline{z}^r t^\gamma \cdot f_{(a,b,\alpha,\beta)} - f_{(a+r,b+r,\alpha+\gamma,\beta+\gamma)} = 0 \right]$$

$$\mapsto \begin{cases} \left(\psi_{v(c)}^{(k)}(a-b) - \psi_{v(\tilde{c})}^{(k)}(a-b) \right) \cdot x^{\sum_\nu (a_\nu + r_\nu)[c^\nu, \eta_0^*(c^\nu)] + (\alpha + \gamma - k)R} \\ \qquad \text{for } k_0 + \gamma \geq k \\ 0 \qquad \text{otherwise} \end{cases}$$

with $k_0 = \alpha + \sum_\nu (a_\nu + r_\nu)\eta_0^*(c^\nu) - \eta_0^*(\tilde{c})$. By definition of $\psi_j^{(k)}$ and $g^{(k)}$ we obtain

$$\psi_{v(c)}^{(k)}(a-b) - \psi_{v(\tilde{c})}^{(k)}(a-b)$$

$$= \sum_\nu (a_\nu - b_\nu) \cdot g^{(k)}\left(\underline{\lambda}(v(c)) + \underline{\mu}^\nu(v(c)) - \underline{\lambda}(v(\tilde{c})) - \underline{\mu}^\nu(v(\tilde{c})), c^\nu \right)$$

$$= \sum_\nu (a_\nu - b_\nu) \cdot g^{(k)}\left(\underline{\lambda}^\nu(v(c)) - \underline{\lambda}^\nu(v(\tilde{c})), c^\nu \right)$$

$$= g_{\underline{d},k}(\underline{t} - t_1) \quad \text{with}$$

$$\underline{d} = \sum_\nu (a_\nu - b_\nu) \cdot \underline{d}\big(\underline{\lambda}^\nu(v(c)) - \underline{\lambda}^\nu(v(\tilde{c})), c^\nu\big)$$

$$= \sum_\nu (a_\nu - b_\nu) \cdot \big(\tilde{\underline{\eta}}(c^\nu) - \underline{\eta}(c^\nu)\big)$$

$$= \sum_\nu (a_\nu - b_\nu) \cdot \big(\tilde{\underline{\eta}}^*(c^\nu) - \underline{\eta}^*(c^\nu)\big),$$

and this completes our proof. Indeed, for relations of type (ii) (i.e. $r = 0$; $\gamma = 1$) we know $c = \tilde{c}$, hence, those relations map onto 0. For relations of type (iii) (i.e. $\gamma = 0$) we can compare the previous formula with the result obtained in Corollary 6.8: The coefficients coincide, and the monomial $\underline{z}^q t^{k_0 - k} \in \mathbb{C}[\underline{z}, t]$ maps onto $x^{\sum_\nu (a_\nu + r_\nu)[c^\nu, \eta_0^*(c^\nu)] + (\alpha + \gamma - k)R} \in A(Y)$. ∎

6.15. It remains to show that the summands $\psi^{(k)}$ of λ^* are indeed surjective maps from $T_Y^2(-kR)^*$ to W_k. We will do so by composing them with auxiliary surjective maps $p^k : \oplus_{\varepsilon < Q} \mathbb{A}^* \otimes_{\mathbb{R}} \mathbb{C} \twoheadrightarrow T_Y^2(-kR)^*$ yielding $\psi^{(k)} \circ p^k = \sum_{\varepsilon < Q} g^{(k)}(\underline{\varepsilon}, \bullet)$. Then the result follows from the first part of Lemma 6.10.

Let us fix some 2-face $\varepsilon < Q$. Assume that $d^1, \ldots, d^m$ are its counterclockwise oriented edges, i.e. the sign vector $\underline{\varepsilon}$ looks like $\varepsilon_i = 1$ for $i = 1, \ldots, m$ and $\varepsilon_j = 0$ otherwise. Moreover, we denote the vertices of $\varepsilon < Q$ by $v^1, \ldots, v^m$ such that d^i runs from v^i to v^{i+1} $(m + 1 := 1)$.

Now p^k maps $[c, z] \in M$ to the linear relation

$$\sum_{i=1}^m \sum_\nu (q_{i,\nu} - q_{i-1,\nu}) \cdot \big[c^\nu, \eta_0^*(c^\nu)\big] + (q_i - q_{i-1}) \cdot [\underline{0}, 1] = 0,$$

where

$$[c, z] = \sum_\nu q_{i,\nu}\big[c^\nu, \eta_0^*(c^\nu)\big] + q_i[\underline{0}, 1]$$

with $\big[c^\nu, \eta_0^*(c^\nu)\big] \in E_i^k \cap E_{i+1}^k$ for every $q_{i,\nu} \neq 0$. This relation is automatically contained in $\ker\big(\oplus_i L(E_i^k) \to L(E)\big)$. Note that only the $c \in \mathbb{L}^*$ is important; choosing another z will not change the differences $q_i - q_{i-1}$. A closer look at the construction and the surjectivity can be taken in [3] sect. 6. Finally, we apply $\psi^{(k)}$ to obtain

$$\psi^{(k)}\big(p^k(c)\big) = \sum_{i=1}^m \sum_\nu (q_{i,\nu} - q_{i-1,\nu}) \cdot g^{(k)}\big(\underline{\lambda}(v^i) - \underline{\lambda}(v(c^\nu)) + \underline{\mu}^\nu(v^i), c^\nu\big)$$

$$= \sum_{i,v} g^{(k)}\big(\underline{\lambda}(v^i) - \underline{\lambda}(v(c^\nu)) + \underline{\mu}^\nu(v^i), q_{i,v}c^\nu\big)$$

$$- \sum_{i,v} g^{(k)}\big(\underline{\lambda}(v^{i+1}) - \underline{\lambda}(v(c^\nu)) + \underline{\mu}^\nu(v^{i+1}), q_{i,v}c^\nu\big)$$

$$= \sum_{i,v} g^{(k)}\big(\underline{\lambda}(v^i) - \underline{\lambda}(v^{i+1}) + \underline{\mu}^\nu(v^i) - \underline{\mu}^\nu(v^{i+1}), q_{i,v}c^\nu\big).$$

We introduce the path ρ^i consisting of the single edge d^i only. Then, if $q_{iv} \neq 0$ and w.l.o.g. $\langle v^i, c^\nu \rangle \geq \langle v^{i+1}, c^\nu \rangle$, the pair of paths $\underline{\mu}^\nu(v^i)$ and $\underline{\mu}^\nu(v^{i+1}) + \rho^i$ meets the assumption of Lemma 6.10(2) (cf. (i)). Hence, we can proceed as follows:

$$\psi^{(k)}\big(p^k(c)\big) = \sum_{i,v} g^{(k)}\big(\underline{\lambda}(v^i) - \underline{\lambda}(v^{i+1}) + \rho^i, q_{iv}c^\nu\big)$$

$$+ \sum_{i,v} g^{(k)}\big(\underline{\mu}^\nu(v^i) - \underline{\mu}^\nu(v^{i+1}) - \rho^i, q_{iv}c^\nu\big)$$

$$= \sum_{i=1}^{m} g^{(k)}\Big(\underline{\lambda}(v^i) - \underline{\lambda}(v^{i+1}) + \rho^i, \sum_{\nu} q_{iv}c^\nu\Big)$$

$$= \sum_{i=1}^{m} g^{(k)}\big(\underline{\lambda}(v^i) - \underline{\lambda}(v^{i+1}) + \rho^i, c\big)$$

$$= g^{(k)}\Big(\sum_{i=1}^{m} \rho^i, c\Big)$$

$$= g^{(k)}(\underline{\varepsilon}, c).$$

Thus, Theorem 6.5 is proven.

7. EXAMPLE

First let us provide a theorem to describe the situation for $\dim \sigma = 3$. We assume σ is smooth in codimension two. Hence, it has an isolated singularity and $\dim T_Y^1 < \infty$, i.e. there are only finitely many $R \in \sigma^\vee \cap M$ with $\dim \big(V(Q)/\underline{1}\big) \neq 0$. The second part of the following theorem provides a combinatorial verification for this fact.

7.1. Theorem. *Let $\sigma \subset \mathbb{R}^3$ be a three dimensional cone with smooth two dimensional faces.*

(i) *Let $R \in \operatorname{int}(\sigma^\vee \cap M)$. We define*

$$Q := \sigma \cap [R = 1] \quad \text{and} \quad Q' := \operatorname{conv}(\text{lattice vertices of } Q).$$

Define $\sigma' := \operatorname{Cone} Q'$. Then we denote by $Y' := \mathbb{TV}(\sigma')$ the associated Gorenstein singularity. If the edge vectors of Q' are primitive (i.e. σ' has smooth two dimensional faces), then Y' has the same deformation theory in degree R^* as Y in degree R.

(ii) *There are only finitely many $R \in \sigma^\vee \cap M$ such that $\dim\left(V(Q)/\underline{1}\right) \neq 0$.*

Proof. (i) This is obvious, since $V(Q) \cong V(Q')$.

(ii) Let $R \in \operatorname{int}(\sigma^\vee \cap M)$. Then $Q := \sigma \cap [R = 1]$ is a two dimensional polytope. We know $\dim T^1_Y(-R) = \dim V(Q) - 1$ by (5.4), hence, to obtain $\dim T^1_Y(-R) \geq 1$ we need $\dim V(Q) \geq 2$. Therefore, Q has to have at least four different components. This is equivalent to Q having at least four lattice vertices. Now for any four generating rays of sigma there are at most one $R \in \operatorname{int}(\sigma^\vee \cap M)$ yielding one on all four of them.

Let us now assume $R \in \partial(\sigma^\vee \cap M)$. If Q has less than three vertices, we immediately obtain $\dim V(Q) \leq 1$. Otherwise Q looks like:

Assume Q has at least two lattice vertices. Then R yields 1 on two rays of σ. Since $R \in \partial(\sigma^\vee \cap M)$, R yields zero on at least one ray of σ. These conditions fully determine R and since σ is spanned by finitely many rays there are only finitely many such R.

Now assume Q has only one lattice vertex. If this lattice vertex is a_1 or a_3, we immediately obtain $\dim V(Q) = 1$. To obtain $\dim V(Q) > 1$ the lattice vertex has to be a_2. Let a_4 be a ray of σ such that $R(a_4) = 0$. By the above observation we know that a_2 and a_4 do not lie in a common two face of σ. There are only two facets of $\sigma^\vee$ that have an infinite intersection with the hyperplane $[a_2 = 1]$, i.e. those defined by a_1 and a_3. Hence, the set $[a_2 = 1] \cap [a_4 = 0] \cap \sigma^\vee$ is bounded and $[a_2 = 1] \cap [a_4 = 0] \cap \sigma^\vee \cap M$ is finite. ∎

Finally we provide an example to illustrate the whole theory and, in particular, Theorem 7.1(i). Let $N = \mathbb{Z}^3$ be a lattice. Define the cone σ by

$$\sigma := \big\langle (0,0,1), (1,0,1), (2,1,1), (1,2,1), (1,4,2), (0,1,2) \big\rangle \subseteq \mathbb{Q}^3 = N_\mathbb{Q}.$$

We choose $R := [0,0,1] \in M = \mathbb{Z}^3$ and obtain the following polygon Q:

$$a^5 = (1/2, 2) \qquad a^4 = (1,2)$$

$$a^6 = (0, 1/2) \qquad\qquad a^3 = (2,1)$$

$$a^1 = (0,0) \qquad a^2 = (1,0)$$

We obtain the following paths:

$$d^1 = \begin{pmatrix} 1 \\ 0 \end{pmatrix}, \quad d^2 = \begin{pmatrix} 1 \\ 1 \end{pmatrix}, \quad d^3 = \begin{pmatrix} -1 \\ 1 \end{pmatrix},$$

$$d^4 = \begin{pmatrix} -\frac{1}{2} \\ 0 \end{pmatrix}, \quad d^5 = \begin{pmatrix} -\frac{1}{2} \\ -\frac{3}{2} \end{pmatrix}, \quad d^6 = \begin{pmatrix} 0 \\ -\frac{1}{2} \end{pmatrix}.$$

Let Q' be the convex hull of the lattice vertices of Q, consisting of d^1, d^2, d^3 and the dashed line in the picture. The associated Gorenstein singularity $Y' = \mathbb{TV}(\sigma')$ with σ' being the cone over Q' equals the affine cone over the Del Pezzo surface of degree 8. It would be interesting to know more about a general geometric relation between the singularities Y and Y', i.e. is there a universal property (depending on R) characterizing the map $Y' \to Y$?

Now we can explicitly describe the ideal $\mathcal{J}$ as defined in (2.4): Q equals its own (and only) 2-face. This yields the following families of polynomials:

$$g_{1,k}(\underline{t}) = t_1^k + t_2^k - t_3^k - \frac{1}{2}t_4^k - \frac{1}{2}t_5^k, \quad k \geq 1$$

and

$$g_{2,k}(\underline{t}) = t_2^k + t_3^k - \frac{3}{2}t_5^k - \frac{1}{2}t_6^k, \quad k \geq 1.$$

Additionally, we have the polynomials $t_4 - t_5$ and $t_5 - t_6$ for the non-lattice vertices. We obtain

$$\mathcal{J} = \big(g_{1,k}(\underline{t}), g_{2,k}(\underline{t}) \mid k \geq 1 \big) + (t_4 - t_5, t_5 - t_6).$$

By Corollary 6.10 we know that it is enough to consider $k \leq 3$. Calculating modulo the the two last equations and hence, only considering t_4, the homogeneous ideal $\mathcal{J}$ defining $\mathcal{M} \subseteq \mathbb{C}^4$ is generated by:

$$\mathcal{J} = \left(t_2 + t_3 - 2 \cdot t_4, t_1 - 2 \cdot t_3 + t_4, t_3^2 - 2 \cdot t_3 t_4 + t_4^2 \right).$$

We introduce the variables $w_{12} := t_1 - t_2$, $w_{23} := t_2 - t_3$ and $w_{34} := t_3 - t_4$ for the differences $t_i - t_j$. Now one can easily see that

$$\mathcal{J} = \left(w_{23} + 2 \cdot w_{34}, w_{12} + w_{23} - w_{34}, w_{34}^2 \right)$$

holds as predicted by Theorem 2.5. Moreover these equations define $\bar{\mathcal{M}} \subseteq \mathbb{C}^3$.

Let us now construct $V(Q)$ as described in (2.2). Since Q has 6 edges we obtain a description of $V(Q)$ as a subspace of $\mathbb{R}^6$.

The polygon Q has two non-lattice vertices, namely a^5 and a^6. These vertices are directly connected by edge d^5 and together they form a component of Q, shown by the dashed line in the picture. This yields the equations $t_4 = t_5$ and $t_5 = t_6$ for $\underline{t} \in V$. From now on we will calculate modulo these equations and hence, only consider t_4.

The remaining two equations are described by the rows of $\sum t_i d^i = 0$. Since all these equations are linearly independent we obtain that V is a two-dimensional subspace of $\mathbb{R}^6$.

The next step is to compute the Hilbert basis E of $\sigma^\vee \cap M$. To do this, we use a program like [5]:

$$E = \big\{ R = [0,0,1], [6,-2,1], [1,0,0], [0,1,0], [2,-1,1],$$

$$[-1,-1,3], [-1,1,1], [0,-1,2], [-1,0,2] \big\}$$

Using the elements of $E \backslash \{R\}$, we can describe $\tilde{C}(Q)^\vee \cap \tilde{M}$. However, since we calculate modulo $t_4 = t_5$ and $t_5 = t_6$ as described above, we will not denote the $\eta^*(c^i)$ as elements of $\mathbb{R}^6$, instead we will consider the evaluation of $\underline{\eta}^*(c^i)$ on components of $\underline{t} \in V$, i.e. in the case of $\eta_4^*(c^i)$, $\eta_5^*(c^i)$, $\eta_6^*(c^i)$ it is only important to know their sum. This means our $\underline{\eta}^*(c^i)$ are built up by the following formula:

$$\underline{\eta}^*(c^i) = \left[\eta_1^*(c^i), \eta_2^*(c^i), \eta_3^*(c^i), \eta_4^*(c^i) + \eta_5^*(c^i) + \eta_6^*(c^i) \right] \in \mathbb{R}^4.$$

i	c^i	$v(c^i)$	$\underline{\lambda}^{c^i}$	$\underline{\eta}^*(c^i)$
1	$[6,-2]$	$(0,1/2)$	$(0,0,0,0,0,-1)$	$[0,0,0,1]$
2	$[1,0]$	$(0,0)$	$(0,0,0,0,0,0)$	$[0,0,0,0]$
3	$[0,1]$	$(0,0)$	$(0,0,0,0,0,0)$	$[0,0,0,0]$
4	$[2,-1]$	$(1/2,2)$	$(0,0,0,0,-1,-1)$	$[0,0,0,1]$
5	$[-1,-1]$	$(2,1)$	$(1,1,0,0,0,0)$	$[1,2,0,0]$
6	$[-1,1]$	$(1,0)$	$(1,0,0,0,0,0)$	$[1,0,0,0]$
7	$[0,-1]$	$(1/2,2)$	$(0,0,0,0,-1,-1)$	$[0,0,0,2]$
8	$[-1,0]$	$(2,1)$	$(1,1,0,0,0,0)$	$[1,1,0,0]$

Using the Hilbert basis E of $\sigma^\vee \cap M$ we want to describe the affine toric variety Y as a subvariety of $\mathbb{C}^{|E|} = \mathbb{C}^9$. To do this, consider the following exact sequence:

$$0 \to L \to \mathbb{Z}^9 \overset{\pi}{\to} M \to 0,$$

where π is defined by mapping the $e_i \in \mathbb{Z}^9$ to the generators of the Hilbert basis of $\sigma^\vee \cap M$, i.e. the matrix

$$\pi = \begin{pmatrix} 0 & 6 & 1 & 0 & 2 & -1 & -1 & 0 & -1 \\ 0 & -2 & 0 & 1 & -1 & -1 & 1 & -1 & 0 \\ 1 & 1 & 0 & 0 & 1 & 3 & 1 & 2 & 2 \end{pmatrix}.$$

Let L be the kernel of this matrix. We build up the so called toric ideal

$$I_L := \left(\underline{x}^{l^+} - \underline{x}^{l^-} \mid l \in L \right) \subseteq k[x_0, \ldots, x_8]$$

and obtain

$$k[\sigma^\vee \cap M] \cong k\left[\pi(\mathbb{N}^n)\right] \cong {}^{k[\underline{x}]}\!/_{I_L}.$$

This yields an inclusion $Y = \mathrm{Spec}\left(k[\sigma^\vee \cap M]\right) \subseteq \mathbb{C}^9$. Now we need to compute the generators of the ideal I_L, which can be easily done by using TORIC.LIB of [6]. The following code will do the calculation needed:

```
LIB "toric.lib";
ring r=0,(t,z1,z2,z3,z4,z5,z6,z7,z8),dp;
intmat pi[3][9]=
0,6,1,0,2,-1,-1,0,-1,
0,-2,0,1,-1,-1,1,-1,0,
1,1,0,0,1,3,1,2,2;
```

```
pi;
ideal I=toric_ideal(pi,"pt");
def L=mstd(I);
I=L[2];
I;
```

Note that we chose the variables of the ring according to the description given in (4.3), i.e. t corresponds to $R \in M$ and z_i corresponds to c^i. We obtain the following polynomials defining Y:

$$
\begin{aligned}
0 \quad & f_{(e^6+e^7,e^8,0,1)} & = \quad & z_6 z_7 - t z_8 \\
1 \quad & f_{(e^3+e^7,e^2+e^8,0,0)} & = \quad & z_3 z_7 - z_2 z_8 \\
2 \quad & f_{(e^5+e^6,2e^8,0,0)} & = \quad & z_5 z_6 - z_8^2 \\
3 \quad & f_{(e^6,e^3+e^8,1,0)} & = \quad & t z_6 - z_3 z_8 \\
4 \quad & f_{(e^3+e^5,e^8,0,1)} & = \quad & z_3 z_5 - t z_8 \\
5 \quad & f_{(e^2+e^5,e^7,0,1)} & = \quad & z_2 z_5 - t z_7 \\
6 \quad & f_{(e^5,e^7+e^8,1,0)} & = \quad & t z_5 - z_7 z_8 \\
7 \quad & f_{(e^3,e^2+e^6,1,0)} & = \quad & t z_3 - z_2 z_6 \\
8 \quad & f_{(0,e^2+e^8,2,0)} & = \quad & t^2 - z_2 z_8 \\
9 \quad & f_{(e^2+2e^7,e^4+e^5,0,0)} & = \quad & z_2 z_7^2 - z_4 z_5 \\
10 \quad & f_{(2e^2+e^8,e^4+e^6,0,0)} & = \quad & z_2^2 z_8 - z_4 z_6 \\
11 \quad & f_{(2e^2+e^7,e^4,0,1)} & = \quad & z_2^2 z_7 - t z_4 \\
12 \quad & f_{(e^2+e^7,e^4+e^8,1,0)} & = \quad & t z_2 z_7 - z_4 z_8 \\
13 \quad & f_{(3e^4,e^1+e^7,0,0)} & = \quad & z_4^3 - z_1 z_7 \\
14 \quad & f_{(2e^2,e^3+e^4,1,0)} & = \quad & t z_2^2 - z_3 z_4 \\
15 \quad & f_{(e^2+2e^4+e^7,e^1+e^5,0,0)} & = \quad & z_2 z_4^2 z_7 - z_1 z_5 \\
16 \quad & f_{(e^2+e^3+2e^4,e^1+e^6,0,0)} & = \quad & z_2 z_3 z_4^2 - z_1 z_6 \\
17 \quad & f_{(2e^2+2e^4,e^1,0,1)} & = \quad & z_2^2 z_4^2 - t z_1 \\
18 \quad & f_{(e^2+2e^4,e^1+e^8,1,0)} & = \quad & t z_2 z_4^2 - z_1 z_8 \\
19 \quad & f_{(4e^2+e^4,e^1+e^3,0,0)} & = \quad & z_2^4 z_4 - z_1 z_3
\end{aligned}
$$

We want to compute the liftings $F_{(a,b,\alpha,\beta)}$ of the $f_{(a,b,\alpha,\beta)}$ in $A(\bar{S})[Z_1, \ldots, Z_w]$. For a given $c \in \mathbb{L}^*$, we have to find a representation

$$[c, \underline{\eta}^*(c)] = \sum_\nu p_\nu [c^\nu, \underline{\eta}^*(c^\nu)],$$

$p_\nu \in \mathbb{Z}_{\geq 0}$. This proves difficult, because we compute the η^* modulo $V^\perp$. It is easier to use Proposition 3.6 instead. If we find a linear combination $[c, \eta_0^*(c)] = \sum_\nu p_\nu [c^\nu, \eta_0^*(c^\nu)]$, we automatically obtain $[c, \underline{\eta}^*(c)] = \sum_\nu p_\nu [c^\nu, \underline{\eta}^*(c^\nu)]$ with the same coefficients $p_\nu \in \mathbb{Z}_{\geq 0}$. Since $\sigma^\vee$ is a pointed cone and we already know a Hilbert basis of $\sigma^\vee \cap M$ this problem is very easy to solve.

Using the equations $t_4 = t_5$ and $t_5 = t_6$ we obtain

$$C(Q)^\vee = \mathbb{R}_{\geq 0}^4 + V^\perp \Big/ V^\perp$$

where $V^\perp$ is generated by $[1, 1, -1, -1]$ and $[0, 1, 1, -2]$ obtained from the edge directions of the polygon Q. As introduced in (4.3) we will use this description of $C(Q)^\vee$ for the liftings of the $f_{(a,b,\alpha,\beta)}$, i.e. the variables $t_1, \ldots, t_4$ correspond to the coordinates of $\mathbb{R}_{\geq 0}^4$. One can easily see that the exponents of the t_i in an $F_{(a,b,\alpha,\beta)}$-term sum up to the exponent of t in the corresponding term of $f_{(a,b,\alpha,\beta)}$.

$$
\begin{aligned}
0 \quad F_{(e^6+e^7,e^8,0,1)} \quad &= \quad Z_6 Z_7 - Z_8 t_1 - Z_8(t_3 - t_1) \\
&= \quad Z_6 Z_7 - Z_8 t_3 \\[6pt]
1 \quad F_{(e^3+e^7,e^2+e^8,0,0)} \quad &= \quad Z_3 Z_7 - Z_2 Z_8 - (t_4^2 - t_1 t_2) \\
&= \quad Z_3 Z_7 - t_4^2 + F_8 \\[6pt]
2 \quad F_{(e^5+e^6,2e^8,0,0)} \quad &= \quad Z_5 Z_6 - Z_8^2 \\[6pt]
3 \quad F_{(e^6,e^3+e^8,1,0)} \quad &= \quad Z_6 t_1 - Z_3 Z_8 - Z_6(t_1 - t_2) \\
&= \quad Z_6 t_2 - Z_3 Z_8 \\[6pt]
4 \quad F_{(e^3+e^5,e^8,0,1)} \quad &= \quad Z_3 Z_5 - Z_8 t_1 - Z_8(t_2 - t_1) \\
&= \quad Z_3 Z_5 - Z_8 t_2 \\[6pt]
5 \quad F_{(e^2+e^5,e^7,0,1)} \quad &= \quad Z_2 Z_5 - Z_7 t_1 - Z_7(t_4 - t_1) \\
&= \quad Z_2 Z_5 - Z_7 t_4 \\[6pt]
6 \quad F_{(e^5,e^7+e^8,1,0)} \quad &= \quad Z_5 t_1 - Z_7 Z_8 - Z_5(t_1 - t_3)
\end{aligned}
$$

$$= Z_5 t_3 - Z_7 Z_8$$

7 $F_{(e^3, e^2 + e^6, 1, 0)}$ $\quad = Z_3 t_1 - Z_2 Z_6$

8 $F_8 := F_{(0, e^2 + e^8, 2, 0)}$ $\quad = t_1^2 - Z_2 Z_8 - (t_1^2 - t_1 t_2)$

$$= t_1 t_2 - Z_2 Z_8$$

9 $F_{(e^2 + 2e^7, e^4 + e^5, 0, 0)}$ $\quad = Z_2 Z_7^2 - Z_4 Z_5$

10 $F_{(2e^2 + e^8, e^4 + e^6, 0, 0)}$ $\quad = Z_2^2 Z_8 - Z_4 Z_6 - Z_2(t_1 t_2 - t_1 t_4)$

$$= Z_2 t_1 t_4 - Z_4 Z_6 - Z_2 F_8$$

11 $F_{(2e^2 + e^7, e^4, 0, 1)}$ $\quad = Z_2^2 Z_7 - Z_4 t_1 - Z_4(t_4 - t_1)$

$$= Z_2^2 Z_7 - Z_4 t_4$$

12 $F_{(e^2 + e^7, e^4 + e^8, 1, 0)}$ $\quad = Z_2 Z_7 t_1 - Z_4 Z_8 - Z_2 Z_7(t_1 - t_3)$

$$= Z_2 Z_7 t_3 - Z_4 Z_8$$

13 $F_{(3e^4, e^1 + e^7, 0, 0)}$ $\quad = Z_4^3 - Z_1 Z_7$

14 $F_{14} := F_{(2e^2, e^3 + e^4, 1, 0)}$ $\quad = Z_2^2 t_1 - Z_3 Z_4 - Z_2^2(t_1 - t_4)$

$$= Z_2^2 t_4 - Z_3 Z_4$$

15 $F_{(e^2 + 2e^4 + e^7, e^1 + e^5, 0, 0)}$ $\quad = Z_2 Z_4^2 Z_7 - Z_1 Z_5$

16 $F_{(e^2 + e^3 + 2e^4, e^1 + e^6, 0, 0)}$ $\quad = Z_2 Z_3 Z_4^2 - Z_1 Z_6 - Z_2^3 Z_4(t_4 - t_1)$

$$= Z_2^3 Z_4 t_1 - Z_1 Z_6 - Z_2 Z_4 F_{14}$$

17 $F_{(2e^2 + 2e^4, e^1, 0, 1)}$ $\quad = Z_2^2 Z_4^2 - Z_1 t_1 - Z_1(t_4 - t_1)$

$$= Z_2^2 Z_4^2 - Z_1 t_4$$

18 $F_{(e^2 + 2e^4, e^1 + e^8, 1, 0)}$ $\quad = Z_2 Z_4^2 t_1 - Z_1 Z_8 - Z_2 Z_4^2(t_1 - t_3)$

$$= Z_2 Z_4^2 t_3 - Z_1 Z_8$$

19 $F_{(4e^2 + e^4, e^1 + e^3, 0, 0)}$ $\quad = Z_2^4 Z_4 - Z_1 Z_3.$

After reformulating the equations one easily notes that the ideal is indeed toric. To achieve positive exponents in the t_i it was necessary to compute modulo $V^\perp$. These liftings together with the equations of $\mathcal{J}$ describe a family contained in

$$\mathbb{C}^9 \times \mathbb{C}^4 \xrightarrow{\mathrm{pr}_2} \mathbb{C}^4 / \mathbb{C} \cdot (\underline{1}).$$

REFERENCES

[1] Altmann, K., The versal deformation of an isolated toric Gorenstein singularity, *Invent. Math.*, **128** (1997), 443–479.

[2] Altmann, K., One-parameter families containing three-dimensional toric Gorenstein singularities, in: *Explicit Birational Geometry of 3-Folds* (ed. Alessio Corti, Miles Reid), p. 21–50, London Mathematical Society Lecture Note Series, 281, Cambridge University Press 2000.

[3] Altmann, K., Infinitesimal deformations and obstructions for toric singularities, *J. Pure Appl. Algebra,* **119** (1997), 211–235.

[4] Arndt, J., *Verselle Deformationen zyklischer Quotientensingularitäten,* Dissertation, Universität Hamburg (1988).

[5] Bruns, W. and Ichim, B., NORMALIZ, Computing normalizations of affine semigroups. Available from http://www.math.uos.de/normaliz.

[6] Greuel, G.-M., Pfister, G. and Schönemann, H., SINGULAR 3.0. A Computer Algebra System for Polynomial Computations. Centre for Computer Algebra, University of Kaiserslautern (2005), http://www.singular.uni-kl.de.

[7] Matsumura, H., *Commutative Algebra,* W. A. Benjamin, Inc., New York, 1970.

Klaus Altmann & Lars Kastner

Fachbereich Mathematik und Informatik,
Institut für Mathematik,
Freie Universität Berlin,
Arnimalle 3,
14195 Berlin,
Germany

e-mails: altmann@math.fu-berlin.de
           kastner@math.fu-berlin.de

BOLYAI SOCIETY
MATHEMATICAL STUDIES, 23

Deformations of
Surface Singularities
pp. 57–97.

Smoothings of Singularities and Symplectic Topology

MOHAN BHUPAL and ANDRÁS I. STIPSICZ

We review the symplectic methods which have been applied in the classification of weighted homogeneous singularities with rational homology disk (QHD) smoothings. We also review the construction of such smoothings and show that in many cases these smoothings are unique up to symplectic deformation. In addition, we describe a method for finding differential topological descriptions (more precisely, Kirby diagrams) of the smoothings and illustrate this method by working out a family of examples.

1. Introduction

Suppose that $(S, o) \subset \mathbb{C}^n$ is a normal surface singularity, and let $L = S \cap S_\varepsilon^{2n-1}(0)$ denote its link. For simplicity, we will always assume in the following that L is a rational homology 3-sphere, which translates to the singularity having a resolution graph being a tree with only rational curves representing the vertices. The singularity induces a contact structure ξ_M (the Milnor fillable contact structure) on L, which according to [4] is unique up to contactomorphism. (For a quick review of some notions of contact and symplectic topology, see Section 2.) Since a smoothing of the singularity provides a Stein (and therefore symplectic) filling of the contact 3-manifold (L, ξ_M), under favourable circumstances one can study differential topological, or even symplectic topological properties of the smoothings of (S, o) by considering symplectic fillings of (L, ξ_M). This line of arguments was fruitfully applied by Ohta and Ono [20, 21] in finding relations of smoothings of simple and simple elliptic singularities and fillings of the corresponding Milnor fillable contact 3-manifolds. Lisca [14] achieved a complete classification (up to diffeomorphism) of symplectic fillings of cyclic quotient singularities, a classification which later was shown in [19] to be identical to smoothings of the singularities at hand. Further results along similar lines were found in [1] for quotient singularities.

In a slightly different direction, we can try to classify singularities which admit smoothings with some fixed topological property. For example, singularities admitting rational homology disk (QHD for short) smoothings played a crucial role in the recent construction of exotic smooth structures on 4-manifolds with small Euler characteristic [10, 24, 27, 30]. In fact, these constructions led to the discovery of simply connected minimal surfaces of general type with $p_g = 0$ and $K^2 = 1, \ldots, 4$ [13, 25, 26]. These developments motivated our study of singularities with QHD smoothings. Combinatorial constraints of such singularities were found in [31], and a complete classification of the resolution graphs have been achieved in [2] for weighted homogeneous singularities.

The common theme of most of these works can be summarized as follows. Suppose that X is a symplectic filling of (L, ξ_M) (for example, X is a smoothing of the singularity) and find a symplectic manifold Z which is a strong concave filling of (L, ξ_M). Then the filling X can be symplectically glued to Z, resulting in a closed symplectic 4-manifold R. Under some restrictions on Z (which in turn pose restrictions on the singularity), it can be shown that R is a rational surface, hence can be given as a repeated blow-up of the complex projective plane $\mathbb{CP}^2$. Conversely, for an appropriate compatible almost-complex structure, in R we can find almost-complex (-1)-spheres, which we can sequentially blow down and arrive at $\mathbb{CP}^2$ at the end of the procedure. Knowledge about the topology of Z (and, in particular, symplectic surfaces and almost-complex curves in it) restricts the possible positions of the (-1)-curves considerably; hence, under favourable circumstances, we can show that X (which is the complement of Z in R) must be the complement of some well-described configuration of curves in some blow-up of the complex projective plane. This method often provides strong constraints, and can sometimes even determine the (symplectic) topology of X.

We will demonstrate the power of this method by recovering a standard result for cyclic quotient singularities, and prove a special case of the main result of [2] (involving less combinatorics). Finally, we will deal with some lose ends left open in [2]. We give a unified way of constructing the QHD smoothings of the weighted homogeneous singularities discussed in [2]. (Constructions of such smoothings were already given in [31, 33]; here we explicitly describe configurations in blown-up projective planes and use this description to identify the fillings.) To demonstrate our method, we show that in many cases the QHD symplectic filling (and hence the QHD smoothing) is 'symplectically unique'. More precisely, we will show

Theorem 1.1. *Suppose that* Γ *is one of the graphs of Figures 1(a), (b), (c), (d), (e), (f) or (g), that is,* Γ *belongs to one of the families* $\mathcal{W}$, $\mathcal{N}$ *or* $\mathcal{M}$. *Suppose furthermore that* W_1, W_2 *are two minimal weak, symplectic QHD fillings of the Milnor fillable contact structure on the link* Y_Γ *of the singularity* (S_Γ, o). *Then there is a diffeomorphism* $f : W_1 \to W_2$ *such that the pull-back* $f^*(\omega_2)$ *of the symplectic form* ω_2 *on* W_2 *is deformation equivalent to the symplectic form* ω_1 *of* W_1.

Finally, using the description of the smoothings as complements in rational surfaces, we show how to derive smooth topological presentations for them. In particular, we describe a method which produces a Kirby diagram for the smoothing at hand. The method is illustrated by a family of examples worked out in detail.

The paper is organized as follows: In Section 2 we review the main ingredients we take from symplectic topology to study topological properties of smoothings. We also reprove some old (and some more recent) results using these techniques. Section 3 is devoted to a unified treatment of the existence of QHD smoothings of all weighted homogeneous singularities admitting such smoothings. In Section 4 we use the results of [21] (and modify some ideas from [2]) to show that in many cases the QHD smoothing is unique up to symplectic deformation. Finally, in Section 5, we provide Kirby diagrams for some of the smoothings found in earlier sections.

Acknowledgements. AS was supported by OTKA NK81203 and by *Lendület program*. Both authors acknowledge support by Marie Curie TOK project BudAlgGeo.

2. SYMPLECTIC TECHNIQUES

2.1. Symplectic and contact preliminaries

Suppose that Y is a closed, oriented 3-manifold. The oriented 2-plane field $\xi \subset TY$ is called a *contact structure* on Y if there is a 1-form $\alpha \in \Omega^1(Y)$ with $\alpha \wedge d\alpha > 0$ satisfying $\xi = \ker \alpha$. The link of a normal surface singularity admits a natural contact structure: consider the 2-plane field of complex tangencies along the link. In fact, by [4] this structure is unique (up to contactomorphism), and is called the *Milnor fillable* contact structure on Y (provided Y can be presented as the link of a singularity). A 2-form $\omega \in \Omega^2(X)$ on an oriented 4-manifold X is called a *symplectic form* if $d\omega = 0$ and $\omega \wedge \omega > 0$. Suppose now that (X, ω) is a compact, symplectic 4-manifold with boundary $\partial X = Y$, and let ξ be a contact structure on Y. We say that (X, ω) is a *(weak) filling* of (Y, ξ) if $\omega|_\xi > 0$. The filling is *strong* if ω is exact near ∂X with $\omega = d\alpha$ such that $\ker \alpha|_Y = \xi$. Obviously, a strong filling is a weak filling. A filling (X, ω) of (Y, ξ) is *minimal* if there is no embedded sphere in X with homological square (self-intersection) being equal to -1. If Y is a rational homology sphere (that is, $H_*(Y; \mathbb{Q}) = H_*(S^3; \mathbb{Q})$) then any weak filling (X, ω) admits a deformation (X, ω') which is a strong filling.

Suppose now that the contact 3-manifold (Y, ξ) is given as the link of a normal surface singularity (S, o). It is not hard to see that a smoothing of (S, o) is a minimal (strong) filling of (Y, ξ). Therefore when studying smoothings of singularities it is sometimes profitable to consider the slightly modified problem of studying the minimal fillings of the Milnor fillable contact structure on the link.

In many cases it is simpler to study closed (symplectic) 4-manifolds as opposed to ones having boundary. By constructing a *symplectic cap* for (Y, ξ), that is, a symplectic 4-manifold (W, ω_W) with the property that $\partial W = -Y$, and that ω and ξ are compatible as in the case of a strong filling, any strong filling (X, ω) can be embedded into a closed symplectic 4-manifold: consider $Z = X \cup_Y W$ and glue the symplectic structures along Y as described in [8]. (The cap is also called a strong *concave* filling of (Y, ξ).)

Remark 2.1. We point out that the gluing scheme given in [8] applies for strong fillings, while no general gluing theorem is known for weak fillings. Another fact is that any contact 3-manifold admits a cap (i.e. a strong concave filling), while the existence of a (convex) filling is an important property of the contact 3-manifold, and in particular implies that it is tight. Notice also that, although in many cases the cap can be chosen to admit a holomorphic structure, even by taking (X, ω) to be a *Stein* filling (provided, eg. by a smoothing of the singularity if (Y, ξ) is a link of a singularity) we cannot glue in general the convex and the concave fillings within the holomorphic category. The flexibility of symplectic and almost-complex structures therefore can be exploited in studying the (symplectic) topology of holomorphic objects such as smoothings of singularities.

Our study of fillings relies on the following fundamental theorem due to McDuff.

Theorem 2.2 (McDuff, [16, Theorem 1.4]). *Let (Z, ω) be a closed symplectic 4-manifold. If Z contains a symplectically embedded 2-sphere S of self-intersection number 1, then Z is a rational symplectic 4-manifold. In particular, Z becomes the complex projective plane after blowing down a finite collection of symplectic (-1)-curves away from S.* ∎

Our strategy for examining the topology of the fillings of (Y, ξ) therefore will be the following. Suppose that we can construct a convenient cap for (Y, ξ) in which we locate some symplectic 2-manifolds, one of which is a sphere S of self-intersection 1. We then glue the filling (X, ω) to the cap symplectically, and apply McDuff's Theorem 2.2 to the resulting closed symplectic 4-manfold Z. It implies that for any almost-complex structure there must be a (-1)-curve in the complement of S. Choosing the almost-complex structure in a way that the symplectic submanifolds in the cap become holomorphic, we arrive at a combinatorial problem as to where the (-1)-curves may be located. Notice that by assuming that the filling (X, ω) is minimal, we require that no (-1)-curve disjoint from the cap can exist in Z. A number of simple observations (listed in [2, Section 2]) narrow the possibilities of the intersection patterns of the potential (-1)-curves, leading to some potential possibilities. When we sequentially blow down all the (-1)-curves, we should arrive at $\mathbb{CP}^2$, with S being a projective line in it. The curves intersecting S (which therefore are *not* in its complement) should become symplectic submanifolds of $\mathbb{CP}^2$, hence their intersection number with S dictates their self-intersections. The resulting combinatorial problem (in some cases) can be solved, providing partial information about the topology of X. Under favourable circumstances this information is enough to show, for example, that certain singularities do not admit fillings which have the rational homology of the 4-disk. This program has been carried out in [2] for weighted homogeneous singularities.

Notice that the method, as it is, only provides obstructions, and one needs constructive methods to show that in some cases certain types of fillings (or, for singularities, certain smoothings) exist. We will show a method for such constructions in the next section.

2.2. A simple example

To demonstrate the effect of the above strategy, we consider a simple (and by now classical) theorem of McDuff [16, Theorem 1.7] (extended by Lisca [14]) and provide a relatively short proof in this language. (We do not claim that this proof relies on ideas different from the ones given by McDuff and Lisca—the presentation of the same ideas is, however, slightly different.)

Let (S_p, o) be the cyclic quotient singularity which can be given by the resolution graph containing a single $(-p)$-framed vertex and no edges. (We assume that $p \geq 2$ is an integer.) Let $(L_p, \xi_{M,p})$ denote the link of (S_p, o), together with its Milnor fillable contact structure. The 3-manifold L_p is then diffeomorphic to the lens space $L(p, 1)$. Suppose that X is a minimal weak symplectic filling of (L, ξ_M). Let D_p denote the total space of the disk bundle of Euler number $-p$ over the sphere S^2. (Notice that D_p is the plumbing 4-manifold defined by the minimal resolution graph of (S_p, o), hence ∂D_p is diffeomorphic to L_p.)

Theorem 2.3 (McDuff, [16]; cf. also Lisca, [14]). *If $p \neq 4$ then X is diffeomorphic to D_p, and for $p = 4$ the filling X is either diffeomorphic to D_4 or to the complement $\mathbb{CP}^2 - Q$ of a quadratic curve Q in the complex projective plane $\mathbb{CP}^2$.*

Proof. First we construct the cap Z_p for the given singularity. Indeed, consider the Hirzebruch surface $\mathbb{F}_p$, admitting a zero-section Σ_0 of self-intersection $-p$ and an infinity section Σ_∞ of self-intersection p. The zero-section Σ_0 therefore can be identified symplectically (after possibly rescaling) with the resolution of (S_p, o), hence the complement of a convex neighbourhood of Σ_0 in $\mathbb{F}_p$ provides a cap for $(L_p, \xi_{M,p})$. Notice that with this choice the cap contains a rational curve of self-intersection p. Repeatedly blowing this curve up $p-1$ times results in Z_p, which contains a chain of rational curves with self-intersections $+1, -1, -2, \ldots, -2$, with exactly $p - 2$

(-2)'s in this sequence. Now gluing X to Z_p we get a closed symplectic 4-manifold which contains a symplectic sphere L of square $+1$, hence by McDuff's Theorem 2.2 we get that $X \cup Z_p$ is the blow-up of $\mathbb{CP}^2$. Next we would like to analyze where the (-1)-curve in $(X \cup Z_p) - L$ can be located. Choose a tame almost complex structure J for which all of the symplectic curves of Z_p listed above are J-holomorphic. Assume that J is generic among such almost complex structures. Since we can perform the blow-downs in the complement of the $(+1)$-sphere L, at the end of this process all (-2)-curves must be eventually blown down, and the (-1)-curve intersecting L (which is therefore not in the complement of L) should turn into a symplectic surface in $\mathbb{CP}^2$ intersecting L exactly in one point. Therefore it will become a projective line, implying that its self-intersection must change from -1 to $+1$ in the course of the blow-downs. Notice first that in order for the (-2)-curves to be blown down, a (-1)-curve must hit the chain of (-2)'s. It is also easy to see that it can happen only at the ends of the chain of (-2)'s, otherwise after three blow-downs we find a symplectic (hence homologically nontrivial) sphere with square 0 and disjoint from L, which would contradict the fact that $b_2^+(X \cup Z_p) = 1$. If the (-1)-curve E hits the chain in its far end (in the end disjoint from the (-1)-curve intersecting L), then it starts a chain reaction of repeated blow-downs, with all (-2)-curves disappearing and the (-1)-curve D intersecting L becoming a 0-curve. Since D eventually becomes a representative of the generator of $H_2(\mathbb{CP}^2; \mathbb{Z})$, a further (-1)-curve must intersect it, which can be blown down and that drives the intersection of D up to 1.

There is, however, another potential choice for the (-1)-curve: we can choose it to intersect the (-2)-curve which intersects D. This choice again starts a chain reaction of blow-downs, and after blowing down all the (-2)-curves the self-intesection of D will go from -1 to $p-3$. Since it must be equal to 1 after all the blow-downs, we either have $p = 4$ (and then we found that X is a QHD), or $p = 3$, in which case the starting and ending curves of the chain of (-2)'s coincide, and we are in the previous case.

Since for $p \neq 4$ the above argument shows uniqueness, and D_p is a filling of $(L_p, \xi_{M,p})$, we get that X and D_p are diffeomorphic. For $p = 4$ the argument given above provides two possibilities (which differ in b_2), and since D_4 and $\mathbb{CP}^2 - Q$ are both fillings of $(L_4, \xi_{M,4})$, the proof is complete.

∎

Remarks 2.4. (a) After locating the (-1)-curves, the diffeomorphism type of X can be determined via a direct Kirby diagrammatic argument as well; this example will be worked out in detail in Section 5.

   (b) There is another way to reduce the self-intersection of the curve Σ_∞ in $\mathbb{F}_p - \Sigma_0$ from p to 1: blow up the curve in $p - 1$ distinct points. The resulting $p - 1$ curves $D_1, \ldots D_{p-1}$ all become $(+1)$-curves intersecting each other in one point after the blow-down procedure. Therefore either there is a (-1)-curve E intersecting all of them, and then for each i there is a further E_i intersecting only D_i (in which case after the blow-downs all D_i will pass through the same point), or there are (-1)-curves E_1, E_2, E_3 intersecting exactly two D_j's, resulting in the exceptional filling for $p = 4$. Of course, in the blow-up procedure we might also mix the two strategies for blowing up Σ_∞, resulting in similar but slightly different combinatorial problems, providing the same result.

2.3. Starshaped graphs and $\mathbb{Q}$HD smoothings

The classification question of singularities admitting $\mathbb{Q}$HD smoothings has a long history. In [32] examples of such singularities have been given, and in fact J. Wahl had a conjectural list of these singularities back in the 80's. (A reference to this 'secret list' can be found in [5].) More recently, relying on Donaldson's diagonalizability theorem and some intricate combinatorics, strong constraints on the resolution graphs of singularities admitting $\mathbb{Q}$HD smoothings (together with many examples) have appeared in [31]. The application of the symplectic topological scheme of Section 2.2 then provided further obstructions for the existence of $\mathbb{Q}$HD smoothings, eventually leading to the complete resolution of the question for weighted homogeneous singularities [2]. (Curiously, this classification turned out to be almost identical to the conjectured list of Wahl.) The combinatorial argument, completing the symplectic topological proof is rather involved in [2]. To indicate the main ideas, below we verify a weaker result with the same techniques, but with significantly simpler combinatorial issues. It already shows an interesting feature of weighted homogeneous singularities with $\mathbb{Q}$HD smoothings: the framing of the node (the central vertex) cannot be very negative.

Theorem 2.5. *Suppose that Γ is a negative definite plumbing graph of spheres, with all framings ≤ -2. If Γ is starshaped with three legs, and* the node (or central vertex) has framing ≤ -10 then the corresponding singularity admits no QHD smoothing.

Proof. Let (S_Γ, o) denote a normal surface singularity with resolution graph Γ. By assumption Γ is a starshaped plumbing graph (with three legs and central framing ≤ -10). Recall the definition of the *dual* graph Γ' of a starshaped graph Γ. To this end, let Γ be a starshaped graph with s legs $\ell_1, \ldots, \ell_s$ and with central framing $-b$. Suppose that the framing coefficients along the leg ℓ_i are given by the negatives of the continued fraction coefficients of $\frac{n_i}{m_i} > 1$. Consider then the 'dual' graph Γ' which is starshaped with s legs $\ell_1', \ldots, \ell_s'$, central framing $b - s$, and the framings along the leg ℓ_i' are given by the negatives of the continued fraction coefficients of $\frac{n_i}{n_i - m_i}$.

Returning to the proof of the theorem, let Γ' denote dual graph of the given graph Γ. Since the central framing of Γ' can be computed from the number of legs and the central framing of Γ, it follows that the central framing of Γ' is at least 7. Blow up each intersection of the central curve with the three legs twice, and then repeatedly blow up the central curve in such an intersection until its framing becomes $+1$. The resulting three-legged starshaped graph Γ'' then has the property that the framing of its central vertex is $+1$, while the framings of the three vertices connected to the central one (corresponding the three curves which will be denoted by D_1, D_2 and D_3) are all -1. In addition, the framings of the curves F_1, F_2, F_3 intersecting D_1, D_2 and D_3 are all equal to -2. The graph Γ'' gives rise to a plumbing manifold Z, which admits a symplectic structure providing a concave filling for (L_{S_Γ}, ξ_M). Suppose now that (S_Γ, o) admits a QHD filling X. As before, symplectically gluing X and Z we get a closed symplectic manifold containing a symplectic sphere L of self-intersection $+1$, hence the repeated blow-down of $X \cup Z$ in the complement of L transforms it to $\mathbb{CP}^2$. Since X is a QHD smoothing, the symplectic 4-manifold $X \cup Z$ (for a tame almost-complex structure generic among those for which the curves in the cap Z are all almost-complex) contains three (-1)-curves disjoint from L and from each other. Let these (-1)-curves be denoted by E_1, E_2, E_3.

Suppose first that one of the (-1)-curves, say E_1, intersects one of the D_i, say D_1. Then E_1 cannot intersect any other curves apart from another D_i, since after blowing it down and the curve it intersects, the self-intersection of D_1 already went up to 1, and this would increase to at least 2 when F_1 collapses to a point. Since the D_i will eventually become lines in $\mathbb{CP}^2$, this would be a contradiction. If E_1 also intersects both D_2 and

D_3, then after blowing it down all the D_i will pass through the same point, hence E_2 and E_3 can intersect only one leg each, hence nothing will start the blow-down sequence on the remaining third leg. If E_1 intersects only one further curve, say D_2, then E_2 and E_3 should intersect two legs each (the first and third and the second and third), but this forces D_1 and D_2 to intersect each other after all the blow-downs in 2 points, a contradiction. Finally, it can happen that E_1 intersects only D_1 and no further curves. In this case E_2 and E_3 are responsible for D_1 to intersect D_2 and D_3 after the blow-downs, hence E_2 should intersect the first and second, while E_3 the first and third leg. Therefore after completing all blow-downs, the D_i will all pass through a single point, and reversing the last step, we have three 0-curves and a (-1)-curve G passing through them. Adding E_1 to D_1, we see that G (which has to be blown up twice to become disjoint from D_2 and D_3) will become F_1. The curve F_1 has framing -2, but already the blow-ups so far drove the framing of G to -3, and in the course of the blow-up sequence none of the framings will increase. This observation provides the desired contradiction.

Assume now that all E_i are disjoint from the D_i. Since the curves D_i will intersect each other after the blow-down procedure is complete, we must have that E_1 intersects two curves in two different legs, and E_2 with the same property. Now E_3 cannot intersect two legs, since then there would be a cycle of holomorphic curves in the complement of L, which contradicts [2, Corollary 2.4]. So E_3 intersects only one leg. Blow down repeatedly E_3 and the further (-1)-curves created in this procedure. It must stop before it reaches the curves intersected by E_1 or E_2, since otherwise by one further blow-down these curves become 0-curves in the complement of L. So at some point we reach a stage when there are only two (-1)-curves present. Since no E_i intersects the D_i, at the end all three D_i must pass through the same point, hence the ultimate blow-down is a (-1)-curve intersecting each D_i, which all have self-intersection 0. Reversing the procedure, each D_i must be blown up one more time, giving three disjoint (-1)-curves, which will persist all the way through, contradicting the fact that this reversed procedure leads to a configuration with only two (-1)-curves. This argument therefore shows that no appropriate collection $\{E_1, E_2, E_3\}$ of (-1)-curves can exist, therefore our assumption on the existence of a $\mathbb{Q}$HD smoothing led to a contradiction. ∎

Remark 2.6. Similar, but more involved analysis verifies the same result as Theorem 2.5 with the bound -10 substituted with -5. Notice that there are examples of singularities with central framing -4 which admit $\mathbb{Q}$HD smoothings (cf. Figure 1(a)), hence the version of Theorem 2.5 with the value -5 is, in fact, sharp.

In fact, a similar technique provides an extension of Lisca's result [14, Corollary 1.2(b)]:

Theorem 2.7. *Suppose that the starshaped plumbing graph Γ with three legs admits the property that all framings are ≤ -5 and the central framing is ≤ -10. Then the Milnor fillable contact structure on the boundary of the plumbing along Γ admits a unique minimal symplectic filling (up to symplectic deformation). In particular, the smoothings of the corresponding singularity are all diffeomorphic.*

Proof (sketch). It is easy to see that the conditions imply that the central framing of the dual graph Γ' is at least 7, and all other framings are either -2 or -3. In addition, between two (-3)-framed curves there are at least two (-2)-framed curves and each leg ends with at least three (-2)-framed curves. As before, blow up the central curve again until its framing becomes $+1$, and all curves intersecting it have framing -1, resulting in the graph Γ''. Let $Z_{\Gamma''}$ denote the corresponding plumbing, and suppose that X is a minimal symplectic filling of the Milnor fillable contact structure at hand. Once again, we try to locate the (-1)-curves in the union $X \cup Z_{\Gamma''}$. Let the (-1)-curves be denoted by $E_1, \ldots, E_k$. (Since now we have no restriction on the topology of X, we do not have any *a priori* restriction on k.) It is easy to see that a curve E_i cannot intersect a (-2)-curve which is between two (-3)-curves, and a (-3)-curve cannot be intersected by two E_i's. In a similar vein, every (-3)-framed curve must be intersected by one of the E_i's, and since no (-1)-curve can intersect three other curves (unless those are the D_i intersecting the $(+1)$-curve L), we get that none of the E_i can intersect two (-3)-framed curves.

This shows that each (-3)-curve is intersected by a unique E_i, and to initiate the blow-down procedure, the (-2)-curves on the ends of the legs should be intersected by some E_j's as well. This means that after all these blow-downs, all curves (except the D_i) are blown down, and the D_i have self-intersection 0. Therefore a final (-1)-curve is needed, which passes through all D_i exactly once.

The above combinatorial analysis shows that there is a single possibility for the position of the E_i, hence there is a single possibility to recover the configuration, and therefore we can construct X (as the complement of the plumbing $Z_{\Gamma''}$), and then the arguments of the proof of Theorem 1.1 (given in Section 4) apply and imply the claimed uniqueness. ∎

Remark 2.8. Once again, by examining more cases, the assumption on the central vertex can be relaxed to ≤ -5.

The main result of [2] provides a complete classification of resolution graphs of weighted homogeneous singularities which admit QHD smoothings:

Theorem 2.9 (Bhupal–Stipsicz, [2]). *Suppose that Γ is a negative definite, starshaped plumbing graph. There is a singularity with resolution graph Γ which admits a QHD smoothing if and only if Γ is one of the graphs in the families QHD_3 and QHD_4 given by Figures 1 and 2.* ∎

The obstruction for the existence of a QHD smoothing follows along the same line outlined in Theorem 2.5 (and these arguments for various subcases are given in detail in [2]), while the existence of the appropriate smoothings can be derived by applying a result of Pinkham, together with appropriate embeddings of curve configurations into rational surfaces. (The existence results were already verified partly by other means in [32, 33].) The next section is devoted to recalling Pinkham's result, and to listing all the necessary embeddings.

3. EXISTENCE OF SMOOTHINGS

The algebro-geometric result guaranteeing the existence of a QHD smoothing rests on an (appropriately modified) result of Pikham, see [29, 31].

Theorem 3.1 ([29, Theorem 6.7], cf. also [31, Theorem 8.1]). *Suppose that Γ is a negative definite starshaped graph. Let Z be a smooth projective rational surface, and $D \subset Z$ a union of smooth rational curves whose intersection graph is Γ', the dual of Γ. Assume*

$$\operatorname{rk} H_2(D; \mathbb{Z}) = \operatorname{rk} H_2(Z; \mathbb{Z}).$$

If Γ is the graph of a rational singularity, then one has a QHD smoothing of a rational weighted homogeneous singularity with resolution dual graph Γ, and the interior of the Milnor fibre is diffeomorphic to $Z - D$. ∎

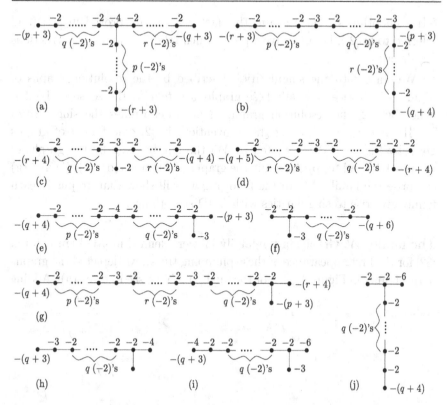

Fig. 1. The graphs defining the class $\mathcal{QHD}_3$ of plumbing graphs. We assume that $p, q, r \geq 0$. Graphs of (a) form the class $\mathcal{W}$, while graphs of (b) and (c) the class $\mathcal{N}$. The graphs given by (d), (e), (f) and (g) form the class $\mathcal{M}$

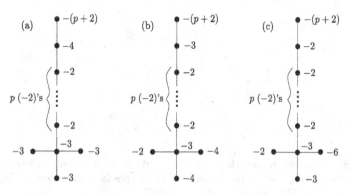

Fig. 2. The graphs (with $p \geq 0$) defining the class $\mathcal{QHD}_4$ of plumbing graphs

In the following (resting on the above theorem) we prove that all the graphs listed in Theorem 2.9 do, in fact, correspond to singularities with

QHD smoothings. Once again, this fact was mainly proved in [31] (and referred to in [2]); here we give a proof which relies on similar constructions for all cases.

We start with the singularities described by the resolution graphs of Figure 1. (Notice that all these graphs are *taut* in the sense of Laufer, hence by [12] the resolution graph, in fact, determines the singularity.) In [31] (and then, following this convention, in [2]) the family of graphs given by Figure 1(a) is denoted by $\mathcal{W}$, the family given by the graphs of Figures 1(b) and (c) by $\mathcal{N}$, while the graphs of Figures 1(d), (e), (f) and (g) comprise the family $\mathcal{M}$. In the following we will show that graphs in each family give rise to singularities with QHD smoothings.

The family $\mathcal{W}$. Graphs in the family $\mathcal{W}$ were defined in [31, Figure 3] (cf. [32] for the first appearance of these plumbing trees); we depicted the graphs of this family in Figure 1(a). For the dual plumbing, see Figure 3(a). Adding

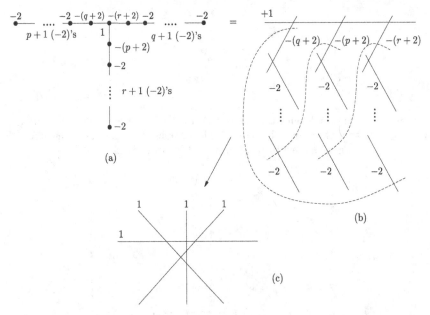

Fig. 3. The dual graphs, the (-1)-curves and the configuration of curves after successively blowing down in the family $\mathcal{W}$. (The graphs in $\mathcal{W}$ are shown in Figure 1(a))

three (-1)-curves to the duals as shown in (b), successive blow-downs results in the configuration shown in Figure 3(c) in the complex projective plane. Since the diagram depicts four generic lines in the complex projective plane, the existence of such a configuration is obvious. Blowing back up we get the dual configuration Γ' in $\mathbb{CP}^2 \# (|\Gamma'|-1)\overline{\mathbb{CP}^2}$, which according to Theorem 3.1 provides the existence of the rational homology disk smoothing. The same statement has been verified in [32] and in [31, Example 8.4].

The family $\mathcal{N}$. Figure 4(a) shows the dual graphs of the triply infinite family of graphs forming $\mathcal{N}$, the family given by the graphs of Figures 1(b) and (c). (Notice that the difference between the $p \geq 0$ and $p = -1$ case,

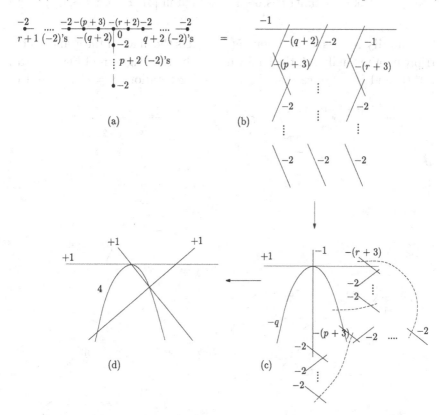

(a) (b) (d) (c)

Fig. 4. The duals, the (-1)-curves and the configuration of curves after successively blowing down in the family $\mathcal{N}$. (The graphs in $\mathcal{N}$ are shown in Figures 1(b) and (c).) When passing from (b) to (c) above, we blow down the horizontal (-1)-curve and one further (-2)-curve to get the $(+1)$-curve depicted by the horizontal line in (c)

which is apparent in the graphs of Figures 1(b) and (c), disappears when we pass to the duals.) The result of one blow-up (shown in Figure 4(b)) and then two blow-downs is shown in (c), where the parabola is tangent to the horizontal line. From this configuration (after successively blowing down the (-1)-curves, starting with the dashed ones) we get a the configuration of a conic and three lines in $\mathbb{CP}^2$. When $p = -1$, the vertical (-1)-curve is missing in (c), and correspondingly the final configuration consists of two lines and a conic. It is elementary to give examples of a conic, a tangent line to it, and two further lines intersecting according to the diagram in Figure 4(d). The reverse of the blow-down procedure, together with Pinkham's Theorem 3.1 shows the existence of the rational homology disk smoothing. Once again, a similar argument for the existence of the rational homology disk smoothing has been presented in [31, Example 8.4].

The family $\mathcal{M}$. As usual, Figure 5(a) depicts the duals of the graphs in the triply infinite family $\mathcal{M}$, the family defined by the diagrams of Figures 1(d), (e), (f) and (g). Notice that the various degenerations $p = -1$, $r = -1$ or

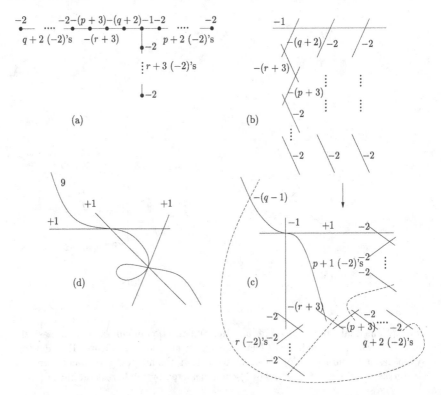

Fig. 5. The dual graphs, the (-1)-curves and the configuration of curves after successively blowing down in the family $\mathcal{M}$. (The graphs in $\mathcal{M}$ are shown by Figures 1(d), (e), (f) and (g).) The horizontal $(+1)$-line is triply tangent to the cubic curve (in (c)) and to the one admitting a self-intersection (in (d))

both (shown in Figure 1) are absorbed by the dual graphs. (For $r = -1$

the vertical -1, together with the (-2)'s hanging off of it are missing, for $p = -1$ the (-2)-curves attached to the horizontal $+1$ are missing, while for $p = r = -1$ both these groups of curves are not there.) Successively blowing down (-1)-curves (starting with the dashed ones) results in the configuration given in (d) (again, for $p = -1$ or $r = -1$ a line is missing, and for $p = r = -1$ two lines are not there). A cubic curve with a transverse double point—for example the one given by $\left\{ y^2 z - x^3 - x^2 z = 0 \right\}$—together with a tangent at one of its inflection points (e.g., $\{z = 0\}$ intersecting it at $[0 : 1 : 0]$) and the two further lines $\{x = 0\}$ and $\{x + y = 0\}$ provide such a configuration. Once again, this argument shows the existence of the rational homology disk filling, which was already verified in [31, Example 8.3].

For future reference, let the cubic curve $\{y^2 z - x^3 - x^2 z = 0\}$ be denoted by C, while the lines $\{z = 0\}$, $\{x = 0\}$ and $\{x + y = 0\}$ by L_z, L_x, L_1, respectively.

The family of Figure 1(h). As before, Figure 6 provides the diagrams for the graphs, their duals, and for the (-1)-curves. The nodal cubic curve C with the tangent L_z at one of its inflection points, together with L'_1 (given by the equation $\{x + z = 0\}$), intersecting the cubic curve once transversally at the point $[0 : 1 : 0]$ and once tangentially at $[-1 : 0 : 1]$), and the line L_y joining $[-1 : 0 : 1]$ with the node $[0 : 0 : 1]$, provide a configuration of curves shown by Figure 6(d). As before, repeated blow-ups then embeds the curves intersecting according to Γ' into $\mathbb{CP}^2 \# \left(|\Gamma'| - 1 \right) \overline{\mathbb{CP}}^2$, which (by Pinkham's Theorem) verifies the existence of a QHD smoothing.

The family of Figure 1(i). The graphs, their duals and the possible locations of the (-1)-curves in the family given by Figure 1(i) are shown in Figure 7. Consider the cubic C and the tangent L_z to it. Let M denote the tangent line $\left\{ y - \left(x + \frac{8}{9} z \right) \sqrt{3} i = 0 \right\}$ passing through another inflection point $\left[-\frac{4}{3} : -i\frac{4}{3\sqrt{3}} : 1 \right]$ of $C = \left\{ y^2 z - x^3 - x^2 z = 0 \right\}$. (It is not hard to see that the further two inflection points of C are $\left[-\frac{4}{3} : \pm i\frac{4}{3\sqrt{3}} : 1 \right]$.) Having these curves in $\mathbb{CP}^2$, the rest of the argument is identical to the previous cases: the appropriate blow-ups embed the dual graphs in the right number of blow-ups of $\mathbb{CP}^2$ and then an application of Pinkham's Theorem 3.1 completes the argument.

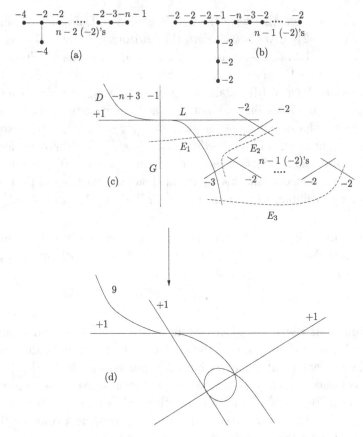

Fig. 6. The one-parameter family of graphs of Figure 1(h), their duals, the blow-down sequence and the final configuration

The family of Figure 1(j). The diagrams of Figure 8 show the graphs, their duals and the locations of the (-1)-curves for the cases given by Figure 1(j). The cubic curve C with a transverse double point and its tangent L_z at one of its inflection points together with the line L_1' (given by the equation $\{x + z = 0\}$) gives the configuration depicted in Figure 8(d). Repeated blow-ups and an application of Theorem 3.1 then verifies the existence of a $\mathbb{Q}$HD smoothing for the singularities given by the graphs under consideration.

With this last case we are finished with the analysis of the graphs of Figure 1, and we start examining starshaped graphs with four legs, given by Figure 2. The existence of $\mathbb{Q}$HD smoothings for singularities with resolution

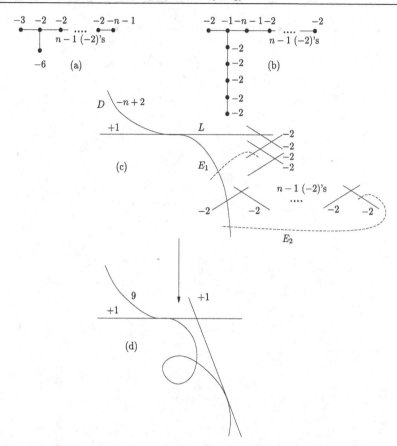

Fig. 7. The one-parameter family of graphs of Figure 1(i), their duals, the location of the (-1)-curves and finally he required configuration of curves in the complex projective plane

graph shown in that Figure has been verified in [31] and (by different methods) in [33].

The family of Figure 2(a). The graphs and their duals, and the two possible locations of (-1)-curves in this case are shown in Figure 9. In order to show that curves intersecting each other according to Γ' can be embedded in $\mathbb{CP}^2 \# \left(|\Gamma'| - 1 \right) \overline{\mathbb{CP}^2}$, we consider the usual singular cubic curve C and its tangent L_z together with the cubic C_1 given by the equation $f_1(x, y, z) = y^2 z + \left(1 - i\sqrt{3} \right) xyz + \frac{4}{9}\left(3 - i\sqrt{3} \right) yz^2 + \frac{1}{2}\left(-1 + i\sqrt{3} \right) x^3 + \left(-2 + i\sqrt{3} \right) x^2 z - \frac{4}{9}\left(-3 + i\sqrt{3} \right) xz^2$. This curve is a rational nodal cubic

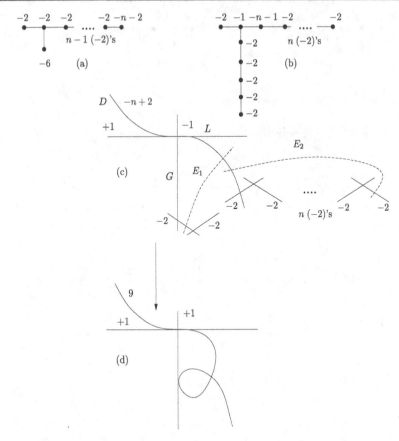

Fig. 8. The one-parameter family of graphs of Figure 1(j), their duals, the (-1)-curves and the final configuration in $\mathbb{CP}^2$

with a node at $\left[-\frac{4}{3} : -\frac{4}{9}i\sqrt{3} : 1\right]$. The line L_z and the curves C and C_1 are pairwise triply tangent at $[0 : 1 : 0]$. Also the curves C and C_1 intersect at each of the points $[0 : 0 : 1]$ and $\left[-\frac{4}{3} : -\frac{4}{9}i\sqrt{3} : 1\right]$ with intersection multiplicity 3. Let N be the line $\left\{y - i\sqrt{3}\left(x + \frac{8}{9}z\right) = 0\right\}$; it is triply tangent to C at $\left[-\frac{4}{3} : -\frac{4}{9}i\sqrt{3} : 1\right]$ and intersects C_1 at the same point with intersection multiplicity 3. Therefore the configuration of curves depicted by Figure 9(d) is verified to exist, from which the appropriate sequence of blow-ups verifies the existence of the embedding of curves with intersection pattern given by Figure 9(b). A simple count of blow-ups shows that the resulting configuration is in $\mathbb{CP}^2 \# \left(|\Gamma'| - 1\right)\overline{\mathbb{CP}^2}$, hence the existence of the smoothing then follows from Pinkham's Theorem 3.1.

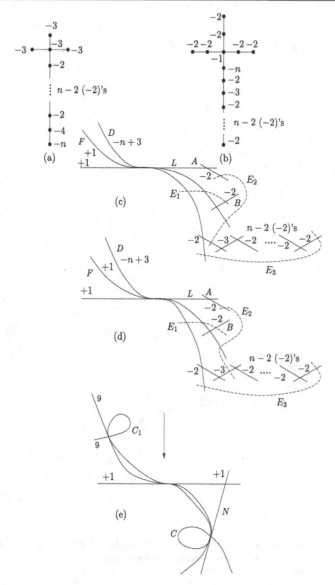

Fig. 9. The one-parameter family of 4-legged graphs of Figure 2(a), the duals, and the two possible configurations of (-1)-curves. In (e) we depict the curve configuration after the blow-downs—this configuration is the same for the two choices of (-1)-curves shown by (c) and (d)

The family of Figure 2(b). The graphs, their duals and the (-1)-curves are shown in Figure 10. The curves L_z and C are as before; let C_2 be the

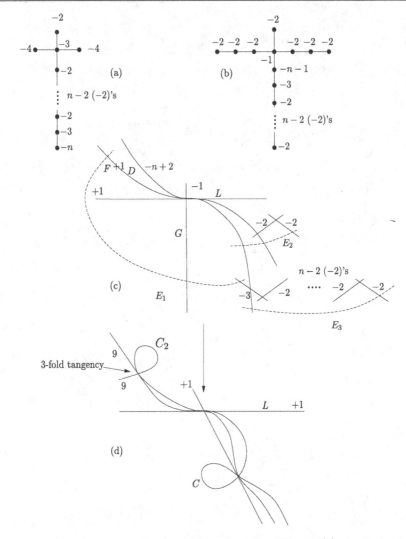

Fig. 10. The one-parameter family of 4-legged graphs of Figure 2(b), the duals, the (-1)-curves and the configuration in $\mathbb{CP}^2$

cubic curve given by the equation $f_2(x, y, z) = y^2 z + 2xyz + 2yz^2 - 2x^3 - 4x^2 z - 2xz^2$. The curve C_2 is a rational nodal cubic with a node at $[-1 : 0 : 1]$, and L_z, C and C_2 are pairwise triply tangent at $[0 : 1 : 0]$. Also, C and C_2 intersect at $[0 : 0 : 1]$ with intersection multiplicity 4 and at $[-1 : 0 : 1]$ with intersection multiplicity 2. Consider furthermore L_1' given by the equation $\{x + z = 0\}$. It passes through the point $[0 : 1 : 0]$ and is tangent to C at $[-1 : 0 : 1]$. Therefore the existence of the configuration of curves depicted by Figure 10(d) is verified, from which the appropriate sequence of blow-ups verifies the existence of the embedding of curves with intersections given by Figure 10(b). The existence of the smoothing of a (weighted homogeneous) singularity with resolution graph given by Figure 10(a) then follows from Pinkham's Theorem 3.1.

The family of Figure 2(c). The graphs, their duals and the (-1)-curves are shown in Figure 11. As before, let L_z be the line $\{z = 0\}$ in $\mathbb{CP}^2$ and let

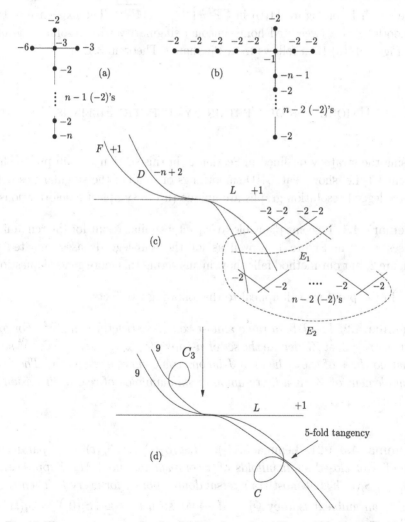

Fig. 11. The one-parameter family of 4-legged graphs of Figure 2(c), the duals, the (-1)-curves and finally the curve configuration in $\mathbb{CP}^2$ we get after the blow-downs

C be the usual cubic given by $f(x, y, z) = y^2 z - x^3 - x^2 z$. Let C_3 denote the cubic given by the equation $f_3(x, y, z) = y^2 z + \frac{1}{2} xyz + yz^2 - \frac{9}{8} x^3 - 2x^2 z - xz^2$. The curves C and C_3 are rational nodal cubics with nodes at $[0 : 0 : 1]$ and $\left[-\frac{2}{3} : -\frac{1}{3} : 1 \right]$, respectively. It is easy to check that both C and C_3 are triply tangent to L_z at the point $[0 : 1 : 0]$ and are also triply tangent to each other at $[0 : 1 : 0]$, and have intersection multiplicity 6 at the point $[0 : 0 : 1]$. Therefore the existence of the configuration of curves depicted by Figure 11(d) is verified, from which the appropriate sequence of blow-ups shows the existence of the embedding of curves with intersections given by the graph Γ' of Figure 11(b) in $\mathbb{CP}^2 \# \left(|\Gamma'| - 1 \right) \overline{\mathbb{CP}^2}$. The existence of the smoothing of a (weighted homogeneous) singularity with resolution graph of Figure 11(a) then follows from Pinkham's Theorem 3.1.

4. On Uniqueness of Certain Symplectic Fillings

Using the strategy outlined in Section 2, in this section we will prove Theorem 1.1, i.e. show that $\mathbb{Q}$HD smoothings of some of the singularities with three-legged resolution graphs are unique (up to symplectic deformations).

Remark 4.1. It seems plausible to expect a similar result for the remaining cases given in Figure 1 as well as for the four-legged cases depicted in Figure 2, but our method falls short in answering this more general question.

In the proof we will appeal to the following two facts.

Lemma 4.2. Let J be a tame almost complex structure on $\mathbb{CP}^2$. For any integer $n \geq 3$ let Z_n denote the set of n-tuples $(z_1, z_2, \ldots, z_n) \in \left(\mathbb{CP}^2 \right)^n$ such that no three of the z_i lie on a J-holomorphic curve of degree 1. Then the complement of Z_n is a finite union of submanifolds of real codimension 2. ∎

Lemma 4.3 ([21, Lemma 5.11]). Let $S_1(t), \ldots, S_k(t)$ be 1-parameter families of closed submanifolds of a compact manifold M. Suppose that $\bigcup_{1 \leq i \leq k} S_i(t)$ has at most transversal double points for every t. Then there exists an ambient isotopy $\Phi_t : M \to M$ such that $\Phi_t \left(S_i(0) \right) = S_i(t)$ for all t. ∎

Proof of Theorem 1.1 in case $\Gamma \in \mathcal{W} \cup \mathcal{N}$. Suppose that Γ is one of the graphs of Figures 1(a), (b) or (c). Let W be a minimal weak symplectic QHD filling of the Milnor fillable contact structure on the link Y_Γ of the singularity (S_Γ, o). After symplectically gluing the symplectic cap Z_Γ to W denote the resulting rational symplectic 4-manifold by R. After repeatedly blowing down all pseudoholomorphic (-1)-curves in R we will arrive at one of the configurations of pseudoholomorphic curves given in Figure 3(c) (if Γ is one of the graphs of Figure 1(a)) or in Figure 4(d) (if Γ is one of the graphs of Figure 1(b) or (c)) in $\mathbb{CP}^2$ for a tame almost complex structure J_1.

First suppose that we end up with a configuration of four pseudoholomorphic curves $\{C_i\}$ of degree 1 in $\mathbb{CP}^2$ as depicted in Figure 3(c). Note that we can specify four pseudoholomorphic curves of degree 1 in general position (that is, no three meet in a point) by specifying four points, no three of which lie on a degree 1 curve, by declaring, say, the curves to be those which pass through the first and second points, the third and fourth points, the first and third points and the second and fourth points. Now suppose that our given configuration $\{C_i\}$ is specified by the four points $\{p_i\}_{1 \leq i \leq 4}$. Let $\{q_i\}_{1 \leq i \leq 4}$ be the four points $[1:0:0]$, $[0:1:0]$, $[0:0:1]$ and $[1:1:1]$, respectively, in $\mathbb{CP}^2$ and let J_0 denote the standard complex structure on $\mathbb{CP}^2$. Let $\{L_i\}$ denote the four J_0-holomorphic curves of degree 1 specified by the four points $\{q_i\}$. We will show that we can find a 1-parameter family of four embedded spheres whose union has at most transversal double points starting with the four embedded spheres $\{L_i\}$ and ending with the four embedded spheres $\{C_i\}$. To this end, let J_t, $t \in [0,1]$, be a family of tame almost complex structures connecting J_0 to J_1. By Lemma 4.2, we can choose a 1-parameter family of four points $\{p_i(t)\}_{1 \leq i \leq 4}$, $t \in [0,1]$ connecting $\{q_i\}$ to $\{p_i\}$ such that for each t no three of the points $\{p_i(t)\}$ lie on a J_t-holomorphic curve of degree 1. For $t \in [0,1]$ let $\{C_{i,t}\}$ be the four J_t-holomorphic curves of degree 1 specified by the four points $\{p_i(t)\}$. Then $\{C_{i,t}\}$ is a 1-parameter family of four embedded spheres whose union has at most transversal double points starting with the four embedded spheres $\{L_i\}$ and ending with the four embedded spheres $\{C_i\}$ as desired. For each t, we now reverse the sequence of blow-downs described in Figure 3 to obtain a 1-parameter family of configurations in $\mathbb{CP}^2 \# (|\Gamma'| - 1) \overline{\mathbb{CP}}^2$ of the type given in Figure 3(a) starting with a configuration Γ'_0 derived from the spheres $\{L_i\}$ and ending with the original dual configuration Γ' in the complement of W. Now, applying Lemma 4.3, we can find an ambient isotopy Φ_t of the pair $\left(\mathbb{CP}^2 \# (|\Gamma'| - 1) \overline{\mathbb{CP}}^2, \Gamma'_0\right)$ with the pair $\left(\mathbb{CP}^2 \# (|\Gamma'| - 1) \overline{\mathbb{CP}}^2, \Gamma'\right)$. The uniqueness of symplectic QHD fillings of Y_Γ up to symplectic deformation equivalence follows at once.

Now suppose that we end up with a configuration of the type given in Figure 4(d) in $\mathbb{CP}^2$. Here we note that given five points in $\mathbb{CP}^2$, no three of which lie on a pseudoholomorphic curve of degree 1, there is a unique nonsingular pseudoholomorphic curve of degree 2 which contains the five points. Thus any set of five points $\{p_i\}_{1\leq i\leq 5}$ in $\mathbb{CP}^2$, no three of which lie on a pseudoholomorphic line, defines a unique configuration of the type given in Figure 4(d). Namely, take the unique pseudoholomorphic curve of degree 2 through the five points together with a pseudoholomorphic curve of degree 1 tangent to the degree two curve at p_1 and the pseudoholomorphic curves of degree 1 through p_1 and p_3, and p_2 and p_3. The rest of the argument is as before. ∎

The proof of Theorem 1.1 in case $\Gamma \in \mathcal{M}$ can be conveniently phrased by using a slightly different compactification from the one we encountered in Section 3. We show the sequence of blow-downs and blow-up in Figure 12 providing this alternative compactification. The proof of Theorem 1.1 in

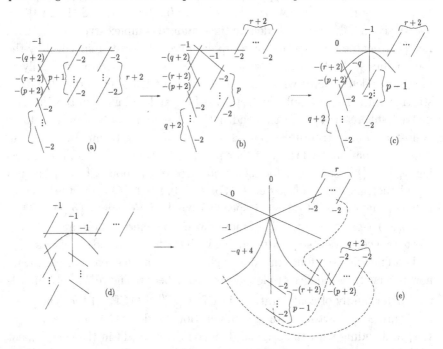

Fig. 12. A sequence of blow-downs and blow-up resulting in another compactifying divisor for a singularity defined by one of the graphs of Figure 1(d), (e), (f) or (g)

this case rests on the following theorem of Ohta and Ono.

Theorem 4.4 (Ohta–Ono, [21, Theorem 5.8]). *Suppose that J_1 is a tame almost complex structure on $\mathbb{CP}^2$. Let D_1 be a J_1-holomorphic cuspidal rational curve which represents the class $3[\mathbb{CP}^1]$. Let q denote the cusp point of D_1. Let D_0 be a cuspidal cubic curve in $\mathbb{CP}^2$ with respect to the standard complex structure J_0 such that the cusp point of D_0 is also at q. Suppose that $\{J_t\}$ is a one-parameter family of tame almost complex structures joining J_0 with J_1. Then there exists a one-parameter family $\phi_t : \mathbb{CP}^1 \to \mathbb{CP}^2$ of pseudoholomorphic maps representing the class $3[\mathbb{CP}^1]$ such that ϕ_t is J_t-holomorphic for each t, $\phi_t(\mathbb{CP}^1)$ connects D_0 to D_1, and each curve $D_t = \phi_t(\mathbb{CP}^1)$ has exactly one nonimmersed point which is of multiplicity 2 and is at q.* ■

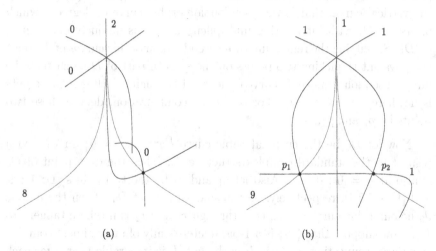

Fig. 13. Blowing up the transversal point of intersection of the cuspidal curve with the $(+2)$-curve on (a) and then blowing down the two (-1)-curves which are the proper transforms of the ruling curves passing through that point, we get (b)

Proof of Theorem 1.1 in case $\Gamma \in \mathcal{M}$. Suppose that Γ is one of the graphs of Figures 1(d), (e), (f) or (g). Let W be a minimal weak symplectic QHD filling of the Milnor fillable contact structure on the link Y_Γ of the singularity (S_Γ, o). After symplectically gluing the symplectic cap Z_Γ, which is a regular neighbourhood of the configuration of solid curves given in Figure 12(e), to W, denote by R the resulting rational symplectic 4-manifold. Since W is a minimal QHD filling, it can be shown that there must be three (-1)-curves in R intersecting the configuration of pseudoholomorphic curves in Z_Γ as indicated by the dashed curves in the Figure 12(e). Blowing down these three (-1)-curves and then successively blowing down other (-1)-curves that are formed in the process, we eventually arrive at the configuration given in Figure 13(a) in $\mathbb{CP}^1 \times \mathbb{CP}^1$. Now blowing up the point where the $(+2)$-curve and the cuspidal curve intersect transversely and then blowing down the two (-1)-curves which are the images of ruling curves passing through that point we arrive at the configuration of pseudoholomorphic curves given in Figure 13(b) in $\mathbb{CP}^2$ for some tame almost complex structure J_1 on $\mathbb{CP}^2$. Denote by D_1 the cuspidal curve and by q its cusp point. By a change of coordinates on $\mathbb{CP}^2$, if necessary, we may assume that $q = [0:0:1]$. Note that we can assume that J_1 is integrable near q since we can assume that the original almost complex structure on R is integrable in a neighbourhood of the compactifying divisor. Now consider the four pseudoholomorphic curves of degree 1 shown in Figure 13(b). Clearly the vertical curve, that is the pseudoholomorphic curve of degree 1 which intersects D_1 with intersection multiplicity 3 at q, is uniquely determined by D_1. Specifying the remaining three pseudoholomorphic curves of degree 1 is equivalent to picking two points on the smooth part of D_1 such that the unique pseudoholomorphic curve of degree 1 through the chosen two points is nowhere tangent to D_1. For our given configuration, denote these two points by p_1 and p_2.

Now let D_0 be the cuspidal cubic curve $\{y^2 z - x^3 = 0\}$ in $\mathbb{CP}^2$ with respect to the standard complex structure J_0. Then the cusp point of D_0 is also at $q = [0:0:1]$. Also let r_1 and r_2 be the two points $[1:1:1]$ and $[1:-1:1]$, respectively, on the smooth part of D_0. Then the unique J_0-holomorphic curve of degree 1 through r_1 and r_2 is nowhere tangent to D_0. Now suppose that $\{J_t\}$ is a 1-parameter family of tame almost complex structures connecting J_0 with J_1 such that J_t is integrable near q for each t. By Ohta and Ono's Theorem 4.4, there exists a one-parameter family $\phi_t : \mathbb{CP}^1 \to \mathbb{CP}^2$ of pseudoholomorphic maps representing the class $3[\mathbb{CP}^1]$ such that ϕ_t is J_t-holomorphic for each t, $\phi_t(\mathbb{CP}^1)$ connects D_0 to D_1, and each curve $D_t = \phi_t(\mathbb{CP}^1)$ has exactly one nonimmersed point which is of multiplicity 2 and is at q. Since the almost complex structure is integrable near q, it follows that the singular point q of D_t is necessarily a $(2,3)$-cusp point for each t (c.f. [21]). By reparametrising the domain, if necessary, we may assume that ϕ_t maps $\infty \in \mathbb{CP}^1 = \mathbb{C} \cup \{\infty\}$ to the cusp point q for each t.

Now consider the 1-parameter family $\{D_t\}$ of cuspidal pseudoholomorphic curves of degree 3. For a point p on the smooth part of D_t, let L_p

denote the unique J_t-holomorphic curve of degree 1 passing through p and tangent to D_t at p. Then either L_p intersects D_t at the point p only (with intersection multiplicity 3) or else it intersects D_t at one further point. Let $\widetilde{p}$ be equal to p in the first case and the intersection point of L_p and D_t different from p in the second case. Now set

$$\Sigma_{1,t} = \left\{ (z_1, z_2) \in \mathbb{C} \times \mathbb{C} \mid \widetilde{\phi_t(z_1)} = \phi_t(z_2) \right\}$$

$$\Sigma_{2,t} = \left\{ (z_1, z_2) \in \mathbb{C} \times \mathbb{C} \mid \widetilde{\phi_t(z_2)} = \phi_t(z_1) \right\}$$

Also let Δ denote the diagonal in $\mathbb{C} \times \mathbb{C}$. Then the real dimension of Δ is 2 and that of $\Sigma_{1,t}$ and $\Sigma_{2,t}$ is also 2 for all t. Thus we may choose a path $(z_1, z_2) : [0,1] \to \mathbb{C} \times \mathbb{C}$ beginning at $\left(\phi_0^{-1}(r_1), \phi_0^{-1}(r_2) \right)$ and ending at $\left(\phi_1^{-1}(p_1), \phi_1^{-1}(p_2) \right)$ such that for all t $\left(z_1(t), z_2(t) \right) \in \mathbb{C} \times \mathbb{C} \backslash (\Delta \cup \Sigma_{1,t} \cup \Sigma_{2,t})$. For such a path, the pair $\left(p_1(t), p_2(t) \right)$, where $p_i(t) = \phi_t \left(z_i(t) \right)$ for $i = 1, 2$, is such that the unique J_t-holomorphic curve of degree 1 through $p_1(t)$ and $p_2(t)$ intersects D_t transversely in 3 points for all t. For each t now consider the cuspidal J_t-holomorphic curve D_t of degree 3, the J_t-holomorphic curve of degree 1 intersecting D_t with multiplicity 3 at q and the three J_t-holomorphic curves of degree 1 passing through each pair of the three points $\left\{ q, p_1(t), p_2(t) \right\}$. Now by reversing the sequence of blow-downs and blow-ups that were used to obtain the configuration in Figure 13(b) from the configuration in Figure 12(a), we can obtain a 1-parameter family of configurations of the type given in Figure 12(a) in $\mathbb{CP}^2 \# \left(|\Gamma'| - 1 \right) \overline{\mathbb{CP}^2}$ starting with a configuration Γ_0' derived from the cuspidal cubic D_0, the line $\{ y = 0 \}$ and the three lines passing through each pair of the three points $\{ q, r_1, r_2 \}$ and ending with the original dual configuration Γ' in the complement of W. The proof is now completed by appealing to Lemma 4.3. ∎

5. The Differential Topology of Some of the QHD Smoothings

The description of a QHD smoothing of a singularity as the complement of the compactifying divisor in some rational surface can be conveniently used to describe the smoothing in differential topological terms. In particular, in this section we will describe a method (and demonstrate it with a few examples) for determining a *Kirby diagram* of the QHD smoothing of a singularity at hand. Kirby diagrammatic descriptions are already available for the QHD smoothings of the singularities given by the diagrams of Figure 1(a), (b) and (c) in [7]; in that paper a method relying on monodromy substitutions and Lefschetz fibrations is used. Here we will use a more direct method. In our subsequent examples we will focus on some of the special cases encountered in the earlier sections.

Recall that a smooth, compact 4-manifold (with possibly non-empty boundary) admits a Morse function with finitely many critical points, and with a unique local minimum. Such a Morse function induces a handle decomposition on the manifold, and Kirby diagrams are designed to record the handle decompositions. Indeed, (4-dimensional) 1-handles can be pictorially presented as dotted and unknotted circles in S^3 (which are unlinked from each other), meaning that if we take a spanning disk properly embedded in D^4 for every dotted circle (disjoint from the others) and delete their open tubular neighbourhoods, then the result will be diffeomorphic to the result of attaching 1-handles to D^4. Then 2-handles are attached along framed circles to the boundary of the union of 0- and 1-handles, hence the handles with indices ≤ 2 can be captured by a diagram in S^3 involving some number of dotted circles (all unknotted and unlinked from the other dotted circles), and some further integer-framed knots. If the 4-manifold has nonempty boundary, and the Morse function has critical points of index ≤ 2, then this picture is complete (this is the case, for example, for all Stein domains). If the 4-manifold is closed, then the 3- and 4-handles can be attached in a unique manner (up to diffeomorphism), hence we do not need to record them.

A fixed 4-manifold, however, can be presented in many different ways, the correspondence among these different presentations is given by the set of Kirby moves in *Kirby calculus*. The moves simply correspond to handle slides and handle cancellations of handles of index n and $n+1$. We will not give a complete treatment of this theory here (cf. [11] for a more thorough discussion), but just highlight the moves we will use later.

A pair of index-n and index-$(n + 1)$ handles can be cancelled if the attaching circle of the $(n + 1)$-handle intersects the belt sphere of the n-handle in a unique point. In the pictorial presentation this means that a 1/2-handle pair cancels if the circle of the 2-handle intersects the spanning disk (in S^3) of the dotted circle corresponding to the 1-handle in a unique point. A 2/3 pair cancels if the circle of the 2-handle can be separated from the rest of the picture, and it comes out as a 0-framed trivial knot. A 2-handle can be slid over any other 2-handle, the circle corresponding to the result is the connected sum of the two circles (along an arbitrarily chosen band connecting the two circles), and the new framing is the sum of the framings modified by (± 2)-times the linking number of the original knots. (The sign depends on whether the band respects or disrespects some chosen orientations on the two knots, choices which also influence the sign of the linking number.)

One specific case of this move can be summarized as follows: if there is a (± 1)-framed unknot such that k further arcs pierce its spanning disk in S^3, then the unknot can be displaced to a disjoint unknot (having the same framing), but the k arcs undergo a full (∓ 1)-twist (with framings changing depending on how many arcs fall into the same connected component of the original link, cf. [11, Figure 5.18] and the text therein). For the pictorial presentation of the above said, see Figure 14 (without the framings).

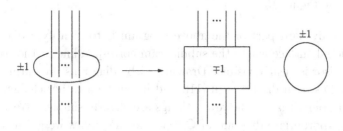

Fig. 14. The diagram describes the effect of blowing up (or down) a sphere of self-intersection ± 1

Recall that in our case the smoothing under consideration is given as the complement of the compactifying divisor K in a suitable rational surface R. Our first aim is to give a handle decomposition of the rational surface, in which the plumbing corresponding to the compactifying divisor K is explicitly visible, and then consider the complement of that plumbing. In the

diagram this means that the part of the diagram building up the plumbing is treated differently: as a collection of handles defining the 3-manifold
on which the handles giving the complement of the plumbing rest. Such
a diagram is called a 'relative Kirby diagram', and is much less popular
than its absolute counterpart (in which case we build the diagram on S^3).
The relative diagram can be turned into an absolute one by 'turning it upside down', which is rather simple for the 2-handles—the attaching circle
of the dual is a 0-framed meridian of the knot corresponding to the original 2-handle, but rather complicated for the 3-handles (since their duals
are 1-handles, which are 'under' the 2-handles). To overcome this difficulty,
we will peel off the 3- and 4-handles of the rational surface R, then turn
the 2-handles in the complement of the compactifying divisor upside down,
and then apply Kirby moves among the original handles (which built up
the 3-manifold $\#_n S^1 \times S^2$ for some n, the boundary of the union of the
3- and 4-handles). We do 3-dimensional handle moves (i.e. we change the
4-manifold, but keep the boundary 3-manifold the same), until the diagram
comprises the n-component 0-framed unlink. This corresponds to n embedded 2-spheres with trivial normal bundle, hence we can perform surgery
along them, arriving at the collection of 1-handles we searched for. Pictorially this last step amounts to trading the 0-framing for a dot on the
corresponding unknot. We also need to reverse the orientation on the result
(which can be done by considering the mirror image of the diagram, with
all framings multiplied by -1) to compensate for the orientation reversal
we introduced when turning the handles upside down. The schematic plan
is given by Figure 15.

The only hard part of the above program is to identify a diagram for
the rational surface where the subdiagram corresponding to the compactifying divisor is clearly visible. Drawing Kirby diagrams presenting specific
surfaces has been discussed in [11], and it can be a rather challenging exercise. In our situation, however, things are slightly simpler. We will start
with an appropriate diagram of $\mathbb{CP}^2$ where the curves we get in the final
blow-down are visible. That step, again, might be complicated in general,
but in our cases we only need to deal with projective lines and cubic curves,
cf. the discussion in [11, Example 6.2.7]. Then the argument for identifying the smoothing provides a blow-up sequence which gives the embedding
of the compactifying divisor in the appropriate rational surface, and which
has exactly the smoothing as its complement. We will start by working out
the description of the fillings given by Theorem 2.3. As given in the proof,
the neighbourhood of the compactifying divisor is a linear plumbing on p

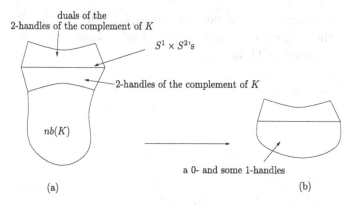

duals of the
2-handles of the complement of K

$S^1 \times S^2$'s

—2-handles of the complement of K

$nb(K)$

a 0- and some 1-handles

(a)

(b)

Fig. 15. The diagrams give a schematic description of the strategy for describing the Kirby picture of a smoothing. $nb(K)$ denotes the neighbourhood of the configuration K for which we take the complement

vertices, the first two with framings $+1$ and -1, and all the others with framing -2. We get this picture by blowing up a pair of lines in the projective plane. In diagrams, we consider a positive Hopf link, with both circles $(+1)$-framed. Indeed, by sliding one off the other, we get a 2-component unlink, with framings 0 and $+1$, so adding a 3- and a 4-handle we get $\mathbb{CP}^2$, i.e. the Hopf link (with the $(+1)$-framings and the 3- and 4-handles) really provides a diagram for the projective plane and the curve configuration (of two generic projective lines) in it. Now blowing up one of the curves, and then repeatedly blowing up the (-1)-curve, and finally blowing up the 0-framed circle (which was one of the knots in the Hopf link) we get a presentation of the compactifying divisor in the blown-up rational surface. In addition we also see the (-1)-spheres we found in the proof of Theorem 2.3 (for $p \neq 4$); in the diagram of Figure 16(a) they are symbolized by dashed circles. Now peel off the 3- and 4-handles, and turn the complement of the compactifying divisor upside down. This amounts to adding 0-framed meridians to the dashed circles. The framings of these new 0-framed meridians now are distinguished by putting them into brackets. The corresponding Kirby diagram is shown by Figure 16(a). After sequentially blowing down the two dashed (-1)-curves and all the original (-2)-curves, we arrive at the diagram of Figure 16(b). Blow down the $(+1)$-curve which is unlinked from the bracketed curves, and then slide the $(p-1)$-framed (bracketed) curve over the other bracketed one. The resulting diagram is given by Figure 16(c). Now converting the (unbracketed) 0-framing to a dot and then cancelling the 1-handle/2-handle pair, and finally reversing the orientation (by taking the mirror image of the unknot, and multiplying its framing by -1), we arrive at the diagram of a single $(-p)$-framed unknot, the Kirby diagram of D_p, as claimed in Theorem 2.3.

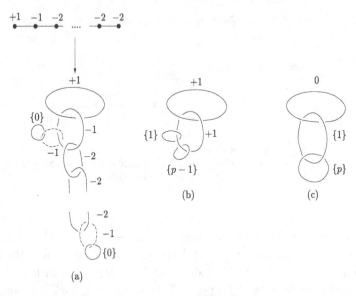

(a)

(b)

(c)

Fig. 16. The blow-down procedure, with the extra (fine) curves, and finally surgery and handle cancellation provides the diagram of the smoothing of the singularities discussed in Theorem 2.3 (for $p \neq 4$)

Recall that there was an exceptional case for $p = 4$. In that case a further possible blow-up sequence was possible: after blowing up one of the unknots in the Hopf link, blow up the intersection of the (-1)-curve with the original projective line, and then blow up this last (-1)-curve. The corresponding blown-up rational surface with the appropriate compactifying divisor and the (-1)-curve in it is shown in Figure 17(a). (We apply the previous convention of drawing the last (-1)-curve dashed, and the dual 0-framed handle to it by fine line, and put the 0-framing into brackets.) Reverse the above blow-up sequence, that is, blow down e_2, e_3 and e_4 of Figure 17(a), and get (b). Blowing down the $(+1)$-framed circle and reversing orientation we get Figure 17(c), which is, after switching the 0-framing for a dot (i.e. surgering the 2-handle into a 1-handle) is exactly the diagram of [11, Figure 8.41] (with $p = 2$ in that convention). In fact, the same diagram also shows that the 4-manifold we get is the complement of the conic in $\mathbb{CP}^2$: It is rather easy to see that if we consider Figure 17(b) with the {0}-framed circle deleted, and we add to it a 3- and a 4-handle, then we get $\mathbb{CP}^2$, and the 4-framed circle corresponds to a conic. Adding a

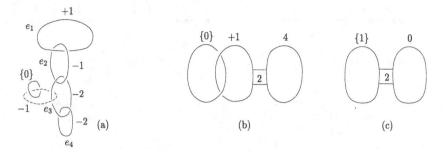

Fig. 17. The extra case of $p = 4$ in Theorem 2.3

dual handle to the $(+1)$-framed circle (representing the complement of the conic) and then blowing down the $(+1)$-framed circle, and finally surgering along the 0-circle and reversing the orientation we get the diagram of the complement, which is identical to the diagram of the smoothing we have found above.

Next we determine the diagram of the $\mathbb{Q}$HD smoothing for the family of graphs shown in Figure 1(f). The corresponding sequence of blow-downs is shown by Figure 5 with the choice $p = r = 0$; the diagrams in this special case are shown in Figure 18. Once again, we start by picturing the end-result, that is, the curves of Figure 18(d) in $\mathbb{CP}^2$. The cubic and the line can be given by a 9-framed trefoil and an $(+1)$-framed unknot linking each other with multiplicity three. (The Seifert surface of the unknot completes the disk in the handle to a sphere, i.e. to a projective line, while an appropriate Seifert surface of the trefoil together with the core disk of the handle gives topologically a smooth cubic.) By taking a Seifert surface with one positive double point, indeed, the same trefoil knot can be used to depict a nodal cubic curve. We need to visualize a cubic and a line which are triply tangent at a point—in our diagrammatic language this corresponds to the fact that the two curves admit a triple linking. Such a triple linking is shown by the box (containing the number three) of Figure 19(a). Indeed, repeatedly blowing it up three times, we can separate the two curves, and this property characterizes the triple tangency. (See also the upper diagram of Figure 19(b) for the three-fold blow-up.) An additional $(+1)$-framed unknot (and two 3-handles and a 4-handle) completes the picture to get a diagram for $\mathbb{CP}^2$. (For this step, disregard the fine curve of Figure 19(a) with bracketed framing.) Indeed, taking the left $(+1)$-framed unknot off, the trefoil and the other circle both become 0-framed unknots, and they form an unlink, hence can be cancelled against the 3-handles. Therefore the

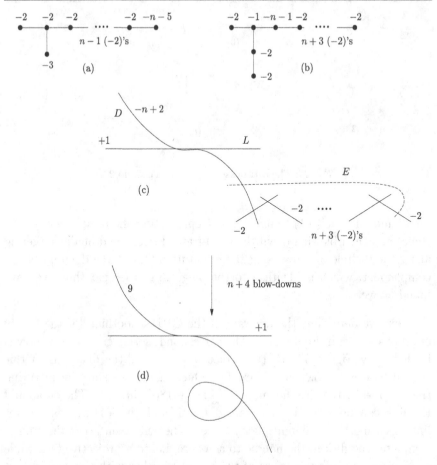

Fig. 18. The special case of Figure 5 for $p = r = 0$

diagram really presents $\mathbb{CP}^2$ with a nodal cubic and a line triply tangent to it. By blowing up the triple tangency three times, and the positive self-intersection of the cubic $(n + 4)$-times (as shown in the diagrams of Figure 19(b)), we get a diagram of a blown-up projective plane, where the plumbing corresponding to the dual graph of the resolution of the singularity at hand (given by Figure 18(b)) is explicitly visible. Once again, at this point we disregard the fine curves with bracketed framings. The diagram contains two curves not in the configuration: the starting $(+1)$-framed unknot, and the last (-1)-framed unknot. In Figures 19(a) and (b) we have already put the meridional 0-framed unknots to these curves. (The framings of these curves, as always, are in brackets.) Recall that our aim is now to transform the curves in the configuration using 3-dimensional Kirby moves

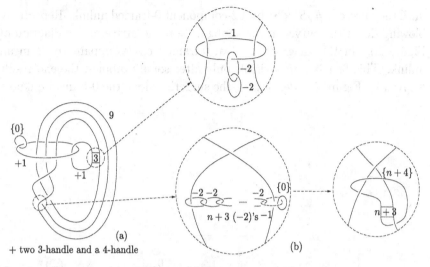

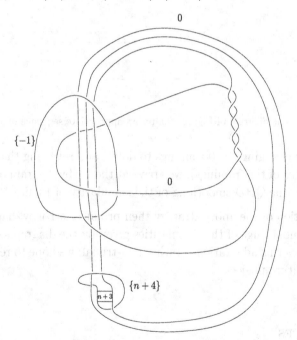

Fig. 19. The diagram (a) of $\mathbb{CP}^2$, and (b) the blow-up of the double point and the triple tangency. In the right we also show the result of blowing the sequence of $(n+3)$ (-2)-circles (and the (-1)-circle) back down

Fig. 20. The diagram we get after blowing down the circles contributing to the configuration K. The two 0-framed circles on the diagram form a 2-component unlink

until they present $\#_2 S^1 \times S^2$ by a 2-component 0-framed unlink. Repeatedly blowing down the curves in the configuration, we arrive at the diagram of Figure 20. In this diagram the two 0-framed circles apparently form an unlink. This fact becomes visible only after some isotopies, the end result is given by Figure 21. Performing the surgeries along the 0-framed unknots

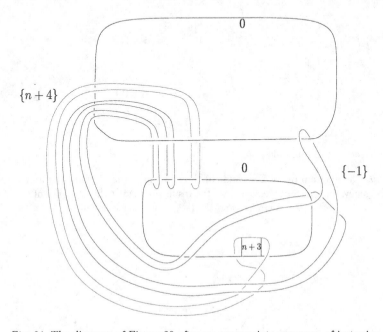

Fig. 21. The diagram of Figure 20 after an appropriate sequence of isotopies

(realized by changing the 0-framings to dots), and reversing the orientation (and the signs of the framings), we arrive at the Kirby diagram of Figure 22, representing the QHD smoothing of the singularity of Figure 1(f).

Adaptation of the above strategy then produces a Kirby diagram for all the QHD smoothings of the singularities given by the diagrams of Figure 1. In some cases this adaptation is rather non-trivial; we hope to return to this issue in a future project.

References

[1] M. Bhupal and K. Ono, *Symplectic fillings of links of quotient surface singularities,* Nagoya Math. J. **207** (2012), 1–45.

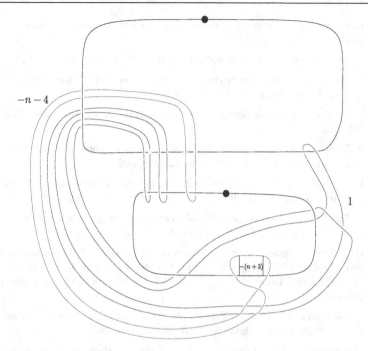

Fig. 22. The Kirby diagram of the smoothing

[2] M. Bhupal and A. Stipsicz, *Weighted homogeneous singularities and rational homology disk smoothings,* Amer. J. Math. **133** (2011), 1259–1297.

[3] A. Casson and J. Harer, *Some homology lens spaces which bound rational homology balls,* Pacific J. Math. **96** (1981), 23–36.

[4] C. Caubel, A. Némethi and P. Popescu-Pampu, *Milnor open books and Milnor fillable contact 3-manifolds,* Topology **45** (2006), 673–689.

[5] T. de Jong and D. van Straten, *Deformation theory of sandwiched singularities,* Duke Math. Journal **95** (1998), 451–522.

[6] Y. Eliashberg, *On symplectic manifolds with some contact properties,* J. Differential Geom. **33** (1991), 233–238.

[7] H. Endo, T. Mark and J. van Horn-Morris, *Monodromy substitutions and rational blowdowns,* J. Topol. **4** (2011), 227–253.

[8] J. Etnyre, *Symplectic convexity in low-dimensional topology,* Symplectic, contact and low-dimensional topology (Athens, GA, 1996). Topology Appl. **88** (1998), 3–25.

[9] R. Fintushel and R. Stern, *Rational blowdowns of smooth 4-manifolds,* J. Diff. Geom. **46** (1997), 181–235.

[10] R. Fintushel and R. Stern, *Double node neighborhoods and families of simply connected 4-manifolds with* $b^+ = 1$, J. Amer. Math. Soc. **19** (2006), 171–180.

[11] R. Gompf and A. Stipsicz, *4-manifolds and Kirby calculus*, Graduate Studies in Mathematics **20** AMS (1999).

[12] H. Laufer, *Taut two-dimensional singularities*, Math. Ann. **205** (1973), 131–164.

[13] Y. Lee and J. Park, *Simply connected surfaces of general type with $p_g = 0$ and $K^2 = 2$*, Invent. Math. **170** (2007), 483–505.

[14] P. Lisca, *On symplectic fillings of lens spaces*, Trans. Amer. Math. Soc. **360** (2008), 765–799.

[15] P. Lisca and A. Stipsicz, *On the existence of tight contact structures on Seifert fibered 3-manifolds*, Duke Math. Journal **148** (2009), 175–209.

[16] D. McDuff, *The structure of rational and ruled symplectic 4-manifolds*, Journal of the Amer. Math. Soc. **3** (1990), 679–712; erratum, **5** (1992), 987–8.

[17] D. McDuff, *The local behaviour of holomorphic curves in almost complex 4-manifolds*, J. Diff. Geom. **34** (1991), 143–164.

[18] D. McDuff, *Singularities and positivity of intersections of J-holomorphic curves*, with Appendix by Gang Liu, in Holomorphic curves in Symplectic Geometry, M. Audin and F. Lafontaine, eds., Progress in Mathematics **117**, Birkhauser (1994), 191–216.

[19] A. Némethi and P. Popescu-Pampu, *On the Milnor fibers of cyclic quotient singularities*, Proc. London Math. Soc. **101** (2010), 554–588.

[20] H. Ohta and K. Ono, *Symplectic fillings of the link of simple elliptic singularities*, J. Reine Angew. Math. **565** (2003), 183–205.

[21] H. Ohta and K. Ono, *Simple singularities and symplectic fillings*, J. Differential Geom. **69** (2005), no. 1, 1–42.

[22] P. Orlik and P. Wagreich, *Isolated singularities of algebraic surfaces with $\mathbb{C}^*$-action*, Ann. Math. **93** (1971), 205–228.

[23] J. Park, *Seiberg–Witten invariants of generalized rational blow-downs*, Bull. Austral. Math. Soc. **56** (1997), 363–384.

[24] J. Park, *Simply connected symplectic 4-manifolds with $b_2^+ = 1$ and $c_1^2 = 2$*, Invent. Math. **159** (2005), 657–667.

[25] H. Park, J. Park and D. Shin, *A simply connected surface of general type with $p_g = 0$ and $K^2 = 3$*, Geom. Topol. **13** (2009), 743–767.

[26] H. Park, J. Park and D. Shin, *A simply connected surface of general type with $p_g = 0$ and $K^2 = 4$*, Geom. Topol. **13** (2009), 1483–1494.

[27] J. Park, A. Stipsicz and Z. Szabó, *Exotic smooth structures on $\mathbb{CP}^2 \# 5\overline{\mathbb{CP}^2}$*, Math. Res. Lett. **12** (2005), 701–712.

[28] H. Pinkham, *Normal surface singularities with $\mathbb{C}^*$-action*, Math. Ann. **227** (1977), 183–193.

[29] H. Pinkham, *Deformations of normal surface singularities with $\mathbb{C}^*$-action*, Math. Ann. **232** (1978), 65–84.

[30] A. Stipsicz and Z. Szabó, *An exotic smooth structure on $\mathbb{CP}^2 \# 6\overline{\mathbb{CP}^2}$*, Geom. Topol. **9** (2005), 813–832.

[31]　A. Stipsicz, Z. Szabó and J. Wahl, *Rational blowdowns and smoothings of surface singularities*, Journal of Topology **1** (2008), 477–517.

[32]　J. Wahl, *Smoothings of normal surface singularities*, Topology **20** (1981), 219–246.

[33]　J. Wahl, *On rational homology disk smoothings of valency 4 surface singularities*, Geom. Topol. **15** (2011), 1125–1156.

Mohan Bhupal

Middle East Technical University
Ankara
Turkey

e-mail: bhupal@metu.edu.tr

András I. Stipsicz

Rényi Institute of Mathematics
Budapest
Hungary

e-mail: stipsicz@renyi.hu

BOLYAI SOCIETY
MATHEMATICAL STUDIES, 23

Deformations of
Surface Singularities
pp. 99–107.

CALCULATING MILNOR NUMBERS AND VERSAL COMPONENT DIMENSIONS FROM P-RESOLUTION FANS

NATHAN OWEN ILTEN

We use Altmann's toric fan description of P-resolutions [1] to formulate a new description of deformation theory invariants for two-dimensional cyclic quotient singularities. In particular, we show how to calculate the dimensions of the (reduced) versal base space components as well as Milnor numbers of smoothings over them.

1. INTRODUCTION

The deformation theory of (two-dimensional) cyclic quotient singularities is fairly well understood. Kollár and Shepherd-Barron have proven a correspondence between so-called P-resolutions and reduced components of the versal base space in [4]. P-Resolutions were further studied using continued fractions by Christophersen and Stevens in [2] and [6] respectively, who both managed to write down explicit equations for the reduced components of the versal deformation. In [1], Altmann uses the continued fractions of Christophersen and Stevens to describe P-resolutions in toric terms, that is, in terms of a fan.

This paper deals with the dimension of the versal base components as well as the Milnor numbers of smoothings over them. A formula for the Milnor numbers is provided in [6]. Furthermore, for the $\mathbb{Q}$-Gorenstein one-parameter smoothing of T-singularities, Altmann has already provided a simple formula in toric terms. The first aim of this paper is to generalize Altmann's formula to smoothings over all components of any cyclic quotient singularity. Our new formula allows the Milnor number of a smoothing to be read directly from the geometry of the fan describing the corresponding P-resolution.

A method for calculating the dimension of the versal base components was first provided by Kollár and Shepherd-Barron in [4]. Even better, the explicit equations in [2] and [6] allow one to write down a simple formula. The second aim of this paper is to translate this component dimension formula into toric language using Altmann's toric description of P-resolutions. Our new formula allows the dimension of a component to be read directly from the geometry of the fans describing the minimal resolution and the P-resolution corresponding to that component. Furthermore, the difference in dimension between two components can be read solely from the two fans describing the two corresponding components. Note also that our proofs of the two formulae, while not differing significantly from those by Stevens, can be easily understood within the context of toric geometry.

In Section 2, we provide necessary definitions and notation. We have chosen notation so as to be completely consistent with [1]; in fact, readers familiar with this paper can probably skip Section 2. In Section 3, we describe Stevens' formula for Milnor numbers and state and prove our new toric formula. Likewise, in Section 4 we present the existing component dimension formula and then state and prove our new toric formula. We finish in Section 5 by providing an example demonstrating the practicality of our formulae.

2. Cyclic Quotients and P-Resolutions

In the following, we recall the notions of cyclic quotients and P-resolutions, as well as fixing notation. References are [3] for toric varieties, and [1] for P-resolutions.

Let n and q be relatively prime integers with $n \geq 2$ and $0 < q < n$. Let ξ be a primitive n-th root of unity. The cyclic quotient singularity $Y_{(n,q)}$ is the quotient $\mathbb{C}^2/(\mathbb{Z}/n\mathbb{Z})$ where $\mathbb{Z}/n\mathbb{Z}$ acts on $\mathbb{C}^2$ via the matrix

$$\begin{pmatrix} \xi & 0 \\ 0 & \xi^q \end{pmatrix}.$$

Every two-dimensional cyclic quotient singularity is in fact a two-dimensional toric variety: Let N be a rank two lattice with dual lattice M; we will identify N with $\mathbb{Z}^2$. Let $\sigma \subset N \otimes \mathbb{R}$ be the cone generated by $(1,0)$ and $(-q, n)$. $Y_{(n,q)}$ is then isomorphic to the toric variety $U_\sigma = \operatorname{Spec} \mathbb{C}[M \cap \sigma^\vee]$. Since by correct choice of basis every singular two-dimensional cone has generators $(1,0)$ and $(-q, n)$ for some q and n as above, every affine singular two-dimensional toric variety is a cyclic quotient singularity.

Introduced by Kollár and Shepherd-Barron in [4], P-resolutions have proven key to understanding the deformation theory of cyclic quotient singularities.

Definition. Let Y be a two-dimensional cyclic quotient singularity. A P-resolution of Y is a partial resolution $f : \widetilde{Y} \to Y$ containing only T-singularities such that the canonical divisor $K_{\widetilde{Y}}$ is ample relative to f. T-singularities are exactly those cyclic quotients admitting a $\mathbb{Q}$-Gorenstein one-parameter smoothing

The following theorem describes the relationship between P-resolutions and reduced components of the base space of the versal deformation for a two-dimensional cyclic quotient singularity Y:

Theorem 2.1. *There is a bijection between the set of P-resolutions $\{\widetilde{Y}_\nu\}$ of Y and the components of the reduced versal base space S, induced by the natural maps* $\operatorname{Def}' \widetilde{Y}_\nu \to S$, *where* $\operatorname{Def}' \widetilde{Y}_\nu$ *is the space of $\mathbb{Q}$-Gorenstein deformations of* $\widetilde{Y}_\nu$.

Proof. See [4], Section 3. ∎

P-Resolutions were described by Christophersen and Stevens in terms of continued fractions in [2] and [6]. In [1], Altmann provides a toric description in terms of a fan; it is this latter description that we shall use in our dimension formula.

Let $c_1, c_2, \ldots, c_k \in \mathbb{Z}$. The continued fraction $[c_1, c_2, \ldots, c_k]$ is inductively defined as follows if no division by 0 occurs: $[c_k] = c_k$, $[c_1, c_2, \ldots, c_k] = c_1 - 1/[c_2, \ldots, c_k]$. Now, if one requires that $c_i \geq 2$ for every coefficient, each continued fraction yields a unique rational number.

Let n and q be relatively prime integers with $n \geq 3$ and $0 < q < n - 1$.[1] We consider the cyclic quotient singularity $Y_{(n,q)}$. Let $[a_2, a_3, \ldots, a_{e-1}]$, $a_i \geq 2$ be the unique continued fraction expansion of $n/(n-q)$. Note that e equals the embedding dimension of $Y_{(n,q)}$. Furthermore, the generators

---

[1]This restriction simply ensures that $Y_{(n,q)}$ isn't a hypersurface, in which case the versal base space is irreducible.

$w^1, \ldots, w^e$ of the semigroup $M \cap \sigma^\vee$ are related to this continued fraction. Indeed, if we order the w^i such that (w^{i+1}, w^i) is positively oriented for all $1 \leq i < e$, then $w^1 = (0,1)$, $w^e = (n,q)$, and $w^{i-1} + w^{i+1} = a_i w^i$.

Likewise, let $[b_1, \ldots, b_r]$ be the unique continued fraction expansion of n/q. The generators of the semigroup $N \cap \sigma$ are related to this continued fraction: $v^0 = (1,0)$, $v^{r+1} = (-q,n)$, and $v^{i-1} + v^{i+1} = b_i v^i$. Drawing rays through the v^i gives a polyhedral subdivision Σ of σ. The corresponding toric variety $\mathrm{TV}(\Sigma)$ is the minimal resolution of Y with self intersection numbers $-b_i$; the number of exceptional divisors in this resolution is r.

For a chain of integers $(k_2, \ldots, k_{e-1})$ define the sequence $q_1, \ldots, q_e$ inductively: $q_1 = 0$, $q_2 = 1$, and $q_{i-1} + q_{i+1} = k_i q_i$. Now define the set

$$K_{e-2} = \left\{ (k_2, \ldots, k_{e-1}) \in \mathbb{N}^{e-2} \,\middle|\, \begin{array}{l} \text{(i) } [k_2, \ldots, k_{e-1}] \text{ is well defined and yields } 0 \\ \text{(ii) The corresponding integers } q_i \text{ are positive} \end{array} \right\}.$$

Further, define the set

$$K\big(Y_{(n,q)}\big) = \big\{ (k_2, \ldots, k_{e-1}) \in K_{e-2} \mid k_i \leq a_i \big\}.$$

Each $\underline{k} \in K\big(Y_{(n,q)}\big)$ determines a fan: $\Sigma_{\underline{k}}$ is built from the rays generating σ and those lying in σ which are orthogonal to $w^i/q_i - w^{i-1}/q_{i-1} \in M_{\mathbb{R}}$ for some $i = 3, \ldots, e-1$. Equivalently, the affine lines $\big[\langle \cdot, w^i \rangle = q_i\big]$ form the "roofs" of the (possibly degenerate) $\Sigma_{\underline{k}}$-cones τ_i. The length in the induced lattice of each roof is $(a_i - k_i)q_i$, and this segment lies in height q_i.

Theorem 2.2. *The P-resolutions of $Y_{(n,q)}$ are in one-to-one correspondence to the elements of $K\big(Y_{(n,q)}\big)$. This correspondence can be realized by the map $\underline{k} \mapsto \mathrm{TV}(\Sigma_{\underline{k}})$, that is, $\underline{k}$ corresponds to the toric variety determined by the fan $\Sigma_{\underline{k}}$.*

Proof. See [6] and [1]. ∎

For each $\underline{k}$, denote by $S_{\underline{k}}$ the versal base component corresponding to the P-resolution $\mathrm{TV}(\Sigma_{\underline{k}})$. We will present examples of several fans corresponding to P-resolutions in Section 5.

Remark. The continued fraction $[1, 2, 2, \ldots, 2, 1] = 0$ always belongs to $K\big(Y_{(n,q)}\big)$. The P-resolution defined by the corresponding fan is the so-called RDP-resolution of $Y_{(n,q)}$. This corresponds to the Artin component of the versal base space, which always has maximal dimension.

3. MILNOR NUMBERS

Altmann notes in the introduction of [1] that T-singularities are exactly those cyclic quotients coming corresponding to a cone σ attained by taking the cone over some line segment of integral length $\mu + 1$ in lattice height 1. In such a case, the Milnor number $b_2(F)$ of the $\mathbb{Q}$-Gorenstein smoothing of the singularity is equal to μ. Note that the corresponding P-resolution is simply the identity. On the other hand, Stevens has proven the following general formula:

Proposition 3.1. *Let Y be a cyclic quotient singularity and let F be the Milnor fiber of a one-parameter smoothing of Y over $S_{\underline{k}}$. Then*

$$b_2(F) = \dim T_Y^1 - 3(e-3) + \#\{2 < i < e - 1 \mid q_i = 1\} + 2.$$

Proof. See lemma 5.3 in [6]. ∎

We will now formulate this in toric terms. Fix a cyclic quotient singularity Y and let $\underline{k} \in K(Y)$. For any two-dimensional cone $\tau \in \Sigma_{\underline{k}}^{(2)}$, let $l(\tau)$ and $h(\tau)$ respectively denote the lattice length and height of its roof. We then have the following theorem:

Theorem 3.2. *Let F be the Milnor fiber of a one-parameter smoothing of Y over $S_{\underline{k}}$. Then*

$$b_2(F) = \left(\sum_{\tau \in \Sigma_{\underline{k}}^{(2)}} l(\tau)/h(\tau) \right) - 1.$$

Proof. The Milnor number $b_2(F)$ can be read from the P-resolution as the sum of the Milnor numbers of the $\mathbb{Q}$-Gorenstein smoothings of each T-singularity plus the total number of exceptional divisors. For any $\tau \in \Sigma_{\underline{k}}^{(2)}$, the Milnor number of the $\mathbb{Q}$-Gorenstein smoothing of the T-singularity $TV(\tau)$ is $l(\tau)/h(\tau) - 1$ by Altmann's result. On the other hand, the number of exceptional divisors is simply the number of internal rays in $\Sigma_{\underline{k}}$, that is, one less than the number of two-dimensional cones $\tau \in \Sigma_{\underline{k}}^{(2)}$. Combining these two facts yields the above formula. ∎

Note that this result generalizes Altmann's result for the special component of a T-singularity.

4. COMPONENT DIMENSION

We now state the explicit formula for computing the dimension of versal base components for a cyclic quotient singularity Y; this comes from Christophersen's and Stevens' description of the versal components.

Proposition 4.1. *The dimension of the versal base component $S_{\underline{k}}$ corresponding to the continued fraction $\underline{k} \in K(Y)$ can be computed as*

$$(4.1) \qquad \dim S_{\underline{k}} = \#\{2 < i < e - 1 \mid q_i = 1\} + \sum_{i=2}^{e-1}(a_i - k_i)$$

where the a_i, k_i, and q_i are as in Section 2.

Proof. This formula follows directly from Christophersen's definition of $V_{[\mathbf{k}]}$ in Section 2.1.1 of [2].[2] ∎

We now translate this formula into toric terms to attain a new dimension formula depending only upon the geometry of the fans of the minimal resolution and the corresponding P-resolution. Let $Y = \text{TV}(\sigma)$ be a cyclic quotient singularity with minimal resolution $\widetilde{Y} = \text{TV}(\Sigma)$ and some P-resolution $\text{TV}(\Sigma_{\underline{k}})$ corresponding to the versal base component $S_{\underline{k}}$. Let v^i be as in Section 2 the generators of the rays in the fan Σ. For any two-dimensional cone $\tau \in \Sigma_{\underline{k}}^{(2)}$, let once again $l(\tau)$ and $h(\tau)$ respectively denote the lattice length and height of its roof. Finally, define

$$\nu = \sum_{i=1}^{r}\left(\det\left(v^{i-1}, v^{i+1}\right)\right),$$

that is, ν is the sum over all r interior rays v^i in Σ of the normed volume of the simplex $\text{Conv}\left\{0, v^{i-1}, v^{i+1}\right\}$. This leads to our formula:

Theorem 4.2. *The dimension of the reduced versal base component $S_{\underline{k}}$ is given by:*

$$\dim S_{\underline{k}} = \nu - 3r + 2 \cdot \sum_{\tau \in \Sigma'^{(2)}} l(\tau)/h(\tau) - 2.$$

---

[2]Note that Christophersen's indices are shifted by one.

Proof. From corollary 3.18 in [7], we have

$$\dim S_{\underline{k}} = h^1(\Theta_{\widetilde{Y}}) + 2b_2(F) - 2r,$$

where F is the Milnor fiber for the component $S_{\underline{k}}$. Now, $h^1(\Theta_{\widetilde{Y}}) = \sum_{i=1}^r (b_i - 1)$, see for example [5]. But each b_i can be computed as $\det\left(v^{i-1}, v^{i+1}\right)$. The desired equation then follows from Theorem 3.2. ■

When comparing the dimension of two components S_1 and S_2 with corresponding fans Σ_1 and Σ_2 we can even forget about ν and r:

Corollary 4.3. *The difference in dimension between S_1 and S_2 is given by:*

$$\dim S_1 - \dim S_2 = 2 \sum_{\tau \in \Sigma_1^{(2)}} l(\tau)/h(\tau) - 2 \sum_{\tau \in \Sigma_1^{(2)}} l(\tau)/h(\tau).$$

Proof. We express $\dim S_1$ and $\dim S_2$ using Proposition 4.2. The term $\nu - 3r - 2$ cancels, leaving the desired expression. ■

5. An Example

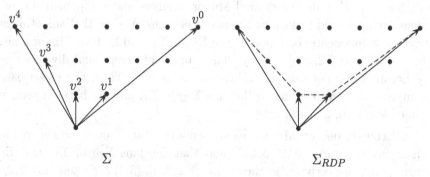

Fig. 1. Minimal and RDP resolution fans for $\sigma = \langle (-2,3), (4,3) \rangle$

Example. Consider the cone $\sigma = \langle (-2,3), (4,3) \rangle$ and let $Y = \mathrm{TV}(\sigma)$. It is quite easy to find the minimal resolution fan Σ; this is done by adding rays through all lattice points on the boundary of $\mathrm{Conv}\left(\sigma \cap N \setminus \{0\}\right)$. Likewise, to get the fan for the RDP resolution, one adds rays through all vertices of

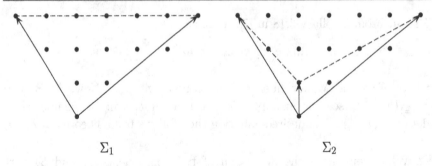

$$\Sigma_1 \hspace{4cm} \Sigma_2$$

Fig. 2. Further P-resolution fans for $\sigma = \langle (-2,3), (4,3) \rangle$

Conv $\left(\sigma \cap N \setminus \{0\} \right)$. These two fans are pictured in Fig. 1; note that the dotted lines represent the roofs of two-dimensional cones in Σ_{RDP}.

Finding further fans corresponding to P-resolutions is slightly more tricky. Of course, one could use the continued fractions followed by Altmann's construction to get them, but it is also possible without them: every P-resolution is dominated by the resolution corresponding to the fan with rays through all lattice points in Conv $\left\{ 0, v^0, v^{r+1} \right\}$. Thus, by removing these rays and checking for T-singularities and convexity of the roofs, one finds all fans corresponding to P-resolutions. The two additional fans Σ_1 and Σ_2 found in this manner are pictured in Fig. 2; note that Y is in fact a T-singularity and the fan Σ_1 corresponds to the space of $\mathbb{Q}$-Gorenstein deformations.

Now, to calculate the desired Milnor numbers and component dimensions, we only need to look at the above pictures. We see that smoothings over the components corresponding to Σ_{RDP}, Σ_1, and Σ_2 have Milnor numbers of 3, 1, and 2, respectively. The number of exceptional divisors r is obviously 3 and one quickly calculates that $\nu = 9$. Thus, the versal base components corresponding to the fans Σ_{RDP}, Σ_1, and Σ_2 have respective dimensions of 6, 2, and 4.

Of course, one could have also calculated that Y corresponds to the chain $(a_2, a_3, a_4, a_5) = (3, 3, 2, 2)$ and that the fans Σ_{RDP}, Σ_1, and Σ_2 correspond, respectively, to the chains $[1, 2, 2, 1]$, $[3, 1, 2, 2]$, and $[2, 3, 1, 2]$. Putting these values into Stevens' formulas yields the same results as with the toric formulas.

REFERENCES

[1] Klaus Altmann, P-resolutions of cyclic quotients from the toric viewpoint, in: *Singularities (Oberwolfach, 1996)*, volume 162 of *Progr. Math.*, pages 241–250. Birkhäuser, Basel, 1998.

[2] Jan Arthur Christophersen, On the components and discriminant of the versal base space of cyclic quotient singularities, in: *Singularity theory and its applications, Part I (Coventry, 1988/1989)*, volume 1462 of *Lecture Notes in Math.*, pages 81–92. Springer, Berlin, 1991.

[3] William Fulton, *Introduction to toric varieties*, volume 131 of *Annals of Mathematics Studies*. Princeton University Press, Princeton, NJ, 1993. The William H. Roever Lectures in Geometry.

[4] J. Kollár and N. I. Shepherd-Barron, Threefolds and deformations of surface singularities, *Invent. Math.*, **91(2)** (1988), 299–338.

[5] Oswald Riemenschneider, Deformationen von Quotientensingularitäten (nach zyklischen Gruppen), *Math. Ann.*, **209** (1974), 211–248.

[6] Jan Stevens, On the versal deformation of cyclic quotient singularities, in: *Singularity theory and its applications, Part I (Coventry, 1988/1989)*, volume 1462 of *Lecture Notes in Math.*, pages 302–319. Springer, Berlin, 1991.

[7] Jonathan Wahl, Smoothings of normal surface singularities, *Topology*, **20(3)** (1981), 219–246.

Nathan Owen Ilten

Department of Mathematics,
Univserity of California, Berkeley
Berkeley, CA 94720

e-mail: nilten@cs.uchicago.edu

BOLYAI SOCIETY
MATHEMATICAL STUDIES, 23

Deformations of
Surface Singularities
pp. 109–162.

Some Meeting Points of Singularity Theory and Low Dimensional Topology

ANDRÁS NÉMETHI

We review some basic facts which connect the deformation theory of normal surface singularities with the topology of their links. The presentation contains some explicit descriptions for certain families of singularities (cyclic quotients, sandwiched singularities).

1. Introduction

The aim of the present paper is to serve as a first introductory guide for those researchers who wish to study the subtle connections between the analytic and deformation invariants of normal surface singularities and topological aspects. Since this subject is rather large, we had to make some selection: here we will say nothing about the connection of singularity theory with the Seiberg–Witten or Heegaard Floer theory, or even with lattice cohomology. Instead, we focus on some aspects regarding deformations, the Milnor fibers of the smoothings, and their connection with the Stein/symplectic fillings in topology. The manuscript can be thought as a complement of the manuscript of A. Stipsicz of this volume, where he concentrates mostly on the topological methods and aspects. Here we wish to collect some facts about the analytic part too.

2. The Link, from Singularity Point of View

2.1. Definition

Let $(X, 0)$ be a complex analytic normal surface singularity embedded in $(\mathbb{C}^N, 0)$, and let B_ε be the corresponding ε-ball. Then, for ε sufficiently small, the intersection $M := X \cap \partial B_\varepsilon$ is a connected compact oriented 3-manifold, whose oriented C^∞ type does not depend on the choice of the embedding and ε. It is called the link of $(X, 0)$ [60]. Moreover, one can verify that $X \cap B_\varepsilon$ is homeomorphic to the cone over M. In particular, M characterizes completely the local topological type of $(X, 0)$. Therefore, if an invariant of $(X, 0)$ can be deduced from M, we say that it is a *topological invariant*.

2.2. M as a plumbed manifold

The bridge between the analytical and topological type of $(X, 0)$ is realized by a resolution, respectively by a resolution graph of $(X, 0)$. Fix a sufficiently small Stein representative X of $(X, 0)$ (e.g. $X \cap B_\varepsilon$ as above) and let $\pi : \tilde{X} \to X$ be a resolution of the singular point $0 \in X$. In particular, $\tilde{X}$ is smooth, and π is a biholomorphic isomorphism above $X \setminus \{0\}$. We can assume that the exceptional divisor $E := \pi^{-1}(0)$ is a normal crossing divisor with smooth irreducible components $\{E_v\}_{v \in \mathcal{V}}$. Such a resolution is called *good*. For a good resolution π, let $\Gamma(\pi)$ be the dual resolution graph associated with π decorated with the self intersection numbers $\{(E_v, E_v)\}_v$ and genera $\{g_v\}_{v \in \mathcal{V}}$ (see [42]). We write $e_v := (E_v, E_v)$. Notice that $H_2(\tilde{X}, \mathbb{Z})$ is freely generated by the fundamental classes $\{[E_v]\}_v$. Let I be the intersection matrix $\{(E_v, E_w)\}_{v,w}$. Since π identifies $\partial \tilde{X}$ with M, the graph $\Gamma(\pi)$ can be regarded as a plumbing graph, and M can be considered as a plumbed manifold whose plumbing graph is $\Gamma(\pi)$ (see e.g. [24, 61] or [62, 65, 89]).

In order to have the plumbing representation, it is necessary to consider resolutions with normal crossing exceptional divisor, that is good resolutions. Among the good resolutions there is a minimal one, called the *minimal good resolution*. On the other hand, in some other considerations, it is preferable to work with the minimal (not necessarily good) resolution. In such a resolution, there is no rational smooth irreducible exceptional divisor E_v with $e_v = -1$.

The link M is called *rational homology sphere* (RHS) if $H_1(M, \mathbb{Q}) = 0$. This is happening if and only if $\Gamma(\pi)$ is a tree and $g_v = 0$ for all $v \in \mathcal{V}$.

2.2.1. $\Gamma(\pi)$ is connected and I is negative definite [61]. The converse of this was proved by Grauert [31]. More precisely, a connected plumbing graph can be realized as a resolution graph of a (complex analytic) normal surface singularity if and only if the associated intersection form I is negative definite. This, by the next statement, gives a complete classification of the possible topological types of (analytic) normal surface singularities.

We say that two plumbing graphs (with negative definite intersection forms) are equivalent if one of them can be obtained from the other by a finite sequence of blow-ups and/or blow-downs along rational (-1)-curves. Obviously, for a given $(X, 0)$, the resolution π, hence the graph $\Gamma(\pi)$ too, is not unique. But different resolutions provide equivalent graphs. By a result of W. Neumann [75], the oriented diffeomorphism type of M determines completely the equivalence class of $\Gamma(\pi)$.

In fact, Neumann's theorem was preceded by the following two key results. First, Mumford proved that the minimal resolution graph of $(X, 0)$ is empty if and only if $M = S^3$ [61]. Later, Orlik and Wagreich generalized this for singularities with a good $\mathbb{C}^*$-action as follows [83]. In this case, M is a Seifert manifold which can be codified by its minimal (star-shaped) plumbing graph, and this minimal plumbing graph coincides with the minimal good resolution graph of $(X, 0)$. In particular, the minimal resolution graph can be recovered from the Seifert invariants of the link.

3. SOME ANALYTIC INVARIANTS OF NORMAL SURFACE SINGULARITIES

3.1.

The analytic type of $(X, 0)$ is characterized by its local analytic ring $\mathcal{O}_{X,0}$ whose maximal ideal will be denoted by $m_0 \subset \mathcal{O}_{X,0}$. It determines the *analytic invariants* of $(X, 0)$. We investigate the topological nature of certain a priori analytical invariants, hence we deal mainly with *discrete* invariants.

3.1.1. The *Hilbert–Samuel function* is defined by

$$f_{HS}(k) = \dim_{\mathbb{C}} \mathcal{O}_{X,0}/m_0^k \quad \text{for any } k \geq 1.$$

Then $f_{HS}(1) = 1$ and $f_{HS}(2) - 1 = \dim m_0/m_0^2$ equals the minimal N for which some embedding $(X,0) \subset (\mathbb{C}^N, 0)$ can be realized, hence is called the *embedding dimension* of $(X,0)$. For $k \gg 1$, $f_{HS}(k) = P_{HS}(k)$ for some polynomial P_{HS} (called the Hilbert–Samuel polynomial)

$$P_{HS}(k) = mk^2/2 + a_1 k + a_2.$$

The integer m above is called the *multiplicity* of $(X,0)$, and it is denoted by $\mathrm{mult}\,(X,0)$. It is not difficult to verify that if $(X,0) \subset (\mathbb{C}^N, 0)$ is an arbitrary embedding and L a generic affine space of codimension 2 (close to the origin), then $\mathrm{mult}\,(X,0) = \#X \cap L$.

3.1.2. The *geometric genus* p_g is defined as $\dim_{\mathbb{C}} H^1(\tilde{X}, \mathcal{O}_{\tilde{X}})$, where $\tilde{X} \to X$ is any resolution as above (see also 3.1.4).

The geometric genus is just one example of the many possible analytic invariants obtained from a fixed resolution $\pi : \tilde{X} \to X$ via sheaf-cohomology. For example, one can replace the structure sheaf by the tangent sheaf $\theta_{\tilde{X}} = \left(\Omega^1_{\tilde{X}}\right)^*$ and obtain $\theta := \dim_{\mathbb{C}} H^1(\tilde{X}, \theta_{\tilde{X}})$. The point is that these invariants can be recovered from the sheaf-cohomology of some one-dimensional spaces as well. Indeed, let $\mathcal{Z}$ (resp. $\mathcal{Z}_{\mathbb{Q}}$) be the set of integral (resp. rational) cycles, i.e. divisors of type $Z = \sum_{v \in \mathcal{V}} m_v E_v$, $m_v \in \mathbb{Z}$ (resp. $m_v \in \mathbb{Q}$) supported on the exceptional divisor E of π. Then, for any $Z \in \mathcal{Z}$, one can consider $h^i(Z) := \dim H^i(Z, \mathcal{O}_Z)$ (for $i = 1, 2$). By the theorem of formal functions (see [34, page 277]) if one takes Z with $m_v \gg 0$ for all v, then, e.g., $h^1(Z) = p_g$.

3.1.3. $\mathcal{Z}$ has a natural partial ordering: $\sum_v m_v E_v \geq \sum_v n_v E_v$ if $m_v \geq n_v$ for all v. If $Z_1 \geq Z_2$ but $Z_1 \neq Z_2$ then we write $Z_1 > Z_2$.

Some cycles $Z > 0$ are of special interest: they describe the possible supports of analytic functions defined on $(X,0)$. More precisely, for any $f \in m_0$ let m_v be the order of vanishing of $f \circ \pi$ along E_v, and set $(f)_E = \sum_v m_v E_v$. Define $\mathcal{Z}_{\mathrm{an}}$ by $\{(f)_E \mid f \in m_0\}$. This yields an ordered lattice (in the sense that if $Z_1, Z_2 \in \mathcal{Z}_{\mathrm{an}}$, then $Z_1 + Z_2$ and $\min\{Z_1, Z_2\}$ are elements of $\mathcal{Z}_{\mathrm{an}}$). In particular, $\mathcal{Z}_{\mathrm{an}}$ has a unique minimal element (corresponding to the "generic linear hyperplane section" in $(\mathbb{C}^N, 0)$). It is called the *maximal ideal cycle* $Z_{\max}$ (of S.S.-T. Yau), or the "fiber cycle". Obviously $Z_{\max} \geq \sum_v E_v$.

By the above definitions, the divisorial part of the sheaf $\pi^*(m_0)$ is given by $Z_{\max}$. More precisely, there exists an ideal $\mathcal{I}$, supported by a finite set $V(\mathcal{I})$, such that $\pi^*(m_0) = \mathcal{I} \cdot \mathcal{O}_{\tilde{X}}(-Z_{\max})$. We say that $\pi^*(m_0)$ (or π) has no basepoint, if $V(\mathcal{I}) = \emptyset$. If π does have some base-points, by some (iterated) blow-ups at the base-points, one can replace π by another resolution which has no base-points. If π has no base-points, then $\mathrm{mult}(X, 0) = -Z_{\max}^2$ (cf. [111]); a fact which illustrates the importance of the maximal cycle and the basepoint freeness of π.

In fact, more generally, one expects an intimate relationship between the pull-back sheaves $\pi^*(m_0^k)$ and the divisorial sheaves $\mathcal{O}_{\tilde{X}}(-kZ_{\max})$ (hence between the Hilbert–Samuel function and the maximal ideal cycle).

3.1.4. Consider now the holomorphic line bundles $\Omega^2_{X \setminus \{0\}}$ of holomorphic 2-forms on $X \setminus \{0\}$. If it is holomorphically trivial then we say that $(X, 0)$ is Gorenstein. If one of its power is holomorphically trivial then $(X, 0)$ is $\mathbb{Q}$-Gorenstein. E.g., complete intersections are Gorenstein.

Similarly, one can also consider the line bundle $\Omega^2_{\tilde{X}}$. By a result of Laufer [43]:

$$p_g = \dim_{\mathbb{C}} H^0\big(\tilde{X} \setminus E, \Omega^2_{\tilde{X}}\big) / H^0\big(\tilde{X}, \Omega^2_{\tilde{X}}\big).$$

This can be read 'from $(X, 0)$' as well: If $H^0_{L^2}$ denotes the global L^2-forms, then

$$p_g = \dim_{\mathbb{C}} H^0\big(X \setminus \{0\}, \Omega^2_X\big) / H^0_{L^2}\big(X \setminus \{0\}, \Omega^2_X\big).$$

4. TOPOLOGICAL CANDIDATES FOR SOME ANALYTIC INVARIANTS

4.1.

One can ask if it is possible to recover a certain analytic invariant from the link M. For a more complete discussion see e.g. [12, 65, 66, 67, 68, 69, 70, 71, 77, 78, 79, 87]. Similarly, we can ask the same question for certain analytic properties which characterize important families of singularities.

4.1.1. The *Hirzebruch–Jung singularities*, by definition, are characterized by the existence of a finite projection $(X, 0) \to (\mathbb{C}^2, 0)$ whose reduced discriminant space is included in the union of the coordinate axes of $(\mathbb{C}^2, 0)$. On the other hand, one can show that $(X, 0)$ is Hirzebruch–Jung if and only if its link is a lens space $L(p, q)$ with $0 < q < p$ and $gcd(p, q) = 1$; or equivalently, if the minimal resolution graph is a straight line graph with all genera zero (here we will use the traditional notation $-b_v$ instead of e_v):

$$\overset{-b_1}{\bullet} \quad\rule[0.5ex]{1.5em}{0.4pt}\quad \overset{-b_2}{\bullet} \qquad \cdots \qquad \overset{-b_s}{\rule[0.5ex]{1.5em}{0.4pt}\bullet}$$

where $-b_1, \ldots, -b_s$ are given by the continued fraction $[b_1, b_2, \ldots, b_s]$:

$$p/q = b_1 - \cfrac{1}{b_2 - \cfrac{1}{\ddots - \cfrac{1}{b_s}}}, \quad b_1, \ldots, b_s \geq 2.$$

Notice also that a Hirzebruch–Jung singularity can also be realized as a *cyclic quotient* singularity $X_{p,q} := (\mathbb{C}^2, 0)/\mathbb{Z}_p$. Here the action is $\xi * (u, v) = (\xi u, \xi^q v)$, where ξ is a primitive p-th root of unity.

For general reference on Hirzebruch–Jung singularities, see e.g. [7] or [89].

4.1.2. Artin defined the *rational singularities* by the vanishing $p_g = 0$. Artin's topological characterization, cf. [3, 4, 21, 91], is reviewed in section 7.

4.1.3. The *rational double points*, or *RDP*-singularities (i.e. rational singularities with multiplicity two) are exactly the simple hypersurface singularities, hence those of type *A-D-E*. Topologically they are characterized by the fact that their minimal resolution graphs are the well-known *A-D-E* (negative definite) graphs. Equivalently, any connected, negative definite graph with $g_v = e_v + 2 = 0$ for all v is the minimal resolution graph of some *RDP* (and it is of type *A-D-E*).

Additionally, the *RDP*'s are exactly the (Kleinian) quotient singularities $\mathbb{C}^2/G$ for finite subgroups $G \subset SL(2, \mathbb{C})$. Fore more details, see e.g. [21, 23].

4.2. (Pseudo)Taut singularities

One can ask even for the possibility to characterize *completely* the analytic type from the topological type. The list of singularities when this is happening was established by Laufer in [44] (completing the list started by Grauert, Brieskorn, Tjurina and Wagreich). These singularities are called *taut,* their topological types carry unique analytical structure. They include all the Hirzebruch–Jung singularities and the rational double and triple points.

In fact, Laufer classified even those resolution graphs which support only countably (equivalently, finitely) many analytic structures. They are called *pseudo-taut singularities* and are distinguished by their topological types and $\theta = h^1(\tilde{X}, \theta_{\tilde{X}})$.

4.3. Topological candidates

Unfortunately, the class of (pseudo)taut singularities is rather restrictive.

On the other hand, in most of our geometrical problems we do not really need this extremely strong, complete topological characterization. In general, what we need is the behavior of *some specific invariant* only, which guides a certain geometric phenomenon. For example, by [92, 112], a deformation of the minimal resolution $\tilde{X}$ blows down to a deformation of $(X, 0)$ if and only if $p_g = h^1(\mathcal{O}_{\tilde{X}})$ remains constant during the deformation.

Therefore, one tries to characterize topologically some of the discrete invariants of $(X, 0)$ only. In general, the analytic invariants listed is section 3 are not topological, but if we restrict our study to some special families, then they might be determined from the graph. In fact, in general, we have 'topological candidates' (or bounds), which in nice cases coincide with their analytical counterparts. In the next paragraphs we will list some of them.

4.3.1. The topological candidate for $\mathcal{Z}_{an}$ is the ordered lattice $\mathcal{Z}_{top} :=$ the set of cycles $Z \in \mathcal{Z}$ with $Z > 0$ and $Z \cdot E_v \leq 0$ for any E_v. The point behind this definition is that any $Z \in \mathcal{Z}_{an}$ satisfies this property, hence $\mathcal{Z}_{an} \subset \mathcal{Z}_{top}$. More precisely, the definition of $\mathcal{Z}_{top}$ contains the topological obstruction which should be satisfied by a cycle Z if it equals $(f)_E$ for some $f \in m_0$. Artin proved that, if $Z_1, Z_2 \in \mathcal{Z}_{top}$, then $\min\{Z_1, Z_2\} \in \mathcal{Z}_{top}$. In particular, $\mathcal{Z}_{top}$ has a unique minimal element Z_{min}, called *Artin's fundamental cycle,* or the *minimal cycle* [3, 4].

$Z_{\min}$ is the topological candidate for $Z_{\max}$. Clearly $Z_{\min} \leq Z_{\max}$. If $Z_{\min} = Z_{\max}$, and $\pi^*(m_0)$ has no base-points, then mult$(X, 0) = -Z_{\min}^2$, in particular, the multiplicity is a topological invariant (a posteriori).

4.3.2. Another 'pair' is the following. The analytic invariant is the canonical divisor $K_{\tilde{X}}$ of $\tilde{X}$. Numerically, it is codified by the *(anti)canonical cycle* Z_K, which is the unique rational cycle in $\mathcal{Z}_{\mathbb{Q}}$, supported by E, satisfying $(Z_K, E_v) = -(K_{\tilde{X}}, E_v)$ for all v. The number $-(K_{\tilde{X}}, E_v)$ can be determined topologically by the adjunction formula:

$$(Z_K, E_v) = E_v \cdot E_v + 2 - 2g_v - 2\delta(E_v) \quad \text{for all} \quad v.$$

(Above $\delta(E_v)$ is the sum of delta–invariants of the singularities of E_v.)

One shows that in the case of the minimal resolution one has $(Z_K, E_v) \leq 0$ for any v. Since the intersection matrix is negative definite, this implies that either $Z_K = 0$ or all the coefficients of Z_K are strictly positive (see e.g. [49]). Moreover, $Z_K = 0$ if and only if $(X, 0)$ is a *RDP*-singularity and π is the minimal resolution.

Note also that the rational number $Z_K^2 + \#\mathcal{V}$ is independent of the choice of π and is an invariant of the link M. On the other hand, Z_K^2 also plays an important role: if π is the *minimal* resolution, Z_K^2 associated with π will be denoted by $Z_{K,m}^2$. Obviously, this number is also an invariant of M.

Finally notice that in the case of the minimal resolution, $(Z_K, E_v) = 0$ for some v if and only if E_v is a smooth, rational (-2)-curve. Sometimes we prefer to avoid such curves (and prefer to have strict inequalities $(K, E_v) > 0$). Therefore, in the minimal resolution we contract all the *RDP*-configurations of curves (cf. 4.1.3). In this way we get a modification $\tilde{X} \to X$ with $\tilde{X}$ in general singular, but only with *RDP*'s. It is called the *minimal resolution with RDP's*.

4.3.3. The main importance of the (anti)canonical cycle Z_K comes from its role in the Riemann–Roch formula. For this, we fix an integral cycle $Z > 0$. Although the individual cohomology dimensions $h^i(Z)$ $(i = 1, 2)$ in general are not topological, the Euler characteristic $\chi(Z) := h^0(Z) - h^1(Z)$ depends only on $\Gamma(\pi)$ by the Riemann–Roch theorem. Indeed, $\chi(Z) = -(Z, Z - Z_K)/2$.

4.3.4. Z_K also characterizes the "topological counterpart" of the set of Gorenstein singularities. The perfect analogue of their definition is the following: we say that $(X, 0)$ is *numerically Gorenstein* if the line bundle $\Omega^2_{X \setminus \{0\}}$ is *topologically* trivial. This happens if and only if $\Omega^1_{X \setminus \{0\}}$ is topologically trivial. Moreover, in terms of a fixed resolution, $(X, 0)$ is numerically Gorenstein if and only if Z_K has integral coefficients. Obviously, Gorenstein singularities are numerically Gorenstein.

Note that the Gorenstein property does not impose any further topological restriction. Indeed, in [90] Popescu-Pampu proved that any numerically Gorenstein topological type can support a Gorenstein analytic structure.

5. Deformations of Normal Surface Singularities

5.1. Deformations

If one tries to clarify how the analytical invariants change in a given topological type, one can start first to understand how they can vary in an infinitesimal deformation. This is the subject of the *(local/formal) deformation theory*.

Deformation theory analyses flat maps $\lambda : (\mathcal{X}, 0) \to (T, 0)$ whose special fibre is identified with $(X, 0)$, up to isomorphisms of maps which respect these identifications. Such maps are called deformations of $(X, 0)$ over $(T, 0)$. In this article, $(T, 0)$ is always supposed to be reduced.

Since $(X, 0)$ has an isolated singularity, one has a *semi-universal deformation* $(\mathcal{X}, 0) \to (B, 0)$. It has the property that all the deformations λ are induced by a map $(T, 0) \to (B, 0)$ which is unique at the level of tangent spaces, see [32, 95]. The space $(B, 0)$ is called the *base space of the semi-universal deformation* of $(X, 0)$. It is smooth if $(X, 0)$ is a complete intersection [107], or if $\operatorname{emb dim}(X, 0) = 4$ [94]. But in general, B can be singular with several irreducible components (with different dimensions), and these components can intersect each other in complicated ways.

The space B, in general, is hardly computable. Usually, one first determines its Zariski tangent space $\left(T^1_{X,0}, 0\right)$, the so-called first order deformations of $(X, 0)$, and then the *obstruction space* $\left(T^2_{X,0}, 0\right)$, with a map $\left(T^1_{X,0}, 0\right) \to \left(T^2_{X,0}, 0\right)$, whose fiber over the origin provides B.

5.1.1. 'Trivial' deformations. Even if (in some cases) we can say very little about B (or $\mathcal{X} \to B$), we would still like to be able to decide when a flat deformation λ is 'trivial'. Here we can think about *analytic triviality,* or *topological triviality.* Analytic triviality means that the map $(T, 0) \to (B, 0)$ is the constant map onto the origin. But, unfortunately, there is no unanimously accepted definition which would answer to all the expectations of the topological triviality. On the other hand, the multitude of the different possibilities explore an extremely rich structure of the deformations; see e.g. the papers of Teissier [104, 106] for different notions (like Whitney conditions, equisingularity, simultaneous resolutions, or μ-constant, resp. μ^*-constant deformations in the case of hypersurfaces).

Here we will emphasize mainly the notion of simultaneous resolution. It also can be defined at different levels of complexity.

5.2. Simultaneous resolutions. Definitions [106, 47, 40].

Let $\lambda : (\mathcal{X}, 0) \to (T, 0)$ be a flat deformation of $(X, 0)$ as above. For any $t \in T$ we write X_t for $\lambda^{-1}(t)$, and we use similar notations for other spaces and invariants as well. Sometimes we fix a section $s : T \to \mathcal{X}$ of λ (which identifies for any t a base-point $s(t) \in X_t$, or a space-germ $(X_t, s(t))$).

Let $\Pi : \tilde{\mathcal{X}} \to \mathcal{X}$ be a proper modification, such that $\lambda \circ \Pi$ is flat. We say that

(i) Π is a *simultaneous RDP resolution* if each $\tilde{X}_t \to X_t$ is a minimal *RDP* resolution.

(ii) Π is a *very weak simultaneous resolution* if each $\tilde{X}_t \to X_t$ is a minimal resolution.

(iii) Π is a *weak simultaneous resolution along* s if it is a very weak resolution near $s(T)$, and the restriction map $\Pi : \Pi^{-1}(s(T))_{\text{red}} \to s(T)$ is simple (i.e. a locally trivial deformation in the Euclidean topology).

(iv) Π is a *strong simultaneous resolution along* s if it is a very weak resolution near $s(T)$ and $\Pi : \Pi^{-1}(s(T)) \to s(T)$ is simple.

We say that λ admits a simultaneous *RDP* (resp. very weak, weak, or strong) resolution, if there exits Π as above with (i) (resp. (ii), (iii) or (iv)).

Notice that (ii) means that $\lambda \circ \Pi$ is an analytic submersion. Also, from (ii) it follows that if T is smooth then Π is a resolution of $\mathcal{X}$.

We emphasize that in (iii) $\Pi^{-1}\big(s(T)\big)_{\text{red}}$ is the topological space with the Euclidean topology, which forgets the analytic space structure of the inverse image; while in (iv) $\Pi^{-1}\big(s(T)\big)$ is the non-reduced analytic space. In particular, in (iv) we require that the inverse images of the maximal ideals of the germs $\big(X_t, s(t)\big)$ behave trivially.

In the next paragraphs, for simplicity, we will take the unit disc D for T.

The literature about simultaneous resolutions and deformations of normal singularities is extremely rich, the interested reader is invited to navigate the articles of Hironaka, Lipman, Schlessinger, Teissier, Lê, Brieskorn, Tjurina, Riemenschneider, Wahl, Looijenga, Arndt, Cristophersen, Behnke, Knörrer, Theo de Jong, van Straten, Pellikaan, Stevens, Vaquié, Kollár, Shepherd-Barron.

5.3.

Our plan is to highlight some topological criteria which assure the existence of some kind of simultaneous resolution, see [106, 40, 47, 48]. Some of these topological descriptions are codified by some numerical invariants which, in general, are *semi-continuous*. If we do not fix a section s, then we also accept multiple singularities in the fiber X_t. If i is an invariant associated with germs of normal surface singularities, then $\sum_{X_t} i$ will denote the sum $\sum i(X_t, p)$ over all the singular points p of the fiber X_t.

Theorem 5.3.1. *Let* $\lambda : (\mathcal{X}, 0) \to (D, 0)$ *be a flat deformation of* $(X, 0)$. *Then*

(i) [110] $\big(Z^2_{K,m}\big)_t$ $(t \in D)$ *is lower semi-continuous; i.e.* $\big(Z^2_{K,m}\big)_0 \leq \sum_{X_t} Z^2_{K,m}$.

(ii) [26] $(p_g)_t$ *is upper semi-continuous; i.e.* $p_g(X, 0) \geq \sum_{X_t} p_g$.

5.4. Very weak simultaneous resolutions

First we mention a result of Brieskorn which says that if a deformation λ admits a simultaneous *RDP* resolution, then by a finite surjective base-change, λ can be transformed onto a deformation which admits a very weak simultaneous resolution, cf. [15, 14], see also the survey of Pinkham: *Résolution simultanée de points doubles rationnels* in [21, pages 179–203].

In particular, if we accept base-changes, then the obstructions to have a *RDP,* or a very weak simultaneous resolution are the same. Notice also that the *RDP* resolution is unique (if it exists, cf. [40]); but the very weak simultaneous resolution, in general, is not.

5.4.1. Topological characterization [47].
(i) Let $\lambda : (\mathcal{X}, 0) \to (D, 0)$ be a flat deformation of a normal *Gorenstein* surface singularity $(X, 0)$. Then $t \mapsto \sum_{X_t} Z^2_{K,m}$ is constant if and only if λ admits a simultaneous *RDP* resolution.

Here one can verify (see e.g. [47], page 12), that if $(X, 0)$ is normal Gorenstein then $\mathcal{X}$ and all the fibers X_t are normal Gorenstein.

(ii) Let λ be a family of normal Gorenstein singularities $\big(X_t, s(t)\big)$. If $Z^2_{K,m}\big(X_t, s(t)\big)$ is constant, then $p_g\big(X_t, s(t)\big)$ is also constant.

Above, the Gorenstein assumption cannot be dropped, cf. [40](2.8), or 6.3 here. Nevertheless, 5.4.1 has a global version without the Gorenstein assumption [40](2.1); see also [40](2.25) for a possible local analog.

5.4.2. Example.
Let $\lambda : (\mathbb{C}^3, 0) \to (\mathbb{C}, 0)$ be a *simple* hypersurface singularity. Then in the minimal resolution graph of $(X, 0) = \big(\lambda^{-1}(0), 0\big)$ one has $Z_K = 0$. On the other hand all the other fibers are smooth. Hence, for any t one has $\sum_{X_t} Z^2_{K,\min} = 0$. The existence of the very weak simultaneous resolution (after a base change) shows that the Milnor fiber of λ is diffeomorphic to $\tilde{X}$, the minimal resolution of $(X, 0)$.

This deformation also exemplifies the necessity of the base change in Brieskorn result 5.4. Indeed, for the miniversal deformation one needs a Galois base change with the corresponding Weyl group W. E.g., in the case of A_1 singularity, $W = \mathbb{Z}_2$, and $\{x^2 + y^2 + z^2 = t^2\}$ admits a very weak simultaneous resolution.

5.4.3. Example.
Let $(X, 0)$ be the hypersurface singularity $\big(\{z^2 = x^3 + y^{12}\}, 0\big) \subset (\mathbb{C}^3, 0)$. Let $\mathcal{X}$ be $\big\{z^2 = x^3 + (y^2 + t)^6\big\}$, a deformation over $t \in D$. The exceptional divisor E of the minimal resolution of $(X, 0)$ has two irreducible components $E_1 \cup E_2$ with $g_1 = 1$, $g_2 = 0$, $E_1^2 = -1$, $E_2^2 = -2$, $E_1 \cdot E_2 = 1$, $Z_K = 2E_1 + E_2$. Hence $Z_K^2 = -2$. On the other hand, in this deformation the singular point splits into two singular points: for $t \neq 0$, X_t has two singular points, each having $Z_K^2 = -1$. Therefore, this deformation admits a very weak (or *RDP*) simultaneous resolution.

5.5. Topological characterization of weak simultaneous resolutions [47, 40]

Let $\lambda : \mathcal{X} \to D$ be a flat family of normal surface singularities, and $s : D \to \mathcal{X}$ a section of λ. Then the germs of $\{X_t\}_{t \in D}$ along s are pairwise homeomorphic (i.e. each X_t has a singularity at $s(t)$ such that $(X_t, s(t))$ is homeomorphic to $(X, 0)$) if and only if λ admits a weak simultaneous resolution along s. Obviously, the topological condition is equivalent with the stability of the links: $M_t \simeq M$ for any t (where $\simeq$ means 'orientation preserving diffeomorphism').

5.5.1. Example. Assume that we are in the situation of 5.4.1 (ii); i.e. λ is a family of normal Gorenstein singularities $(X_t, s(t))$. Then if the link is constant, then evidently $Z^2_{K,m}(X_t, s(t))$ is constant too, hence $p_g(X_t, s(t))$ is also constant.

If, additionally, each $(X_t, s(t))$ has a smoothing, with Milnor number μ_t, then using the above statement and 6.5(2), one gets that if the link is constant then μ_t is also constant.

5.6. Strong simultaneous resolutions

We assume that λ is a flat deformation of isolated *hypersurface* singularities $(X_t, s(t))$ $(t \in D)$.

We recall (see [104]), that for any isolated hypersurface singularity $(X, 0)$, Teissier defined an invariant μ^*. In dimensional two $\mu^* = (\mu^{(3)}, \mu^{(2)}, \mu^{(1)})$ has the following significance: $\mu^{(3)}$ is the Milnor number μ (see next section) $\mu^{(1)} = \text{mult}(X, 0) - 1$, while $\mu^{(2)}$ is the Milnor number of the plane curve singularity obtained by cutting $(X, 0)$ by a generic hyperplane through the origin.

In [106] Teissier proved that the existence of a strong simultaneous resolution implies the Whitney condition along $s(D)$. Moreover, by another result of Teissier [104], and respectively of Briançon and Speder [13], the Whitney condition is equivalent to μ_t^* constant. Finally, the circle is completed by Laufer in [48], where he proved that μ_t^* constant implies strong simultaneous resolution for λ.

In particular, λ admits a strong simultaneous resolution if and only if $(\mu_t, \mu_t^{(2)}, \text{mult}(X_t))$ is constant.

5.7.

If one tries to compare the strong, respectively the weak simultaneous resolutions of deformations of isolated hypersurface singularities, then one realizes that the difference which really separates them is codified in the invariants $\mu_t^{(2)}$ and mult (X_t). Both depend essentially on the properties of the generic hyperplane sections. Let m_t be the maximal ideal of $\mathcal{O}_{X_t, s(t)}$. Then we recall (cf. 3.1.3) that $\Pi_t^*(m_t) = \mathcal{I}_t \cdot \mathcal{O}_{\tilde{X}_t}(Z_{\max,t})$. If μ_t and $\Pi_t^*(m_t)$ is constant then λ admits a strong simultaneous resolution. If for a deformation λ (with M_t and μ_t constant) a strong simultaneous resolution does not exist, then we can expect that either one of the coefficients of $Z_{\max,t}$ jumps at $t = 0$, or at the embedded points $V(\mathcal{I}_t)$ the structure of $\mathcal{I}_t$ is not constant.

For a deformation which admits a weak simultaneous resolution the existence of a strong simultaneous resolution is not guaranteed, even if the stable link is a rational homology sphere.

5.7.1. Example. Consider the deformation of hypersurface (even Newton non–degenerate) isolated singularities $f_t = z_3^3 + z_2^4 z_1 + z_1^{10} + t z_2^3 z_3$ [12]. Then the link is stable under the deformation, nevertheless $\mu^{(2)}(f_1) = 7$ and $\mu^{(2)}(f_0) = 8$. Note that in this case the stable link is a rational homology sphere (similar examples without this additional property were constructed by Briançon and Speder [13]).

6. SMOOTHINGS AND THEIR INVARIANTS

6.1.

A (flat) deformation $\lambda : (\mathcal{X}, 0) \to (D, 0)$ of $(X, 0)$ is called a *one parameter smoothing* of $(X, 0)$ if $X_t = \lambda^{-1}(t)$ is smooth for $t \neq 0$. We will write F for (the C^∞ type of) X_t ($t \neq 0$), and we will call it the *Milnor fiber* of λ. F is an oriented real connected 4-manifold with boundary M. As the fibers of λ are Stein, F has the homotopy type of a finite CW-complex of dimension ≤ 2. (For the choice of the 'good' representatives of the local Milnor balls, Milnor fibers and fibrations, see e.g. the book of Looijenga [56].)

We write $\mu := \operatorname{rank} H_2(F, \mathbb{Z})$, and we call it the *Milnor number* of the smoothing λ. In fact, $H_1(F, \mathbb{Q}) = 0$ (fact conjectured by Wahl [115], and proved by Greuel and Steenbrink [33]), hence the topological Euler characteristic $\chi_{\text{top}}(F)$ of F is $1 + \mu$. The intersection form in $H_2(F, \mathbb{Z})$ is symmetric. Let (μ_+, μ_-, μ_0) be its Sylvester invariants and $\sigma := \mu_+ - \mu_-$ the *signature* of the Milnor fiber F. Notice that the Milnor fiber F, hence all its invariants, depend on the choice of the smoothing λ.

Consider the semi-universal deformation $(\mathcal{X}, 0) \to (B, 0)$ of $(X, 0)$. An irreducible component B_i of B is called a *smoothing component* of $(X, 0)$ if the general fiber over B_i is smooth. Any one parameter smoothing λ of $(X, 0)$ lies on a unique smoothing component.

6.1.1. Not every singularity $(X, 0)$ admits a smoothing, and, in general, it is rather difficult to decide if a specific singularity has any smoothing at all. For different obstructions the reader is invited to search in the papers of Laufer, Wahl, Looijenga, Pinkham [46, 115, 58, 55, 88] (and the references therein).

For example, Dolgachev's triangle singularity D_{pqr} (with Dolgachev numbers p, q, r, cf. 8.3.2) admits a smoothing if $p + q + r < 22$ [55], cannot be smoothed at all if $p + q + r > 22$ [115], and the remaining cases $p + q + r = 22$ are decided in [88]: all but $D_{2,10,10}$ can be smoothed.

6.2. Example

For rational singularities all the components of B are smoothing components. One of them is distinguished. It can be constructed as follows [5]: let $\tilde{X} \to X$ be the minimal resolution of $(X, 0)$. Then all the deformations of $\tilde{X}$ come from deformations of $(X, 0)$ (since p_g is constantly zero). These deformations provide a component, called the Artin component. It is smooth [113].

6.3. Example

In general, even in the case of rational singularities, there are more (smoothing) components. The first example was constructed by Pinkham [86]: the cone over the rational normal curve of degree 4 has two smoothing components. In this case, the minimal resolution graph has only one vertex with $e = -4$ and $g = 0$ (hence evidently $(X, 0)$ is rational, cf. 7.3). B has two irreducible components; one of them is the Artin component B_1, which has dimension 3. The other component B_2 has dimension 1. In this case both components are smooth and intersect each other transversally (cf. also with the corresponding results of section 7 and 10).

The Milnor number corresponding to the Artin component is 1, while the other Milnor number is 0. This means that the Milnor fiber F above B_2 is a rational homology ball (with boundary the lens space $M = L(4,1)$).

The smoothing above B_2 can be realized as follows, see e.g. [115], (5.9.1). Consider the hypersurface singularity $\{f = 0\} = \{xy - z^2 = 0\}$ in $(\mathbb{C}^3, 0)$ with the $\mathbb{Z}_2$-diagonal action $(\pm 1) * (x, y, z) = (\pm x, \pm y, \pm z)$. The quotient singularity $\{f = 0\}/\mathbb{Z}_2$ is exactly $(X, 0)$, and the smoothing $\{f = t\}$ of f induces a one parameter smoothing of $(X, 0)$. The Milnor fiber F_f of f has the homotopy type of S^2, and $\mathbb{Z}_2$ acts on F_f freely; hence $\chi_{\text{top}}(F) = 1$, which shows that $\mu = 0$.

By [113], for any rational singularity, the semi-universal deformation restricted above the Artin component admits a very weak simultaneous resolution. This is not true for the other components. In the example of Pinkham, the deformation above B_2 does not admit a very weak simultaneous resolution.

Notice also that in this example if one takes the semi-universal deformation restricted above the Artin component, then the invariant $\sum_{X_t} Z^2_{K,m}$ is not constant. Indeed, for $t = 0$ it is -1, otherwise it is zero. Nevertheless, a very weak simultaneous resolution exists, hence the Gorenstein assumption in 5.4.1 is needed.

6.4.

There are interesting relations which connect the topological invariants $Z^2_K + \#\mathcal{V}$, $b_1(M)$ (the first Betti number of the link M), the smoothing invariants μ and σ, and the analytic invariant p_g.

6.5. Theorem [22, 46, 99, 115, 58]

Consider a smoothing of the normal surface singularity $(X, 0)$. Then

(1) $4p_g = \mu + \sigma + b_1(M)$.

In addition, if $(X, 0)$ is Gorenstein, then

(2) $\mu = 12p_g + Z^2_K + \#\mathcal{V} - b_1(M)$.

In particular, for a smoothing of a Gorenstein singularity, (1) and (2) give

(3) $\sigma + 8p_g + Z_K^2 + \#\mathcal{V} = 0$.

This shows that modulo the link-invariants $Z_K^2 + \#\mathcal{V}$ and $b_1(M)$, there are two independent relations connecting p_g, μ and σ, provided that $(X,0)$ is Gorenstein. Moreover, in the Gorenstein case, μ and σ are independent of the smoothing.

6.5.1. Wahl in [115] added to this list some other relations. He also explained a recipe by which one can create even more identities, modulo a conjectural globalization assumption. This assumption was proved by Looijenga in [57].

We will recall in the next paragraphs only one such identity which is strongly related to deformations.

Let $\lambda : (\mathcal{X},0) \to (D,0)$ be a smoothing, and consider the relative derivations $\theta_{\mathcal{X}/D}$. Then $coker(\theta_{\mathcal{X}/D} \otimes \mathcal{O}_X \to \theta_X)$ has finite dimension, which will be denoted by β. Note that β is an a priori analytic invariant of the smoothing λ. Recall that θ denotes $h^1(\tilde{X}, \theta_{\tilde{X}})$.

6.6. Theorem [115]

Consider the minimal resolution of $(X,0)$ with $\#\mathcal{V}_{\min}$ irreducible exceptional components and (anti)canonical cycle $Z_{K,m}$. Then, with the above notations, one has:

(1) β equals the dimension of the irreducible component B_i of B on which the smoothing λ occurs.

(2) $\beta - \theta + 14p_g = 2(\mu + b_1(M) - \#\mathcal{V}_{\min})$.

If $(X,0)$ is Gorenstein, then this (via 6.5(2)) transforms into

(3) $\beta = \theta + 10p_g + 2Z_{K,m}^2$.

Notice that (by (2)) β depends only on $(X,0)$ and the Milnor number of the smoothing λ; in fact, $\beta - 2\mu$ is independent of the smoothing. If $(X,0)$ is Gorenstein, then β itself is independent of the smoothing.

6.6.1. Example. If $(X, 0)$ is rational then $p_g = b_1(M) = 0$. Then, for any smoothing $\sigma = -\mu$ (i.e. the intersection form on $H_2(F)$ is negative definite, cf. 6.5(1)), $0 \le \mu \le \#\mathcal{V}_{\min}$, and

$$\beta = \theta - 2(\#\mathcal{V}_{\min} - \mu).$$

For smoothings supported by the Artin component B_1 one has $\mu = \#\mathcal{V}_{\min}$ and $\dim B_1 = \beta = \theta$.

Assume that $(X, 0)$ is rational, but not *RDP*. Then Wahl proved (cf. [8]) that

$$\dim T^1_{X,0} \ge \theta + \mathrm{emb\,dim}\,(X, 0) - 4,$$

which gives a lower bound for the codimension of the Artin component in $T^1_{X,0}$. For many cases the equality holds (see e.g. [8], cf. also with 7.7 here). But Wahl have found an example when the inequality is strict (cf. [8]).

For the ultimate formula for the codimension of the Artin component of a rational singularity see [19].

6.7. Remark

In general it is extremely difficult to determine the number of irreducible components of B and their dimensions. Hence, it is really remarkable if in some cases this can be done, and it is even more remarkable if this can be done only from the topology of the link; see the next section.

7. RATIONAL SINGULARITIES

7.1.

In this section we exemplify for rational singularities that the topology indeed carries a lot of analytic and smoothing information. Recall that, by definition, $(X, 0)$ is rational if $p_g = 0$. In the sequel we fix a resolution π. It is easy to see that $p_g = 0$ if and only if $h^1(Z) = 0$ for any $Z > 0$. In particular, all the genera g_v should vanish, and $\Gamma(\pi)$ should be a tree.

The main point is that Artin succeeded to replace the vanishings of $h^1(Z)$'s by a criterion formulated in terms of $\chi(Z)$. In fact, the beauty of the second part of the next characterization is that it is enough to consider only one cycle, namely the fundamental cycle $Z_{\min}$. It is instructive to recall that for *any* normal surface singularity $h^0(Z_{\min}) = 1$, hence $\chi(Z_{\min}) \leq 1$.

7.2. Topological characterizations of rational singularities [3, 4]

(a) $p_g = 0$ if and only if $\chi(Z) \geq 1$ for all cycles $Z > 0$.

(b) $p_g = 0$ if and only if $\chi(Z_{\min}) = 1$.

Notice that the characterizations (a) and (b) are independent of the choice of the resolution π. If a resolution graph satisfies (a) or (b), we say that it is a *rational graph*. Note also that any resolution of a rational graph is automatically good.

7.3. Examples

(a) If a rational singularity $(X, 0)$ is numerically Gorenstein, then in the minimal resolution $Z_K = 0$. In particular, $(X, 0)$ is a *RDP*.

(b) Let Γ be an arbitrary tree. For any vertex v set δ_v to be the number of edges with endpoint v. Consider the decorations $g_v = 0$ and

$$
e_v = \begin{cases} -\delta_v & \text{if } \delta_v \neq 1 \\ -2 & \text{if } \delta_v = 1 \end{cases} \quad \text{for any } v.
$$

Then the intersection matrix I is automatically negative definite, hence Γ is the minimal resolution graph of some singularity. It is not difficult to show that $Z_{\min} = \sum_v E_v$, $\chi(Z_{\min}) = 1$, hence a singularity with such a resolution graph Γ is rational. Moreover, $Z_{\min}^2 = -\#\{v : \delta_v = 1\}$.

(c) Rational surface singularities with reduced fundamental cycle (i.e. with $Z_{\min} = \sum E_v$) are also called *minimal singularities* (cf. [40], section 7). E.g., the cyclic quotients are minimal.

The class of rational graphs is closed while taking subgraphs and decreasing self-intersections. The graphs of minimal singularities are obtained from those considered in (b) by decreasing the decorations.

(d) There is another important subclass of rational singularities, the so-called *sandwiched singularities*, studied by Zariski, Lipman, Hironaka, Spivakovsky (who invented the name of the family), Theo de Jong, van Straten and others, see e.g. [98, 39]. A sandwiched singularity is, by definition, a normal surface singularity which is analytically isomorphic to the germ of an algebraic surface that admits a birational morphism to $(\mathbb{C}^2, 0)$. If we consider a resolution $\tilde{X} \to X$ of such a singularity, then we get a diagram $(\tilde{X}, E) \to (X, 0) \to (\mathbb{C}^2, 0)$. In particular, X is sandwiched between two smooth spaces via birational maps.

They are also characterized by their (minimal) resolution graphs [98]. Consider a plane curve singularity $(C, 0) \subset (\mathbb{C}^2, 0)$, and let $\phi : Y \to \mathbb{C}^2$ be a (in general, non-minimal) embedded resolution of it. Consider the collection E of those irreducible exceptional divisors which are not (-1)-curves (and assume that they form a connected curve). If one contracts E then one gets a sandwiched singularity, and any sandwiched singularity can be obtained in this way (although the choice of $(C, 0)$ and ϕ is not unique). Notice that this can be reformulated in terms of the combinatorics of the graph as well.

The next theorem targets some of the analytic invariants introduced in section 3.

7.4. Theorem [3, 4]

Assume that $(X, 0)$ is rational and $k \geq 1$. Then:

(a) $Z_{an} = Z_{top}$, *in particular* $Z_{min} = Z_{max}$;

(b) $\pi^* m_0^k = \mathcal{O}(-kZ_{min})$, *in particular* $\mathrm{mult}\,(X, 0) = -Z_{min}^2$;

(c) $\dim_{\mathbb{C}} m_0^k / m_0^{k+1} = -kZ_{min}^2 + 1$, *in particular* $\mathrm{emb\,dim}\,(X, 0) = -Z_{min}^2 + 1$;

(d) $f_{HS}(k) = -k(k-1)/2 \cdot Z_{min}^2 + k$.

7.5.

Next we discuss the base space B of the semi-universal deformation of $(X, 0)$. Although there is a very general conjecture of Kollár about B, valid for any rational singularity [41], we present the case of cyclic quotient singularities only.

Notice that the cyclic quotient singularities are *taut* singularities. Hence, the next discussion shows that even in those cases when a priori we know that some invariant is topological, its representation from the graph can be really complicated.

For cyclic quotient singularities, Riemenschneider in [93] determined the infinitesimal deformations $T^1_{X,0}$, and later Arndt gave the equations of the base space B [2]. Here we reproduce the combintorial part of some results obtained by J. Christophersen, J. Stevens and J. Kollár and N. I. Shepherd-Barron (after [101]).

7.6. Combinatorial characterization of B

Assume as in subsection 4.1.1 that $p/q = [b_1, \ldots, b_s]$. We write $p/(p-q) = [a_2, \ldots, a_{e-1}]$. (Here we will follow the traditional index notation.) These two expansions are dual; one finds the first one from the second one with Riemenschneider's point diagram: place in the i-th row $a_i - 1$ dots, the first one under the last dot of the $(i-1)$-st row; then column j contains $b_j - 1$ dots. From this it is easy to see that the length of the sequence a_* (i.e. $e-2$) is $1 + \sum_j (b_j - 2)$. Hence, the integer e is exactly $\mathrm{emb\,dim}\,(X_{p,q})$. Indeed, since $Z_{\min} = \sum E_v$, 7.4 gives $\mathrm{emb\,dim} = 3 + \sum_j (b_j - 2)$.

If $e = 2$, then $(X, 0)$ is smooth. If $e = 3$ then $(X, 0)$ is a hypersurface singularity A_s; in particular B is smooth and irreducible of dimension $\mu = s$.

In the sequel we will assume that $e \geq 4$. Let K_{e-2} be the set of tuples of positive integers $[\mathbf{k}] = (k_2, \ldots, k_{e-1})$, $(k_i \geq 1)$, such that the continued fraction $[k_2, \ldots, k_{e-1}]$ is zero and $[\mathbf{k}]$ is admissible in the sense of (10.2). The cardinality of this set is $\frac{1}{e-2}\binom{2(e-3)}{e-3}$, the Catalan number C_{e-2}. This number also describes the number of ways to subdivide an $(e-1)$-gon in triangles using only diagonals which intersect at most at vertices. The correspondence between the two methods of counting is the following. Mark one of the vertices of the $(e-1)$-gon by $*$ (this is called the distinguished vertex), and the others by $v_2, \ldots, v_{e-1}$. Given a subdivision, we define k_i as the number of triangles having v_i as vertex $(2 \leq i \leq e - 1)$. Then the integers $[\mathbf{k}] = (k_2, \ldots, k_{e-1})$ determine an element in K_{e-2}.

In a similar way we also define the integer k_* associated with the distinguished vertex $*$. It satisfies

$$k_* + \sum k_i = 3(e - 3).$$

Then one has the following numerical identities (for details, see [101]):

7.7. Theorem [Riemenschneider, Christophersen, Stevens]

- $\dim T^1_{X,0} = \sum_j (b_j - 1) + e - 4 = 2e - 7 + s$.
- $\theta = \sum_j (b_j - 1) = e - 3 + s$.
- There is a one-to-one correspondence between the irreducible components of B and continued fractions $[\mathbf{k}] \in K_{e-2}$ with $k_i \le a_i$ for all i. For each $[\mathbf{k}]$ we denote the corresponding component by $B_{[\mathbf{k}]}$, its dimension by $\beta_{[\mathbf{k}]}$, and the corresponding Milnor number by $\mu_{[\mathbf{k}]}$.
- Each $B_{[\mathbf{k}]}$ is smooth.
- $\beta_{[\mathbf{k}]} = \sum_j (b_j - 1) - 2(e - 3 - k_*) = s + k_* - (e - 3 - k_*)$.
- $\mu_{[\mathbf{k}]} = s - (e - 3 - k_*) = \beta_{[\mathbf{k}]} - k_* \ge 0$.
- For any $[\mathbf{k}]$ clearly $k_* \le e - 3$, hence $\beta_{[\mathbf{k}]} \le \theta$ and $\mu_{[\mathbf{k}]} \le s$.
- The Artin component corresponds to $[\mathbf{k}] = [1, 2, \ldots, 2, 1]$. This is the unique $[\mathbf{k}]$ with $k_* = e - 3$. In particular, the Artin component is the unique component with $\beta_{[\mathbf{k}]} = \theta$ and $\mu_{[\mathbf{k}]} = s$.

7.7.1. The above data follow from a very detailed analysis of the first order deformation space $T^1_{X,0}$, and the corresponding equations of B in it. Of course, from this one can obtain even more information about B (e.g. about the intersection properties of the components). On the other hand, there is another, completely different approach of Kollár and Shepherd-Barron, which finds all the components of B, and all their dimensions, *without* computing $T^1_{X,0} \to T^2_{X,0}$ (but provides no information e.g. about their intersections) [40]. This gives possibilities for remarkable generalizations (see [41]) to situations when $T^1_{X,0} \to T^2_{X,0}$ is out of reach.

This construction depends essentially on a very special subclass of cyclic quotient singularities.

7.8. The class T of cyclic quotient singularities [58](5.9), [114], cf. also with [40]

For a cyclic quotient singularity $X_{p,q}$, the following facts are equivalent:

(i) $Z^2_K \in \mathbb{Z}$;

(ii) $(q + 1)^2 / p \in \mathbb{Z}$;

(iii) *For some positive integers r, l and d with $d \leq r$ and $\gcd(d,r) = 1$ one has $p = r^2 l$ and $q = drl - 1$.*

In fact, the integers r, l, d can be recovered from p and q by

$$l := \gcd\left(q + 1, p, (q + 1)^2/p\right); \quad rl := \gcd(q + 1, p); \quad drl := q + 1.$$

Notice that $r = 1$ implies $d = 1$, $p = l$ and $q = l - 1$; hence in this case $(X, 0)$ is the RDP A_{l-1}.

(iv) *The minimal resolution graph can be recognized as follows: the graph is either an A_k graph, or can be obtained by starting with one of the graphs*

$$\overset{-4}{\bullet} \quad \text{or} \quad \overset{-3}{\bullet}\!-\!\!\!\overset{-2}{\bullet}\!-\!\!\!\overset{-2}{\bullet}\!- \quad \cdots \quad -\overset{-2}{\bullet}\!-\!\!\!\overset{-3}{\bullet}$$

and iterating a few times (in an arbitrary order) the next step. By one step, change $[b_1, \ldots, b_k]$ to either $[2, b_1, \ldots, b_{k-1}, b_k + 1]$ or $[b_1 + 1, b_2, \ldots, b_k, 2]$ and take the corresponding graph.

7.8.1. Definitions. A cyclic quotient singularity satisfying one of the above properties is called of *class* T. For any such singularity one defines $def'(X_{p,q})$ to be l if $r > 1$, respectively to be $l - 1$ if $r = 1$.

In [40], the key object which provides valuable information about the irreducible components of B is the set of P-resolutions.

7.9. Definition

A *P-resolution* of a cyclic quotient singularity $(X, 0)$ is a partial resolution $f : Y \to X$ (i.e. a modification which is an isomorphism above $X \setminus \{0\}$) such that K_Y is ample relative to f (cf. next paragraph), and Y has only singularities of class T.

7.9.1. Fix a modification $f : Y \to X$. Let F_j $(1 \leq j \leq k)$ be the set of irreducible exceptional divisors of f, and set $\text{Sing}(Y) = \{P_1, \ldots, P_l\}$. Let $\rho : \tilde{X} \to Y$ be the minimal resolution of Y, and set $f \circ \rho = \pi$. Denote the strict transform (via ρ) of F_j by F'_j. Let Γ_i $(1 \leq i \leq l)$ be the subgraph of the resolution graph $\Gamma(\pi)$ of π corresponding to the singular points P_i.

Let $Z_K(\Gamma(\pi))$, respectively $Z_K(\Gamma_i)$ be the (anti)canonical cycle of $\Gamma(\pi)$, respectively of Γ_i.

Then K_Y is ample relative to f (i.e. $K_Y \cdot F_j > 0$ for all j) if and only if

$$\left(Z_K(\Gamma(\pi)) - \textstyle\sum_i Z_K(\Gamma_i)\right) \cdot F'_j < 0 \quad \text{for any } 1 \le j \le k.$$

In particular, the set of P-resolutions can be determined combinatorially from the plumbing graph of the link of $(X,0)$. The set of P-resolutions is not empty. For example, the minimal RDP resolution of $(X,0)$ is a P-resolution.

If $f : Y \to X$ is a P-resolution, set $def'(f) := \sum_{i=1}^{l} def'(Y, P_i)$ (cf. 7.8.1).

The set of exceptional divisors F'_j $(1 \le j \le k)$ form h connected, pairwise disjoint exceptional curves $C_1, \ldots, C_h$, each determining a cyclic quotient singularity (X_j, Q_j) (by contracting C_j into the point Q_j). The dimension of the Artin component of (X_j, Q_j) is denoted by $\theta(X_j, Q_j)$ (cf. 7.7). Set $\theta(f) := \sum_{j=1}^{h} \theta(X_j, Q_j)$.

7.9.2. [40, (3.14)]. When one searches for all the P-resolutions, the next fact is very helpful: Any P-resolution is dominated by the *maximal* resolution of $(X,0)$.

Here, a resolution $\pi : \tilde{X} \to X$ is *maximal* if all the coefficients of $Z_K(\pi)$ are strictly positive, and π is not dominated by any other resolution with this property. Any cyclic quotient singularity has a unique maximal resolution.

Now, we can state (3.9) of [40] for the particular case of cyclic quotients.

7.10. Theorem [Kollár–Shepherd-Barron]

For any cyclic quotient singularity $(X,0)$, there is a one-to-one correspondence between the set of P-resolutions of $(X,0)$ and the set of irreducible components of the base space B of the semi-universal deformation of $(X,0)$. If the component $B(f)$ corresponds to the P-resolution f, then $\dim B(f) = def'(f) + \theta(f)$.

Under this correspondence, the Artin component corresponds to the minimal RDP resolution.

The correspondence between the pictures 7.7 and 7.10 is realized in [101].

7.10.1. Example (Cf. with 6.3). Assume that the minimal resolution graph of $(X,0)$ has only one vertex with $g_v = 0$ and $e_v = -4$. Then $(X,0)$ has two P-resolution: the minimal resolution and the identity of $(X,0)$.

7.10.2. Example [40]. Assume that the minimal resolution $\tilde{X}$ of $(X,0)$ has the dual graph:

$$\overset{-3}{\bullet}\!\!-\!\!-\!\!-\!\!-\!\!\overset{-4}{\bullet}\!\!-\!\!-\!\!-\!\!-\!\!\overset{-2}{\bullet}$$

Consider also the resolution $\tilde{X}' \to X$ with graph:

$$\overset{-4}{\bullet}\!\!-\!\!\overset{-1}{\bullet}\!\!-\!\!\overset{-5}{\bullet}\!\!-\!\!-\!\!-\!\!\overset{-2}{\bullet}$$

Then $(X,0)$ has three P-resolutions Y_1, Y_2 and Y_3. Y_1 is the minimal RDP resolution obtained from $\tilde{X}$ by contracting the (-2)-curve. Y_2 is obtained from $\tilde{X}$ by contracting the (-4)-curve. Finally, Y_3 is obtained from $\tilde{X}'$ by contracting all the curves except the (-1)-curve.

The corresponding dimensions of the irreducible components of B are 6, 4 and 2.

7.11. Remark

If $(X,0)$ is rational with $\text{mult}\,(X,0) \leq 3$ then the base space B is smooth. See [37] and [102] for the description of the deformation of rational quadruple points, [38] for minimal rational singularities, and [39] for sandwiched singularities. All these descriptions are topological.

7.12. Kollár's Conjectures

As we already mentioned, Kollár formulated a conjecture about the irreducible components of B, in terms of P-resolutions, for any rational singularity [41]. He also stated a simplified version of this rather complex conjecture as follows. Let $(X,0)$ be a rational singularity. Suppose that any irreducible exceptional divisor of the minimal resolution of $(X,0)$ has self-intersection at most -5. Then B has only one component, namely the Artin component. This second conjecture was verified for cyclic quotients [101] and minimal rational singularities [39].

8. ELLIPTIC SINGULARITIES

8.1.

Recall that for any fixed resolution graph $\Gamma(\pi)$ of a normal surface singularity $(X,0)$ one has $\chi(Z_{\min}) \leq 1$. Similarly, $\min_{Z>0} \chi(Z) \leq 1$. By 7.2, $(X,0)$ is rational if and only if $\chi(Z_{\min}) = 1$, or equivalently, $\min_{Z>0} \chi(Z) = 1$. Similarly, $\chi(Z_{\min}) = 0$ if and only if $\min_{Z>0} \chi(Z) = 0$. Laufer in [45, page 1281] proved this via the theory of analytic deformations. For a combinatorial proof see [62, (4.3)].

8.2. Definition (Wagreich [111], Laufer [45])

$(X,0)$ is called *elliptic* if $\Gamma(\pi)$ is elliptic. $\Gamma(\pi)$ is elliptic if $\min_{Z>0} \chi(Z) = 0$, or equivalently, $\chi(Z_{\min}) = 0$. The definition is independent of the choice of the resolution.

Laufer in [45] identified topologically a subclass of elliptic singularities for which p_g is topological (in fact, $p_g = 1$). It is the set of *minimally elliptic singularities*. They are defined via their *minimally elliptic cycle*. Let π be the minimal resolution. A cycle $Z > 0$ is called *minimally elliptic* if $\chi(Z) = 0$ and for any $0 < D < Z$ one has $\chi(D) > 0$. Laufer proved that if $\chi(Z_{\min}) = 0$ then there is a unique minimally elliptic cycle. It will be denoted by C. Clearly, by its minimality, $C \leq Z_{\min}$.

Notice also that if $(X,0)$ is numerically Gorenstein, then Z_K associated with the minimal resolution is in $\mathcal{Z}_{\text{top}}$, hence $Z_{\min} \leq Z_K$.

8.3. Theorem. Minimally elliptic singularities [45]

Consider the minimal resolution π of $(X,0)$. Then the following statements are equivalent. If a singularity satisfies them, it is called minimally elliptic.

(i) $(X,0)$ is numerically Gorenstein and $Z_K = Z_{\min}$.

(ii) $\chi(Z_{\min}) = 0$ and any proper subgraph of $\Gamma(\pi)$ supports a rational singularity.

(iii) $(X,0)$ is numerically Gorenstein and Z_K is a minimally elliptic cycle.

(iv) $Z_{\min}$ is a minimally elliptic cycle.

(v) $p_g = 1$ and $(X, 0)$ is Gorenstein.

In particular, for minimally elliptic singularities $Z_{\min} = Z_K = C$.

8.3.1. Example. Assume that $\Gamma(\pi)$ consists of only one vertex with $g_v = 1$, and arbitrary self-intersection e_v. Then it is elliptic, and is called *simple elliptic* of degree $-e_v$. If $e_v \in \{-1, -2, -3\}$ then they are realized by the hypersurface singularities $(\{x^a + y^b + z^c + \lambda xyz = 0\}, 0)$ with $(a, b, c) = (2, 3, 6)$ $(\tilde{E}_8)$, $(2, 4, 4)$ $(\tilde{E}_7)$ and $(3, 3, 3)$ $(\tilde{E}_6)$. The case $e_v = -4$ is realized by the complete intersection $(\tilde{D}_5)$ $\{x^2 + y^2 + \lambda zw = xy + z^2 + w^2 = 0\}$.

8.3.2. Example. Dolgachev's triangle singularity $D_{-e_1, -e_2, -e_3}$**.** Assume that the resolution graph Γ has the following form with $g_v = 0$ for all v:

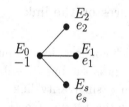

The intersection matrix is negative definite if and only if $1 + \sum_{v>0} 1/e_v > 0$. Note that $Z_{\min} \neq Z_K$. But in this case Γ is not minimal: E_0 is a rational (-1)-curve, so it should be contracted. Then in the new minimal graph, $Z_K = Z_{\min} = \sum E_v$.

The next theorem was obtained by Laufer and Reid; M. Tomari also proved similar results for elliptic singularities in the presence of some additional assumptions [108].

8.4. Theorem [45, 91]

Assume that $(X, 0)$ is a minimally elliptic singularity. Let $Z_{\min}$ be the fundamental cycle in the minimal resolution π. Then:

(a) $Z_{\min} = Z_{\max}$;

(b) If $Z_{\min}^2 \leq -2$, then $\pi^*(m_0) = \mathcal{O}(-Z_{\min})$, hence $\mathrm{mult}\,(X, 0) = -Z_{\min}^2$;

(c) If $Z_{\min}^2 = -1$, then $\pi^*(m_0) = m_Q \mathcal{O}(-Z_{\min})$ for a smooth point Q of E, and $\mathrm{mult}\,(X,0) = 2$;

(d) $\mathrm{emb\,dim}\,(X,0) = \max\left(3, -Z_{\min}^2\right)$;

(e) If $Z_{\min}^2 \leq -3$ then $\dim \mathcal{O}_{X,0}/m_0^k = \chi(\mathcal{O}_{kZ_{\min}}) + 1$ and $\dim m_0^k/m_0^{k+1} = -kZ_{\min}^2$ $(k \geq 1)$.

For more about elliptic singularities, see [45, 63]. For deformation invariants of minimally elliptic singularities, see Wahl's paper [114] and the references therein.

9. Links and Fillings form Contact Geometry Point of View

9.1. Contact structures on the link

One of the objects which connects local complex analytic singularities with low-dimensional topology is the link of isolated singularities. One defines a contact structure on a singularity link as follows. Fix an embedding $(X,0) \subset (\mathbb{C}^N, 0)$. Then the distance function $\rho := \sum_{k=1}^N |z_k|^2$ is strict pseudosubharmonic, and $\xi := TM \cap J(TM)$, the J-invariant subspace of TM, defines a contact structure on M. Its isotopy class is called the *canonical contact structure* of M induced by the analytic structure of $(X,0)$. (Here J is the almost complex structure of TX.)

This definition imposes several questions/problems for any fixed M:

1. *Classify all the possible contact structures induces by different analytic structures supported by the topological type determined by M.*

2. *How is the subclass from (1) related with the class of all contact structures of M?*

A partial answer to (1) was given in [16, 17]: all the possible (canonical) contact structures induces by different analytic structures are *contactomorphic*. Notice again that this fact is not valid in higher dimensions, see Ustilovsky's work [109].

In fact, we conjecture that all the possible (canonical) contact structures induces by different analytic structures are even *isotopic*. But, definitely, in order to prove this (or even to state this) one needs first to provide a canonical construction which identifies the link M up to an isotopy (and independently of the supported analytic structure); the existing plumbing construction identifies M only up to an orientation preserving diffeomorphism.

Although there is an intense activity in classification of contact structures, and a considerably impressive list of positive results finishing the classification for lens spaces, torus bundles over circles, circle bundles over surfaces, some Seifert manifolds (thanks to the work of Giroux, Etnyre, Honda, Lisca, Stipsicz and others, see e.g. [28, 29, 35, 36, 54] and the references listed therein), Part (2) is mainly open. Recall that a contact structure is either *overtwisted* or *tight,* and all overtwisted structure are characterized by the homotopy of their underlying oriented plane field (by a result of Eliashberg). Hence, the difficulty appears in the classification of tight structures. The canonical structure is one of them. Surprisingly, even the finiteness of the possible tight structures on M is not guaranteed in general (as was shown by Colin, Giroux and Honda [20]): although the number of corresponding homotopy classes of plane fields is finite, the number of isotopy classes of tight contact structures is finite if and only if M is atoroidal (i.e. the minimal resolution graph is star-shaped with at most three 'legs', and all genus decorations are zero). It is also not clear at all how one can identify in this multitude of structures the canonical one.

In the sequel the canonical contact structure will be denoted by (M, ξ_{can}).

9.2. Fillings of (M, ξ_{can})

Although in the literature there are many different versions of fillability (holomorphic, Stein, strong/weak symplectic), here we will deal only with Stein one: any Stein manifold whose contact boundary is contactomorphic to (M, ξ_{can}) is a Stein filling of (M, ξ_{can}). For a singularity link, the following questions/problems are natural:

1. *Is any (M, ξ_{can}) Stein fillable?*

2. *Classify all the Stein fillings of (M, ξ_{can}).*

3. *Determine all the Stein fillings 'coming from singularity theory'.*

The answer to (1) is yes: Consider the minimal resolution of $(X, 0)$. It is a holomorphic, non-Stein filling of (M, ξ_{can}), but by a theorem of Bogomolov and de Oliveira [11] this holomorphic structure can be deformed into a Stein one. This construction already provides an example for (3); another one is given by the Milnor fibers of the smoothings of different analytic realizations $(X, 0)$ of the topological type fixed by M (if there are any). More precisely, the existence (and uniqness) of the miniversal deformation of isolated singularities is guaranteed by results of Schlessinger [95] and Grauert [32]. In general, its base space has many irreducible components. A component is called *smoothing component* if the generic fiber over it (the so-called Milnor fiber) is smooth. In general, different analytic structures might have different smoothings; or for a fixed $(X, 0)$, different smoothing components might produce diffeomorphic Milnor fibers. It might also happen that smoothings do not exist at all (for the last two situation see e.g. the case of some simple elliptic singularities [58, 80]).

In fact, in the literature basically only these two constructions are present regarding (3); it is a high desire to find some other general constructions too. We notice that rational singularities are always smoothable, moreover, the Milnor fiber of one of the smoothing component (the Artin component) is diffeomorphic with the space of the minimal resolution.

The list regarding part (2) starts with a result of Eliashberg [25] showing that $(\mathbb{S}^3, \xi_{can})$ has only one Stein filling (up to diffeomorphism), namely the ball. For links of simple and simple elliptic singularities the classification was finished by Ohta and Ono [80, 81]. Moreover, in all these cases all possible Stein fillings are provided by the minimal resolution or Milnor fibers. This parallelism sometimes is really striking. E.g. for simple elliptic singularities with degree $k > 0$, Ohta and Ono proved that the existence of a Stein filling with vanishing first Chern class imposes $k \leq 9$. This can be compared with the fact that in the case of a Milnor fiber the Chern class is vanishing (by [96]) and the smoothability condition is the same $k \leq 9$ (cf. [86]).

Fillings of links of lens spaces $L(p, q)$ were classified by Lisca [52, 53] (as a generalization a result of McDuff [59] valid for the spaces $L(p, 1)$, for all $p \geq 2$), and of quotient surface singularities by Bhupal and Ono [10]. We will return to Lisca's result in the next section showing that Lisca's list agrees perfectly with the list of Milnor fibers (or, with the smoothing/all deformation components).

We would like to notice that this phenomenon, namely that all the Stein fillings are obtained either by minimal resolution or Milnor fibers – at some point of the singularity complexity – might stop. This emphasizes the importance of the research in direction (3) even more.

Moreover, in general, the finiteness of the Stein fillings might fail too: Ohta and Ono produced on some singularity links infinitely many symplectic fillings [82], also Ozbagci and Stipsicz in [85], and independently Smith in [97], have shown that certain contact structures have infinitely many Stein fillings (although they are not singularity links, one expects that at some moment similar fact will be established for some singularity links as well). For similar result see also the recent article [1] too.

9.3. (M, ξ_{can}) and open books

By a result of Giroux [30], there is a one-to-one correspondence between open book decompositions of M (up to isotopy and stabilization) and positive (that is, which induces the ambient orientation) contact structures on M (up to isotopy).

The link M of a normal surface singularity $(X, 0)$ admits a natural family of open book decompositions, the so-called *Milnor open books*. They are cut out by analytic germs $f : (X, 0) \to (\mathbb{C}, 0)$ which define isolated singularities. If $L_f \subset M$ denotes the (transversal) intersection of $f^{-1}(0)$ with M, then the Milnor fibration of f defines an open book decomposition of M with binding L_f. By [17], all the Milnor open book decompositions support the same contact structure on M, namely the canonical contact structure (M, ξ_{can}). Section 12 contains a more detailed discussion about Milnor open books and their invariants.

10. FILLINGS. THE CASE OF CYCLIC QUOTIENT SINGULARITIES

10.1.

In this section we assume that $(X, 0) = (X_{p,q}, 0)$ is a *cyclic quotient* (or *Hirzebruch–Jung*) singularity. Its oriented link is the oriented lens space $L(p, q)$. Recall that $(X_{p,q}, 0)$ is taut and rational.

For cyclic quotient singularities the classification of the Stein fillings of (M, ξ_{can}) and their connection with Milnor fibers and Milnor open books is completely understood. In this section we review this problem; the presentation follows [72].

10.2. Notations

If $\underline{x} = (x_1, \ldots, x_n)$ are variables, the *Hirzebruch–Jung continued fraction* $[x_1, \ldots, x_n]$ can be defined by induction on n through the formulae: $[x_1] = x_1$ and $[x_1, \ldots, x_n] = x_1 - 1/[x_2, \ldots, x_n]$ for $n \geq 2$. One shows that:

$$[x_1, \ldots, x_n] = \frac{Z_n(x_1, \ldots, x_n)}{Z_{n-1}(x_2, \ldots, x_n)},$$

where the polynomials $Z_n \in \mathbb{Z}[x_1, \ldots, x_n]$ satisfy the inductive formulae:

(10.2.1)
$$Z_n(x_1, \ldots, x_n) = x_1 \cdot Z_{n-1}(x_2, \ldots, x_n) - Z_{n-2}(x_3, \ldots, x_n) \text{ for all } n \geq 1,$$

with $Z_{-1} \equiv 0$, $Z_0 \equiv 1$ and $Z_1(x) = x$. In fact, $Z_n(\underline{x})$ equals the determinant of the $n \times n$-matrix $M(\underline{x})$, whose entries are $M_{i,i} = x_i$, $M_{i,j} = -1$ if $|i - j| = 1$, $M_{ij} = 0$ otherwise. Hence, besides (10.2.1), they satisfy many 'determinantal relations' too; e.g. $Z_n(x_1, \ldots, x_n) = Z_n(x_n, \ldots, x_1)$. Following [84], $\underline{x} \in \mathbb{N}^n$ is *admissible* if the matrix $M(\underline{x})$ is positive semi-definite of rank $\geq n - 1$. Denote by $\mathrm{adm}\,(\mathbb{N}^n)$ the set of admissible n-tuples.

If $\underline{x}$ is admissible and $n > 1$, then each $x_i > 0$. Moreover, if $[x_1, \ldots, x_n]$ is admissible then $[x_n, \ldots, x_1]$ is admissible too. For any $r \geq 1$, denote:

$$K_r := \left\{ \underline{k} = (k_1, \ldots, k_r) \in \mathrm{adm}\,(\mathbb{N}^r) \mid [k_1, \ldots, k_r] = 0 \right\}.$$

For $\underline{k} = (k_1, \ldots, k_r) \in K_r$ set $\underline{k}' := (k_r, \ldots, k_1) \in K_r$.

For p, q as above and HJ-expansion $\frac{p}{p-q} = [a_1, \ldots, a_r]$, set:

$$K_r\left(\frac{p}{p-q}\right) = K_r(\underline{a}) := \{\underline{k} \in K_r \mid \underline{k} \leq \underline{a}\} \subset K_r.$$

Here, $\underline{k} \leq \underline{a}$ means that $k_i \leq a_i$ for all i.

10.3. Lisca's Conjecture

We recall briefly the classification of fillings of lens spaces (endowed with their canonical contact structure) established by Lisca. He provides by surgery diagrams a list of compact oriented 4-manifolds $W_{p,q}(\underline{k})$ with boundary $L(p,q)$. They are parametrized by the set $K_r\left(\frac{p}{p-q}\right)$ of sequences of integers $\underline{k} \in \mathbb{N}^r$ (cf. (10.2)). He showed that each manifold $W_{p,q}(\underline{k})$ admits a structure of Stein surface, filling $\left(L(p,q), \xi_{\text{can}}\right)$, and that any symplectic filling of $\left(L(p,q), \xi_{\text{can}}\right)$ is orientation-preserving diffeomorphic to a manifold obtained from one of the $W_{p,q}(\underline{k})$ by a composition of blow-ups. In general, the oriented diffeomorphism type of the boundary and the parameter $\underline{k}$ does not determine uniquely the (orientation-preserving) diffeomorphism type of the fillings: for some pairs the corresponding types might coincide (see subsection (10.8) for this 'ambiguity').

Lisca also noted that, following the works of Christophersen [18] and Stevens [101], $K_r\left(\frac{p}{p-q}\right)$ parametrizes also the irreducible components of the reduced miniversal base space of deformations of the cyclic quotient singularity $X_{p,q}$. Since each component of the miniversal space is in this case a smoothing component, Lisca conjectured in [53, page 768] that *the Milnor fiber of the irreducible component of the reduced miniversal base space of the cyclic quotient singularity $X_{p,q}$, parametrized in [101] by $\underline{k} \in K_r\left(\frac{p}{p-q}\right)$ is diffeomorphic to $W_{p,q}(\underline{k})$.*

On the other hand, in [39], de Jong and van Straten studied by an approach completely different from Christophersen and Stevens the deformation theory of cyclic quotient singularities (as a particular case of sandwiched singularities). They also parametrized the Milnor fibers of $X_{p,q}$ using the elements of the set $K_r\left(\frac{p}{p-q}\right)$. Hence, one can formulate the previous conjecture for their parametrization as well.

10.4. The answers to the (improved) conjecture

[72] answers positively these questions. Its main results are the following:

1. *One defines an additional structure associated with any (non-necessarily oriented) lens space: the 'order'. Its meaning is the following: geometrically it is a (total) order of the two solid tori separated by the (unique) splitting torus of the lens space; in plumbing language, it is an order of the two ends of the plumbing graph (provided that*

this graph has at least two vertices). Then one shows that the oriented diffeomorphism type and the order of the boundary, together with the parameter $\underline{k}$ determines uniquely the filling.

2. *One endows in a natural way all the boundaries of the spaces involved (Lisca's fillings $W_{p,q}(\underline{k})$, Christophersen–Stevens' Milnor fibers $F_{p,q}(\underline{k})$, and de Jong–van Straten's Milnor fibers $F'_{p,q}(\underline{k})$) with orders (this extra structure is denoted by $*$). Then one proves that all these spaces are connected by orientation-preserving diffeomorphisms which preserve the order of their boundaries: $W_{p,q}(\underline{k})^* \simeq F_{p,q}(\underline{k})^* \simeq F'_{p,q}(\underline{k})^*$. This is an even stronger statement than the result expected by Lisca's conjecture since it eliminates the ambiguities present in Lisca's classification.*

3. *In fact, [72] even provides a fourth description of the Milnor fibers constructed by a minimal sequence of blow ups of the projective plane which eliminates the indeterminacies of a rational function which depends on $\underline{k}$. This is in the spirit of Balke's work [6].*

4. *As a byproduct it follows that both Christophersen–Stevens and de Jong–van Straten parametrized the components of the miniversal base space in the same way.*

5. *Moreover, one obtains that the Milnor fibers corresponding to the various irreducible components of the miniversal space of deformations of $X_{p,q}$ are pairwise non-diffeomorphic by orientation-preserving diffeomorphisms whose restrictions to the boundaries preserve the order.*

In the sequel we will not say more about Lisca's construction, instead, we will describe briefly the Milnor fibers associated with the different smoothing components, with a special emphasis on the construction (10.4)(3). This will be compatible with the description of Christophersen and Stevens on the structure of the reduced miniversal base space of cyclic quotients [9, 18, 103] (cf. also with [2]). The de Jong–van Straten construction will be reviewed in the next section.

10.5.

First we concentrate on $X_{p,q}$. It can be embedded into $\mathbb{C}^{r+2}$ by some regular functions $z_0, \ldots, z_{r+1}$. *Some* of the equations of the embedding are

$$(10.5.1) \qquad z_{i-1}z_{i+1} - z_i^{a_i} = 0 \quad \text{for all } i \in \{1, \ldots, r\}.$$

Using equations (10.5.1) and induction, one shows that the restriction of each z_i to $X_{p,q}$ is a rational function in (z_0, z_1) of the form

$$z_i = z_1^{Z_{i-1}(a_1, \ldots, a_{i-1})} \cdot z_0^{-Z_{i-2}(a_2, \ldots, a_{i-1})} \quad \text{for } i \in \{1, \ldots, r+1\}.$$

In particular, the restriction pr_{01} of the projection $(z_0, \ldots, z_{r+1}) \mapsto (z_0, z_1)$ to $X_{p,q}$ is birational, i.e. it is a 'sandwiched representation' of $X_{p,q}$ (cf. next section for the terminology). (Here a comment is in order. Recall that $X_{p,q}$ is minimal, hence, by a general construction, it admits a canonical sandwiched representation $pr : X_{p,q} \to \mathbb{C}^2$, see e.g. (11.1). We wish to emphasize that the two birational maps pr_{01} and pr are different capturing two different geometrical aspects about $X_{p,q}$.)

The equations of $X_{p,q}$ are weighted homogeneous, however the weights $w_i := w(z_i)$ are not unique. With the choice $w_0 = w_1 = 1$ one has $w_i = Z_{i-1}(a_1, \ldots, a_{i-1}) - Z_{i-2}(a_2, \ldots, a_{i-1})$ for all $i \geq 1$, and $1 = w_0 = w_1 \leq w_2 \leq \cdots \leq w_{r+1} = q$.

10.6. The deformations

Next, we fix $\underline{k} \in K_r(\underline{a})$, and we denote by $S_{\underline{k}}^{CS}$ the corresponding deformation component (as it is described by Christophersen and Stevens). Then, we consider a special 1-parameter deformation with equations $\mathcal{E}_{\underline{k}}^t$ of $X_{p,q}$. This deformation is determined by the deformed equations of (10.5.1) (cf. [2], [101, (2.2)]). These are:

$$(10.6.1) \qquad z_{i-1}z_{i+1} = z_i^{a_i} + t \cdot z_i^{k_i} \quad \text{for all } i \in \{1, \ldots, r\},$$

where $t \in \mathbb{C}$. Let $\mathcal{X}_{\underline{k}}^t$ be the affine space determined by the equations $\mathcal{E}_{\underline{k}}^t$ in $\mathbb{C}^{r+2}$. One proves that *the deformation $t \mapsto \mathcal{X}_{\underline{k}}^t$ has negative weight and is a smoothing belonging to the component $S_{\underline{k}}^{CS}$*. Hence, $\mathcal{X}_{\underline{k}}^t$ is a smooth affine space for $t \neq 0$ (and by [115, (2.2)]) it is *diffeomorphic to the Milnor fiber of $S_{\underline{k}}^{CS}$*.

10.7. $\mathcal{X}_{\underline{k}}^t$ as a rational surface

Similarly as for $X_{p,q}$, using (10.6.1), on $\mathcal{X}_{\underline{k}}^t$ all the coordinates z_i can be expressed as rational functions in (z_0, z_1). Indeed, for each $i \in \{1, \ldots, r+1\}$, on $\mathcal{X}_{\underline{k}}^t$ one has:

$$z_i = z_0^{-Z_{i-2}(a_2,\ldots,a_{i-1})} P_i$$

for some $P_i \in \mathbb{Z}[t, z_0, z_1]$. The polynomials P_i satisfy the inductive relations:

$$(10.7.1) \qquad P_{i-1} \cdot P_{i+1} = P_i^{a_i} + t P_i^{k_i} \cdot z_0^{(a_i - k_i) \cdot Z_{i-2}(a_2,\ldots,a_{i-1})}$$

with $P_1 = z_1$ and with the convention $P_0 = 1$.

Consider the application $\pi : \mathbb{C}^2 \setminus \{z_0 = 0\} \longrightarrow \mathcal{X}_{\underline{k}}^t$ given by

$$(10.7.2)$$
$$(z_0, z_1) \mapsto (z_0, z_1, P_2, \ldots, z_0^{-Z_{i-2}(a_2,\ldots,a_{i-1})} P_i, \ldots, z_0^{-(p-q)} P_{r+1}) \in \mathbb{C}^{r+2}.$$

We are interested in the birational map $\mathbb{C}^2 \dashrightarrow \mathcal{X}_{\underline{k}}^t$, still denoted by π, and its extension $\overline{\pi} : \mathbb{P}^2 \dashrightarrow \overline{\mathcal{X}_{\underline{k}}^t}$, where $\overline{\mathcal{X}_{\underline{k}}^t}$ is the closure of $\mathcal{X}_{\underline{k}}^t$ in $\mathbb{P}^{r+2}$.

Let $\rho'_{\underline{k}} : B'\mathbb{P}^2 \to \mathbb{P}^2$ be the minimal sequence of blow ups such that $\overline{\pi} \circ \rho'_{\underline{k}}$ extends to a regular map $B'\mathbb{P}^2 \to \overline{\mathcal{X}_{\underline{k}}^t}$. Let $L_\infty \subset \mathbb{P}^2$ be the line at infinity and by L_0 the closure in $\mathbb{P}^2$ of $\{z_0 = 0\}$. We use the same notations for their strict transforms via blow ups of $\mathbb{P}^2$. Since the projection $pr : \overline{\mathcal{X}_{\underline{k}}^t} \to \mathbb{C}^2$ is regular and $pr \circ \pi$ is the identity, one gets that $\overline{\pi} \circ \rho'_{\underline{k}}$ sends L_0 and the total transform of L_∞ in the curve at infinity $\overline{\mathcal{X}_{\underline{k}}^t} \setminus \mathcal{X}_{\underline{k}}^t$.

Therefore, let us modify $\rho'_{\underline{k}}$ into $\rho_{\underline{k}} : B\mathbb{P}^2 \to \mathbb{P}^2$, the *minimal* sequence of blow ups which resolve the indeterminacies of $\overline{\pi}$ *sitting in* $\mathbb{C}^2$ (hence $\rho'_{\underline{k}}$ and $\rho_{\underline{k}}$ over $\mathbb{C}^2$ coincide). Denote by E_π its exceptional curve and by C_π the union of those irreducible components of E_π which are sent to $C_{\underline{k}}^\infty$. Set $B\mathbb{C}^2 := B\mathbb{P}^2 \setminus L_\infty$. Summing up all the above discussions, one obtains:

Theorem 10.7.1. *The restriction of $\overline{\pi} \circ \rho_{\underline{k}}$ induces an isomorphism $B\mathbb{C}^2 \setminus (L_0 \cup C_\pi) \to \mathcal{X}_{\underline{k}}^t$. In particular, the Milnor fiber can be realized as the complement of the projective curve $L_\infty \cup L_0 \cup C_\pi$ in $B\mathbb{P}^2$.*

The point is that the indeterminacies of $\overline{\pi}$ above $\mathbb{C}^2$, hence the modification $\rho_{\underline{k}}$ too, can be described precisely. This leads to the following description of the Milnor fiber.

Corollary 10.7.2. *Consider the lines L_∞ and L_0 on $\mathbb{P}^2$ as above. Blow up $r - 1 + \sum_{i=1}^{r}(a_i - k_i)$ infinitely closed points of L_0 in order to get the dual graph in Figure 1 of the configuration of the total transform of $L_\infty \cup L_0$ (this procedure topologically is unique, and its existence is guaranteed by the fact that $\underline{k} \in K_r(\underline{a})$). Denote the space obtained by this modification by $B\mathbb{P}^2$. Then the Milnor fiber $\mathcal{X}_{\underline{k}}^t$ of $S_{\underline{k}}^{CS}$ is diffeomorphic to $B\mathbb{P}^2 \setminus (\cup_{j=0}^{r} V_j)$.*

Moreover, let T be a small open tubular neighbourhood of $\cup_{j=0}^{r} V_j$, and set $F_{p,q}(\underline{k}) = B\mathbb{P}^2 \setminus T$. Then $F_{p,q}(\underline{k})$ is a representative of the Milnor fiber of $S_{\underline{k}}^{CS}$ as a manifold with boundary whose boundary is $L(p,q)$.

Furthermore, the marking $\{V_i\}_i$ as in the Figure 1, defines on the boundary of $F_{p,q}(\underline{k})$ an order; denote this supplemented space by $F_{p,q}(\underline{k})^*$. Then its ordered boundary is $L(p,q)^*$ endowed with the preferred order.

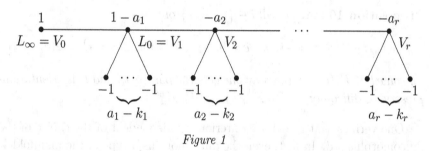

Figure 1

Remark 10.7.3. In fact, $\rho_{\underline{k}}$ serves also as the minimal modification which eliminates the indeterminacy of the last component of π from (10.7.2), namely of the rational function $z_{r+1} = P_{r+1}/z_0^{p-q}$. In particular, we find the following alternative description of the Milnor fiber $F_{p,q}(\underline{k})$:

For each $\underline{k} \in K_r(\underline{a})$, define the polynomial P_{r+1} via the inductive system (10.7.1). Let $\rho_{\underline{k}} : B\mathbb{P}^2 \to \mathbb{P}^2$ be the minimal modification of $\mathbb{P}^2$ which eliminates the indeterminacy points of P_{r+1}/z_0^{p-q} sitting in $\mathbb{C}^2$. Then the dual graph of the total transform of $L_\infty \cup L_0$ has the form indicated in Figure 1, and $F_{p,q}(\underline{k})$ is orientation-preserving diffeomorphic to $B\mathbb{P}^2 \setminus (\cup_{j=0}^{r} V_j)$.

10.8. An identification criterion of the Milnor fibers

The next criterion generalizes Lisca's criterion [53, §7] to recognize the fillings of $L(p,q)$, it is valid for the spaces with *ordered* boundaries. It also implies immediately part (2) of (10.3), namely that $W_{p,q}(\underline{k})^* \simeq F_{p,q}(\underline{k})^*$.

Associate to the sequence $\underline{a}$ the string $G(\underline{a})$ decorated with the entries $a_1, \ldots, a_r$ (this might also serve as the minimal resolution graph of $X_{p,p-q}$). Regarded $G(\underline{a})$ as a plumbed graph, it determines the plumbed 4-manifold $\Pi(\underline{a})$ (which is diffeomorphic to the minimal resolution space of $X_{p,p-q}$, whose oriented boundary is $-L(p,q)$ ($L(p,q)$ with opposite orientation).

Let F be a Stein filling of $(L(p,q), \xi_{\mathrm{can}})$, e.g. one of the Milnor fibers considered above. Set V for the closed 4-manifold obtained by gluing F and $\Pi(\underline{a})$ via an orientation preserving diffeomorphism $\phi : \partial F \to \partial(-\Pi(\underline{a}))$ of their boundaries. Denote by $\{s_i\}_{1 \leq i \leq r}$ the classes of 2-spheres $\{S_i\}_{1 \leq i \leq r}$ in $H_2(\Pi(\underline{a}))$ (listed in the same order as $\{a_i\}_{1 \leq i \leq r}$), and also their images via the monomorphism $H_2(\Pi(\underline{a})) \to H_2(V)$ induced by the inclusion.

Lisca's criterion (implied also by the results of (10.7)) reads as follows:

Proposition 10.8.4. *For all $i \in \{1, \ldots, r\}$ one has*

$$\#\{e \in H_2(V) \mid e^2 = -1,\ s_i \cdot e \neq 0,\ s_j \cdot e = 0 \text{ for all } j \neq i\} = 2(a_i - k_i)$$

for some $\underline{k} \in K_r(\underline{a})$. In this way one gets the pair $(\underline{a}, \underline{k})$ and F is orientation-preserving diffeomorphic to $F_{p,q}(\underline{k})$ $\left(\simeq F_{p,q'}(\underline{k}') \right)$.

One verifies that the above criterion is independent of the choice of the diffeomorphism ϕ. In fact, even the diffeomorphism type of the manifold V is independent of the choice of ϕ.

Notice that $\{S_i\}_{1 \leq i \leq r}$ and $\{S_{r-i}\}_{1 \leq i \leq r}$ cannot be distinguished, hence the above algorithm does not differentiate $(\underline{a}, \underline{k})$ from $(\underline{a}', \underline{k}')$, or $F_{p,q}(\underline{k})$ from $F_{p,q'}(\underline{k}')$. On the other hand, these are the only ambiguities. (In fact, if $r = 1$, or even of $r > 1$ but $\underline{a}$ and $\underline{k}$ are symmetric, then there is no ambiguity, since $(p, q, \underline{k}) = (p, q', \underline{k}')$.)

Using the notion of order of the boundaries, one can eliminate the above ambiguity. Notice that any diffeomorphism $F_{p,q}(\underline{k}) \to F_{p,q'}(\underline{k}')$ (whenever $(p, q, \underline{k}) \neq (p, q', \underline{k}')$) does not preserve any fixed order of the boundary. One has:

Theorem 10.8.5. *All the spaces $F_{p,q}(\underline{k})^*$ are different, hence their boundaries $L(p,q)^*$ and $\underline{k} \in K_r(\underline{a})$ determine uniquely all the Milnor fibers up to orientation-preserving diffeomorphisms which preserve the order of the boundary.*

In order to prove this, the criterion (10.8.4) is modified as follows. Let F^* be a Stein filling of $\left(L(p,q), \xi_{\mathrm{can}}\right)$ with an order on its boundary. Consider

$\Pi(\underline{a})^*$ with its preferred order (provided a well-determined order of the s_i's, cf. [72]). Construct V as in (10.8.4), and consider the two pairs $(q,\underline{k})$ and $(q',\underline{k}')$ provided (but undecided) by (10.8.4).

Proposition 10.8.6. *If ϕ preserves (resp. reverses) the orders of the boundary then F^* is orientation and order preserving diffeomorphic to $F_{p,q}(\underline{k})^*$ (resp. to $F_{p,q'}(\underline{k}')^*$).*

11. FILLINGS. THE CASE OF SANDWICHED SINGULARITIES

11.1.

For the definition of minimal rational and sandwiched singularities see 7.3.

In [39] de Jong and van Straten related the theory of sandwiched surface singularities with *decorated plane curve singularities*.

Consider a reduced germ of plane curve $(C,0) \subset (\mathbb{C}^2,0)$ with *numbered* branches (irreducible components) $\{C_i\}_{1 \leq i \leq r}$. Fix the minimal (not necessarily good) embedded resolution of $(C,0)$ obtained by a sequence of blow ups. The *multiplicity sequence* associated with C_i is the sequence of multiplicities on the successive strict transforms of C_i, starting from C_i itself and not counting the last strict transform. The *total multiplicity* $m(i)$ of C_i with respect to C is the sum of the sequence of multiplicities of C_i.

A *decorated germ* of plane curve is a weighted germ (C,l) such that $l = (l_i)_{1 \leq i \leq r} \in (\mathbb{N}^*)^r$ and $l_i \geq m(i)$ for all $i \in \{1, \ldots, r\}$.

Starting from a decorated germ, one can blow up iteratively points infinitely near 0 on the strict transforms of C, such that the sum of multiplicities of the strict transform of C_i at such points is exactly l_i. Such a composition of blow-ups is determined canonically by (C,l). If l_i is sufficiently large (in general, larger than $m(i)$) then the union of the exceptional components which do not meet the strict transform of C form a *connected* configuration of (compact) curves $E(C,l)$. After the contraction of $E(C,l)$, one gets a sandwiched singularity $X(C,l)$, determined uniquely by (C,l) (for details see [39]). This follows from the fact that the collection of the irreducible components of the exceptional curve E not contained in $E(C,l)$ are exactly the (-1) curves involved in the definition of sandwiched singularities. Conversely, for any sandwiched singularity X one can find (C,l) such that X can be represented as $X(C,l)$.

11.2. Markings of the links [73]

By a result of Neumann [75], the information codified in the link of a surface singularity up to an orientation-preserving homeomorphism and in the
weighted dual graph of the minimal good resolution are equivalent. In [89,
Theorems 9.1 and 9.7] this is generalized as follows. From the resolution,
the abstract link inherits a *plumbing structure*, that is, a family of pairwise disjoint embedded tori whose complement is fibered by circles, and
such that on each torus the intersection number of the fibers from each side
is ±1. (The tori correspond to the edges of the dual graph and the connected components of their complement – the "pieces of M" – correspond
to the 'un-numbered vertices'.) Then, by [89, Theorem 9.7], the plumbing structure corresponding to the minimal normal crossings resolution is
determined up to an isotopy by the oriented link.

Consider again $X(C, l)$ associated with some (C, l), and let M be its link.
Using the notations of the previous subsection, write $E = E(C, l) + \sum_{i=1}^{r} E_i$,
where $(E_i)_{1 \leq i \leq r}$ is numbered such that E_i is the unique irreducible component which intersects the strict transform of C_i. Denote by F_i the unique
irreducible component of $E(C, l)$ which intersects E_i. To F_i corresponds
a well-defined "piece" of M. In this way, one gets a map from the set
$\{1, \ldots, r\}$ to the set of pieces of M.

Definition 11.2.1. A map from $\{1, \ldots, r\}$ (the index set of the numbered
branches of C) to the set of "pieces" of M obtained as above is called *a
marking* of M.

Hence, each realization of M as the link of some $X(C, l)$, where (C, l) is
a decorated curve with numbered branches, provides a well-defined marking
of M.

11.3. Deformation of sandwiched singularities after de Jong and van Straten

The point is that the above correspondence between sandwiched singularities and decorated plane curves extends to their deformation theory as well.

The total multiplicity of C_i with respect to C may be encoded also as
the unique subscheme of length $m(i)$ supported on the preimage of 0 on the
normalization of C_i. This allows to define the *total multiplicity scheme* $m(C)$
of any reduced curve contained in a smooth complex surface, as the union

of the total multiplicity schemes of all its germs. Given a smooth complex analytic surface Σ, a pair (C, l) consisting of a reduced curve $C \hookrightarrow \Sigma$ and a subscheme l of the normalization $\tilde{C}$ of C is called *a decorated curve* if $m(C)$ is a subscheme of l ([39, (4.1)]). The deformations of (C, l) considered by de Jong and van Straten are:

Definition 11.3.1 ([39, p. 476]). (i) A 1-*parameter deformation* of a decorated curve (C, l) over a germ of smooth curve $(S, 0)$ consists of:

(1) a δ-constant deformation $C_S \to S$ of C;

(2) a flat deformation $l_S \subset \tilde{C}_S = \tilde{C} \times S$ of the scheme l, such that:

(3) $m_S \subset l_S$, where the *relative total multiplicity scheme* m_S of $\tilde{C}_S \to C_S$ is defined as the closure $\overline{\bigcup_{s \in S \setminus 0} m(C_s)}$.

(ii) A 1-parameter deformation (C_S, l_S) is called a *picture deformation* if for generic $s \neq 0$ the divisor l_s is reduced.

The singularities of $C_{s \neq 0}$ are only ordinary multiple points. Hence, it is easy to draw a real picture of a deformed curve, which motivates the terminology.

Theorem 11.3.2 ([39, (4.4)]). *All the 1-parameter deformations of $X(C, l)$ are obtained by 1-parameter deformations of the decorated germ (C, l). Moreover, picture deformations provide all the smoothings of $X(C, l)$.*

Next, let us fix a decorated germ (C, l) and one of its picture deformations (C_S, l_S). Fix a closed Milnor ball B for the germ $(C, 0)$. For $s \neq 0$ sufficiently small, C_s will have a representative in B, denoted by D, which meets ∂B transversally. It is a union of immersed discs $\{D_i\}_{1 \leq i \leq r}$ canonically oriented by their complex structures (and whose set of indices correspond canonically to those of $\{C_i\}_{1 \leq i \leq r}$). The singularities of D consist of ordinary multiple points.

Denote by $\{P_j\}_{1 \leq j \leq n}$ the set of images in B of the points in the support of l_s. It is a finite set of points which contains the singular set of D (because $m_s \subset l_s$ for $s \neq 0$), but it might contain some other 'free' points as well. There is a priori no preferred choice of their ordering. (Hence, the matrix introduced next is well-defined only up to permutations of columns.)

The Milnor fiber of the smoothing associated with the fixed picture deformation has the following description. Let $\beta : (\tilde{B}, \tilde{D}) \to (B, D)$ be the simultaneous blow-up of the points P_j of D. Here $\tilde{D} := \cup_{1 \leq i \leq r} \tilde{D}_i$, where $\tilde{D}_i$ is the strict transform of the disc D_i by the modification β. Let T_i be a sufficiently small open tubular neighbourhood of $\tilde{D}_i$ in $\tilde{B}$ (with pairwise disjoint closures).

Proposition 11.3.3 ([39, (5.1)]). *Let (C, l) be a standard decorated germ. Then the Milnor fiber of the smoothing of $X(C, l)$ corresponding to the picture deformation (C_S, l_S) is orientation-preserving diffeomorphic to the compact oriented manifold with boundary $F' := \tilde{B} \setminus \left(\bigcup_{1 \leq i \leq r} T_i \right)$ (whose corners are smoothed).*

Moreover, by this presentation (for a fixed (C, l)) one can also canonically identify the boundaries of all the Milnor fibers with the link.

Finally, one reads from the above deformation a combinatorial object too, which will be crucial in the sequel:

Definition 11.3.4 ([39, page 483]). The *incidence matrix* of a picture deformation (C_S, l_S) is the matrix $\mathcal{I}(C_S, l_S)$ with r rows and n columns whose entry at the i-th row and the j-th column is the multiplicity of P_j as a point of D_i.

11.4. The characterization of the Milnor fibers by the incidence matrix

First, let us start with some general remarks. Obviously, the main goal would be to extend for sandwiched singularities the statements of (10.4), valid for cyclic quotients. This is obstructed seriously in both sides of the correspondence. First, at this moment there is no classification of the Stein fillings of the canonical contact structures of sandwiched singularity links. Second, there is no classification of the smoothing components either, or of the possible Milnor fibers. (Recall that in principle it might happen that several different smoothing components have the same Milnor fiber. Also, it is also still open the characterization of those 'combinatorial' picture deformations which can be analytically realized.)

Nevertheless, we have the following characterization result, which provides a homological/combinatorial method to 'separate' some Milnor fibers associated with different smoothing components. In the next paragraphs we follow [73].

Theorem 11.4.1. (a) *The incidence matrix $\mathcal{I}(C_S, l_S)$ associated to a picture deformation of a decorated germ (C, l) is determined (up to a permu-*

tation of its columns) by the associated Milnor fiber and the marking of the link.

(b) Consider two topologically equivalent decorated germs of plane curves, and for each one of them a picture deformation. If their incidence matrices are different up to permutation of columns, then their associated Milnor fibers are not diffeomorphic by an orientation preserving diffeomorphism which preserves the markings of the boundaries.

The proof is based on a construction which glues a special universal "cap" to each Milnor fiber. This is explained in the next subsection.

11.5. Closing the boundary of the Milnor fiber

Let (C_S, l_S) be a picture deformation of the decorated germ (C, l).

As the disc-configuration D is obtained by deforming C, its boundary $\partial D := \cup_{1 \leq i \leq r} D_i \hookrightarrow \partial B$ is isotopic as an oriented link to $\partial C \hookrightarrow \partial B$. Therefore, we can isotope D outside a compact ball containing all the points P_j till its boundary coincides with the boundary of C. From now on, D will denote the result of this isotopy. Let (B', C') be a second copy of (B, C), and define:

$$(V, \Sigma) := (B, D) \cup_{id} (-B', -C').$$

Here V is the oriented 4-sphere obtained by gluing the boundaries of B and $-B'$ by the tautological identification, and $\Sigma := \cup_{i=1}^{r} \Sigma_i$, where Σ_i is obtained by gluing D_i (perturbed by the above isotopy) and $-C_i'$ along their common boundaries. Moreover, one also glues $(-B', -C')$ with $(\tilde{B}, \tilde{D})$ in such a way that the blow-up morphism β may be extended by the identity on $-B'$, yielding $\beta : (\tilde{V}, \tilde{\Sigma}) \longrightarrow (V, \Sigma)$. Here $\tilde{\Sigma} := \cup_{i=1}^{r} \tilde{\Sigma}_i$, where $\tilde{\Sigma}_i$ denotes the strict transform of Σ_i, i.e. $\tilde{\Sigma}_i = \tilde{D}_i \cup -C_i'$. Since C_i' is a topological disc, and D_i is an immersed disc, $\tilde{\Sigma}_i$ is a topologically *embedded* 2-sphere in $\tilde{V}$. Write $T := \cup_{1 \leq i \leq r} T_i$ and set

$$U := -B' \cup T.$$

Since $F' = \tilde{B} \backslash T$ (cf. 11.3.3), the closed oriented 4-manifold $\tilde{V}$ is obtained by closing the boundary of F' by the cap U. The point is that U *is independent of the chosen picture deformation and it is always glued in the same way (up to an isotopy) to the boundaries of all the involved Milnor fibers.*

In particular, the homology of $\tilde{V}$ serves as an invariant of the Milnor fiber F'. More precisely one has the next statement (which generalizes the 'identification criterion' (10.8.4), and implies (11.4.1) too):

Proposition 11.5.1. (a) *Up to permutations of columns, there exists a unique basis $(e_j)_{1 \leq j \leq n}$ of $H_2(\tilde{V})$ such that $e_j^2 = -1$ for all j, and the matrix*

$$N(C_S, l_S) := \left([\tilde{\Sigma}_i] \cdot e_j \right)_{1 \leq i \leq r, 1 \leq j \leq n}$$

has only non-negative entries.

(b) *For any picture deformation (C_S, l_S), the incidence matrix $\mathcal{I}(C_S, l_S)$ is equal to $N(C_S, l_S)$, up to a permutations of the columns.*

Corollary 11.5.2. *Let (M, ξ_{can}) be a link of a sandwiched singularity endowed with its canonical contact structure. Fix the topological type of a defining decorated germ. Then there are at least as many Stein fillings (up to diffeomorphisms fixed on the boundary) of (M, ξ_{can}) as there are incidence matrices (up to permutation of columns) realised by the picture deformations of all the decorated germs with the given topology.*

The above corollary captures all the Milnor fibers associated with all the analytic structures supported by (the cone over) M. Notice that it might happen that some Milnor fibers of one of the analytic structure cannot be realized by another analytic structure.

12. Milnor Open Books

12.1. Notations

As we already mentioned in (9.3), (M, ξ_{can}) can be studied via Milnor open books. On the other hand, Milnor open books can also be studied via the combinatorics of the resolutions. In this section we assume that M is a RHS. For any fixed resolution (graph) we consider the lattice $L := \left(\mathbb{Z}^{|\mathcal{V}|}, (\cdot, \cdot) \right)$, and in order to emphasize the graph Γ, we write $\mathcal{Z}_{\mathrm{an}}^{\Gamma}$ and $\mathcal{Z}_{\mathrm{top}}^{\Gamma}$ instead of $\mathcal{Z}_{\mathrm{an}}$ and $\mathcal{Z}_{\mathrm{top}}$.

Recall that for rational singularities $\mathcal{Z}_{\mathrm{an}}^{\Gamma} = \mathcal{Z}_{\mathrm{top}}^{\Gamma}$, hence $Z_{\max} = Z_{\min}$ too. But, in general, these equalities do not hold: the structure of $\mathcal{Z}_{\mathrm{an}}^{\Gamma}$ (hence the identification of $Z_{\max}$ too) can be very difficult, it depends essentially on the analytic structure of $(X, 0)$. (Maybe the most general result in this direction is the combinatorial description of $\mathcal{Z}_{an}^\Gamma$ for any splice-quotient singularity, a family which includes all the weighted homogeneous singularities as well, cf. [66].)

12.2. (Milnor) open books

One has the following facts for the Milnor open book decompositions of M, cf. 9.3.

1. For any such $f \in \mathcal{O}_{X,0}$, consider an embedded good resolution π of the pair $(X, f^{-1}(0))$. Then the strict transform $S(f) := \operatorname{div}(\pi^* f) - (f)_E$ intersects E transversally, and the number of intersection points $(S(f), E_v)$ is exactly $-((f)_E, E_v)$. Since the intersection form is negative definite, the collection of binding components $\{(S(f), E_v)\}_v$ and $(f)_E \in \mathcal{Z}_{an}^\Gamma$ determine each other.

   Moreover, by classical results of Stallings and Waldhausen, the (topological type of the) binding $L_f \subset M$ determines completely the open book up to an isotopy, provided that M is a rational homology sphere. (For counterexamples for these statement in the general case, see e.g. [64].)

   In particular, all the Milnor open book decompositions associated with a fixed *analytic* type $(X, 0)$ are completely described by the semi-groups $\{\mathcal{Z}_{an}^\Gamma\}_\Gamma$ (associated with all the possible resolutions and natural identifications of elements of them). Hence, the classification of all the Milnor open books associated with a fixed analytic type $(X, 0)$ is, in fact, as difficult as the determination of the semi-groups $\mathcal{Z}_{an}^\Gamma$ of $\mathcal{Z}_{top}^\Gamma$.

2. Therefore, from topological points of view, it is more natural to consider the open books of all the analytic germs associated with *all the analytic structures* supported by the topological type of some $(X, 0)$.

   As we already mentioned, for a fixed topological type of $(X, 0)$, in any (negative definite) plumbing graph Γ of M one can also define the cone $\mathcal{Z}_{top}^\Gamma$. The point is that for any $D \in \mathcal{Z}_{top}^\Gamma$ there is a convenient analytic structure on $(X, 0)$ and an analytic germ f, such that the plumbing graph can be identified with a dual embedded resolution graph (for the pair $(X, f^{-1}(0))$), and D is the compact part $(f)_E$ [76]. Hence, modifying the analytic structure of $(X, 0)$, we fill by the collections $\mathcal{Z}_{\text{an}}^{\Gamma}$ all the semi-group $\mathcal{Z}_{\text{top}}^{\Gamma}$.

In particular, the collection of all the Milnor open book decompositions associated with a fixed topological type M is described by the collection of semi-groups $\{\mathcal{Z}_{\text{top}}^{\Gamma}\}_{\Gamma}$ (considered in all the possible resolution graphs with a natural identifications of elements of them).

3. For any fixed analytic type $(X, 0)$, the open book associated with $Z_{\max}$ is the Milnor fibration of the generic hyperplane section, in particular this open book is (resolution) graph-independent. Similarly, for a fixed topological type of $(X, 0)$, the open book associated with $Z_{\min}$ is also graph-independent, it depends only on the topology of the link.

12.3. Invariants of Milnor open books

From the above correspondence, any property defined for elements of $\mathcal{Z}_{\text{top}}^{\Gamma}$ (or $\mathcal{Z}_{\text{an}}^{\Gamma}$) can be translated in the language of open books. This is true also in the opposite direction, invariants of open books can be studied via the lattice L. The aim of this section is to emphasise exactly this second direction applied for some key invariants of open books.

Let us fix M, a plumbing (or, a dual resolution) graph Γ. Let us consider a Milnor open book associated with an element $Z \in \mathcal{Z}_{\text{top}}^{\Gamma}$, cf. 12(2). In the sequel we will consider the following numerical invariants of the open book and also their description in terms of L:

1. The number of binding components $\text{bn}(Z)$ is given by $-(Z, E)$.

2. Let F_f be the fiber of the open book. It is an oriented connected surface with $-(Z, E)$ boundary components. Let $g(Z)$ be its genus (the so-called page-genus of the open book) and $\mu(Z)$ be the first Betti-number of F_f (the so-called Milnor number). Clearly $\mu(Z) = 2g(Z) - 1 - (Z, E) \geq 2g(Z)$.

What is less obvious is the following identity, which connects open books with the Riemann–Roch formula: for any $Z \in \mathcal{Z}_{\text{top}}^{\Gamma}$ one has

$$g(Z) = 1 + (Z, E) + \chi(-Z), \quad \text{and}$$

$$\mu(Z) = 1 + (Z, E) + 2 \cdot \chi(-Z) = g(Z) + \chi(-Z).$$

12.4. 'Monotone increasing' invariants

This description allows to prove the next property of these invariants.

Definition 12.4.1. Assume that for any resolution π with graph Γ one has a map $I_\Gamma : \mathcal{Z}_{\mathrm{top}}^\Gamma \to \mathbb{Z}_{\geq 0}$. We say that $I = \{I_\Gamma\}_\Gamma$ is 'monotone increasing' if for any two cycles $Z_i \in \mathcal{Z}_{\mathrm{top}}^\Gamma$ $(i = 1, 2)$ with $Z_1 \leq Z_2$ one has $I_\Gamma(Z_1) \leq I_\Gamma(Z_2)$ for any Γ.

Remark 12.4.2. Assume that the collection of invariants $\{I_\Gamma\}_\Gamma$ can be transformed into (or comes from) an invariant I which associates with any Milnor open book $\mathfrak{m}$ of the link an integer. For any fixed analytic type, let $\mathfrak{m}_{\max}$ be the Milnor open book associated with $Z_{\max}$ (considered in any resolution). Similarly, for any topological type, let $\mathfrak{m}_{\min}$ be the Milnor open book associated with $Z_{\min}$ (in any resolution of an analytic structure conveniently chosen); cf. 12(3). Then, *whenever $\{I_\Gamma\}_\Gamma$ is monotone increasing*, one has automatically the next consequences:

1. Fix an analytic singularity $(X, 0)$ and consider all the Milnor open books associated with all isolated holomorphic germs $f \in \mathcal{O}_{X,0}$. Then the minimum of integers $I(\mathfrak{m})$ of all these Milnor open books $\mathfrak{m}$ is realized by the generic hyperplane section, i.e. by $I(\mathfrak{m}_{\max})$.

2. Fix a topological type of a normal surface singularity, and consider the open books associated with *all* the isolated holomorphic germs of *all* the possible analytic structures supported by the fixed topological type. Then the minimum of all integers $I(\mathfrak{m})$ of all these Milnor open books $\mathfrak{m}$ is realized by the open book associated with the Artin cycle, i.e. by $I(\mathfrak{m}_{\min})$.

The above definition is motivated by the following result:

Theorem 12.4.3 ([74]). *Both invariants $Z \mapsto g(Z)$ and $Z \mapsto \mu(Z)$ are monotone increasing.*

12.5. Invariants of the canonical contact structure

In [27] Etnyre and Ozbagci consider three invariants associated with fixed contact structure defined in terms of all open book decomposition supporting it:

- the *support genus* sg (ξ) is the minimal possible genus for a page of an open book that supports ξ;
- the *binding number* bn (ξ) is the minimal number of of binding components for an open book supporting ξ and that has pages of genus sg (ξ);
- • the *norm* $\mathfrak{n}(\xi)$ of ξ is the negative of the maximal (topological) Euler characteristic of a page of an open book that supports ξ.

Now, we consider the above invariants restricted on the realm of *Milnor* open books (i.e. for all those open books which might appear in the analytic context). In particular, ξ will be the *canonical contact structure* ξ_{can}. Let us denote the corresponding invariants by $sg_{an}(\xi_{can})$, $bn_{an}(\xi_{can})$ and $\mathfrak{n}_{an}(\xi_{can})$. Then Theorem 12.4.3 has the following consequence:

$$sg_{an}(\xi_{can}) = g(Z_{min});$$

$$bn_{an}(\xi_{can}) = bn(Z_{min});$$

$$\mathfrak{n}_{an}(\xi_{can}) = \mu(Z_{min}) - 1.$$

In particular,

$$\mathfrak{n}_{an}(\xi_{can}) - bn_{an}(\xi_{can}) = 2 \cdot sg_{an}(\xi_{can}) - 2.$$

These facts answer some of the questions of [27], section 8, at least in the realm of Milnor open books. Since $\chi(-Z) + \chi(Z) + Z^2 = 0$, one also has

$$g(Z_{min}) = 1 + (Z_{min}, E - Z_{min}) - \chi(Z_{min}).$$

Therefore, if $(X, 0)$ is rational (i.e. $\chi(Z_{min}) = 1$) then $g(Z_{min}) = (Z_{min}, E - Z_{min})$ (this can be strict positive, e.g. for the E_8-singularity it is 1); if $(X, 0)$ is elliptic (i.e. $\chi(Z_{min}) = 0$) then $g(Z_{min}) = 1 + (Z_{min}, E - Z_{min}) \geq 1$. In general,

$$g(Z_{min}) \geq 1 - \chi(Z_{min}),$$

hence it can be arbitrarily large.

REFERENCES

[1] Akhmedov, A., Etnyre, J. B., Mark, T. E. and Smith, I., A note on Stein fillings of contact manifolds, *Math. Res. Lett.*, **15**, no. 6 (2008), 1127–1132.

[2] Arndt, J., Verselle Deformationen zyklischer Quotientensingularitäten, Diss. Hamburg, 1988.

[3] Artin, M., Some numerical criteria for contractibility of curves on algebraic surfaces, *Amer. J. of Math.*, **84** (1962), 485–496.

[4] Artin, M., On isolated rational singularities of surfaces, *Amer. J. of Math.*, **88** (1966), 129–136.

[5] Artin, M., Algebraic construction of Brieskorn's resolutions, *J. of Algebra*, **29** (1974), 330–348.

[6] Balke, L., Smoothings of cyclic quotient singularities from a topological point of view, arXiv:math/9911070.

[7] Barth, W., Peters, C. and Van de Ven, A., Compact Complex Surfaces, Springer-Verlag, 1984.

[8] Behnke, K. and Knörrer, H., On infinitesimal deformations of rational surface singularities, *Comp. Math.*, **61** (1987), 103–127.

[9] Behnke, K. and Riemenschneider, O., Quotient surface singularities and their deformations, in: *Singularity theory*, D. T. Lê, K. Saito & B. Teissier eds., World Scientific, 1995, 1–54.

[10] Bhupal, M. and Ono, K., Symplectic fillings of links of quotient surface singularities, arXiv:0808.3794.

[11] Bogomolov, F. A. and de Oliveira, B., Stein Small Deformations of Strictly Pseudoconvex Surfaces, *Contemporary Mathematics*, **207** (1997), 25–41.

[12] Braun, G. and Némethi, A., Surgery formula for Seiberg–Witten invariants of negative definite plumbed 3-manifolds, *J. reine angew. Math.*, **638** (2010), 189–208.

[13] Briançon J. and Speder, J., Les conditions de Whitney impliquent "μ^* constant", *Ann. Inst. Fourier (Grenoble)*, **26** (1976), 153–164.

[14] Brieskorn, E., Die Auflösung der rationalen Singularitäten holomorpher Abbildungen, *Math. Ann.*, **178** (1968), 255–270.

[15] Brieskorn, E., Singular elements in semi-simple algebraic groups, *Proc. Int. Con. Math. Nice*, **2** (1971), 279–284.

[16] Caubel, C., and Popescu-Pampu, P., On the contact boundaries of normal surface singularities, *C. R. Acad. Sci. Paris*, Ser. I **339** (2004), 43-48.

[17] Caubel, C., Némethi, A. and Popescu-Pampu, P., Milnor open books and Milnor fillable contact 3-manifolds, *Topology*, **45** (2006), 673–689.

[18] Christophersen, J. A., On the components and discriminant of the versal base space of cyclic quotient singularities; in: *Singularity theory and its applications*, Warwick 1989, Part I, D. Mond, J. Montaldi eds., LNM **1462**, Springer, 1991.

[19]   Christophersen, J. A. and Gustavsen, T. S., On infinitesimal deformations and obstructions for rational surface singularities, *J. Algebraic Geometry,* **10** (1) (2001), 179–198.

[20]   Colin, V., Giroux, E., Honda, K., Finitude homotopique et isotopique des structures de contact tendues, *Publ. Math. Inst. Hautes Études Sci.,* **109** (2009), 245–293.

[21]   Demazure, M., Pinkham, H., Teissier, B. (editors), Séminaire sur les Singularités des Surfaces, *Lecture Notes of Math.,* **777**, Springer-Verlag, 1980.

[22]   Durfee, A., The Signature of Smoothings of Complex Surface Singularities, *Math. Ann.,* **232** (1978), 85–98.

[23]   Durfee, A., Fifteen characterizations of rational double points and simple critical points, *L'enseignement Math.,* **25** (1979), 131–163.

[24]   Eisenbud, D. and Neumann, W., *Three-Dimensional Link Theory and Invariants of Plane Curve Singularities,* Ann. of Math. Studies, **110**, Princeton University Press, 1985.

[25]   Eliashberg, Y., Filling by holomorphic discs and its applications, Geometry of low-dimensional manifolds, 2 (Durham, 1989), 45–67, *London Math. Soc. Lecture Note Ser.,* **151**, Cambridge Univ. Press, 1990.

[26]   Elkik, R., Singularités rationelles et Déformations, *Inv. Math.,* **47** (1978), 139–147.

[27]   Etnyre, J. and Ozbagci, B., Invariants of contact structures from open books, *Trans. AMS,* **360** (6) (2008), 3133–3151.

[28]   Giroux, E., Structures de contact en dimension trois et bifurcations des foilletages de surfaces, *Invent. Math.,* **141** (2000), 615–689.

[29]   Giroux, E., Structures de contact sur les variétés fibrées en cercles au-dessus d'une surface, *Comment. Math. Helv.,* **76** (2001), 218–262.

[30]   Giroux, E., Géometrie de contact: de la dimension trois vers les dimensions supérieures, *Proc. ICM, Beijing 2002,* **Vol. II.**, 405–414.

[31]   Grauert, H., Über Modifikationen und exceptionelle analytische Mengen, *Math. Annalen,* **146** (1962), 331–368.

[32]   Grauert, H., Über die Deformationen Isolierten Singularitäten Analytischer Mengen, *Inv. Math.,* **15** (1972), 171–198.

[33]   Greuel, G.-M. and Steenbrink, J., On the topology of smoothable singularities, *Proc. of Symp. in Pure Math.,* **40**, Part 1 (1983), 535–545.

[34]   Hartshorne, R., *Algebraic Geometry,* Graduate Texts in Math., **52**, Springer-Verlag 1977.

[35]   Honda, K., On the classification of tight contact structures I., *Geom. Topol.,* **4** (2000), 309–368.

[36]   Honda, K., On the classification of tight contact structures II., *J. Differential Geom.* **55** (2000), 83–143.

[37]   de Jong, T. and van Straten, D., On the base space of a semi-universal deformation of rational quadruple points, *Annals of Math.,* **134** (2) (1991), 653–678.

[38]   de Jong, T. and van Straten, D., On the deformation theory of rational surface singularities with reduced fundamental cycle, *J. Alg. Geom.,* **3** (1994), 117–172.

[39] de Jong, T. and van Straten, D., Deformation theory of sandwiched singularities, *Duke Math. J.,* **95** (3) (1998), 451–522.

[40] Kollár, J. and Shepherd-Barron, N. I., Threefolds and deformations of surface singularities, *Invent. Math.,* **91** (1988), 299–338.

[41] Kollár, J., Flips, flops, minimal models, etc., *Surveys in Diff. Geom.,* **1** (1991), 113–199.

[42] Laufer, H. B., Normal two–dimensional singularities, *Annals of Math. Studies,* **71**, Princeton University Press, 1971.

[43] Laufer, H. B., On rational singularities, *Amer. J. of Math.,* **94** (1972), 597–608.

[44] Laufer, H. B., Taut two–dimensional singularities, *Math. Ann.,* **205** (1973), 131–164.

[45] Laufer, H. B., On minimally elliptic singularities, *Amer. J. of Math.,* **99** (1977), 1257–1295.

[46] Laufer, H. B., On μ for surface singularities, *Proceedings of Symposia in Pure Math.,* **30** (1977), 45–49.

[47] Laufer, H. B., Weak simultaneous resolution for deformations of Gorenstein surface singularities, *Proc. of Symp. in Pure Math.,* **40**, Part 2 (1983), 1–29.

[48] Laufer, H. B., Strong Simultaneous Resolution for Surface Singularities, *Adv. Studies in Pure Math.,* **8** (1986), 207–214. *Complex Analytic Singularities.*

[49] Laufer, H. B., The multiplicity of isolated two-dimensional hypersurface singularities, *Transactions of the AMS,* **302**, Number 2 (1987), 489–496.

[50] Lê Dũng Tráng, Topologie des singularités des hypersurfaces complexes, *Astérisque,* **7–8** (1973), 171–182.

[51] Lipman, J., Double point resolutions of deformations of rational singularities, *Compositio Math.,* **38** (1979), 37–42.

[52] Lisca, P., On lens spaces and their symplectic fillings, *Math. Res. Letters,* **1**, vol. 11 (2004), 13–22.

[53] Lisca, P., On symplectic fillings of lens spaces, *Trans. Amer. Math. Soc.,* **360** (2008), 765–799.

[54] Lisca, P. and Stipsicz, A. I., On the existence of tight contact structures on Seifert fibered 3-manifolds, *Duke Math. J.,* **148**m no. 2, (2009), 175–209.

[55] Looijenga, E., The smoothing components of a triangle singularity. I, *Proc. of Symp. in Pure Math.,* **40**, Part 2, (1983), 173–183.

[56] Looijenga, E. J. N., Isolated Singular Points on Complete Intersections, *London Math. Soc. Lecture Note Series,* **77**, Cambridge University Press 1984.

[57] Looijenga, E., Riemann–Roch and smoothing of singularities, *Topology,* **25** (3) (1986), 293–302.

[58] Looijenga, E. and Wahl, J., Quadratic functions and smoothing surface singularities, *Topology,* **25** (1986), 261–291.

[59] McDuff, D., The structure of rational and ruled symplectic 4-manifolds, *J. Amer. Math. Soc.,* **3**, no. 3, (1990), 679–712.

[60] Milnor, J., Singular points of complex hypersurfaces, *Annals of Math. Studies,* **61**, Princeton University Press, 1968.

[61] Mumford, D., The topology of normal singularities of an algebraic surface and a criterion for simplicity, *IHES Publ. Math.,* **9** (1961), 5–22.

[62] Némethi, A., Five lectures on normal surface singularities, lectures delivered at the Summer School in *Low dimensional topology,* Budapest, Hungary, 1998; Bolyai Society Math. Studies, **8** (1999), 269–351.

[63] Némethi, A., "Weakly" Elliptic Gorenstein Singularities of Surfaces, *Inventiones Math.,* **137** (1999), 145–167.

[64] Némethi, A., The resolution of some surface singularities, I., (cyclic coverings); Proceedings of the AMS Conference, San Antonio, 1999; *Contemporary Mathematics,* **266**, 89–128.

[65] Némethi, A., Invariants of normal surface singularities, Proceedings of the Conference: *Real and Complex Singularities,* San Carlos, Brazil, August 2002; *Contemporary Mathematics,* **354** (2004), 161–208.

[66] Némethi, A., The cohomology of line bundles of splice-quotient singularities, arXiv:0810.4129.

[67] Némethi, A. and Nicolaescu, L. I., Seiberg–Witten invariants and surface singularities, *Geometry and Topology,* Volume **6** (2002), 269–328.

[68] Némethi, A. and Nicolaescu, L. I., Seiberg–Witten invariants and surface singularities II (singularities with good $\mathbf{C}^*$-action), *Journal of London Math. Soc.* (2), **69** (2004), 593–607.

[69] Némethi, A. and Nicolaescu, L. I., Seiberg–Witten invariants and surface singularities III (splicings and cyclic covers), *Selecta Mathematica,* New series, Vol. **11**, Nr. 3–4 (2005), 399–451.

[70] Némethi, A. and Okuma, T., On the Casson invarint conjecture of Neumann–Wahl, *J. of Algebraic Geometry,* **18** (2009), 135–149.

[71] Némethi, A. and Okuma, T., The Seiberg–Witten invariant conjecture for splice-quotients, *J. of London Math. Soc.,* **28** (2008), 143–154.

[72] Némethi, A. and Popescu-Pampu, P., On the Milnor fibers of cyclic quotient singularities, *Proc. London Math. Soc.,* **101**(2) (2010), 497–553.

[73] Némethi, A. and Popescu-Pampu, P., On the Milnor fibers of sandwiched singularities, *Int. Math. Res. Not.,* **6** (2010), 1041–1061.

[74] Némethi, A. and Tosun, M., Invariants of open books of links of surface singularities, *Studia Sc. Math. Hungarica,* **48**(1) (2011), 135–144.

[75] Neumann, W. D., A calculus for plumbing applied to the topology of complex surface singularities and degenerating complex curves, *Transactions of the AMS,* **268**, Number 2, (1981), 299–344.

[76] Neumann, W. D. and Pichon, A., Complex analytic realization of links, *Intelligence of low dimensional topology 2006,* 231-238, Ser. Knots Everything, 40, World Sci. Publ., Hackensack, NJ, 2007.

[77] Neumann, W. and Wahl, J., Complex surface singularities with integral homology sphere links, *Geometry and Topology,* **9** (2005), 757–811.

[78] Neumann, W. and Wahl, J., Complete intersection singularities of splice type as universal abelian covers, *Geometry and Topology,* **9** (2005), 699–755.

[79] Okuma, T., The geometric genus of splice-quotient singularities, *Transaction AMS,* **360** (2008), 6643–6659.

[80] Ohta, H. and Ono, K., Symplectic fillings of the link of simple elliptic singularities, *J. reine angew. Math.,* **565** (2003), 183–205.

[81] Ohta, H. and Ono, K., Simple singularities and symplectic fillings, *J. Differential Geom.,* **69** (2005), 1–42.

[82] Ohta, H. and Ono, K., Examples of isolated surface singularities whose links have infinitely many symplectic fillings, *J. Fixed Point Theory Appl.,* **3** (2008), 51–56.

[83] Orlik, P. and Wagreich, Ph., Isolated singularities of algebraic surfaces with $\mathbb{C}^*$ action, *Ann. of Math.* (2), **93** (1971), 205–228.

[84] Orlik, P. and Wagreich, P., Algebraic surfaces with k^*-action, *Acta Math.,* **138** (1977), 43–81.

[85] Ozbagci, B. and Stipsicz, A., Contact 3-manifolds with infinitely many Stein fillings, *Proc. AMS,* **132** (2004), 1549–1558.

[86] Pinkham, H., Deformations of algebraic varieties with G_m action, *Astérisque,* **20** (1974), 1–131.

[87] Pinkham, H., Normal surface singularities with $\mathbb{C}^*$ action, *Math. Ann.,* **117** (1977), 183–193.

[88] Pinkham, H., Smoothing of the D_{pqr} singularities, $p + q + r = 22$, *Proc. of Symp. in Pure Math.,* **40**, Part 2, (1983), 373–377.

[89] Popescu-Pampu, P., The geometry of continued fractions and the topology of surface singularities, in *Singularities in Geometry and Topology 2004,* Advanced Studies in Pure Mathematics, **46** (2007), 119–195.

[90] Popescu-Pampu, P., Numerically Gorenstein surface singularities are homeomorphic to Gorenstein ones, *Duke Math. Journal,* **159**, No. 3, (2011), 539–559.

[91] Reid, M., Chapters on Algebraic Surfaces, in: *Complex Algebraic Geometry,* IAS/Park City Mathematical Series, Volume **3** (J. Kollár editor), 3–159, 1997.

[92] Riemenschneider, O., Bemerkungen zur Deformationstheorie Nichtrationaler Singularitäten, *Manus. Math.,* **14** (1974), 91–99.

[93] Riemenschneider, O., Deformationen von Quotintensingularitäten (nach zyklischen Gruppen), *Math. Ann.,* **209** (1974), 211–248.

[94] Schaps, M., Deformations of Cohen-Macauley Schemes of codimension 2 and Non-Singular Deformations of Space Curves, *Am. J. Math.,* **99** (1977), 669–685.

[95] Schlessinger, M., Functors of Artin Rings, *Trans. AMS,* **130** (1968), 208–222.

[96] Seade, J. A., A cobordism invariant for surface singularities, *Proc. of Symp. in Pure Math.,* **40**(2) (1983), 479–484.

[97] Smith, I., Torus fibrations on symplectic four-manifolds, *Turkish J. Math.,* **25**, no. 1, (2001), 69–95.

162 A. Némethi

[98] Spivakovsky, M., Sandwiched singularities and desingularization of surfaces by normalized Nash transformations, *Annals of Math.*, **131** (1990), 411–491.

[99] Steenbrink, J. H. M., Mixed Hodge structures associated with isolated singularities, *Proc. Symp. Pure Math.*, **40**, Part 2 (1983), 513–536.

[100] Stevens, J., Elliptic Surface Singularities and Smoothings of Curves, *Math. Ann.*, **267** (1984), 239–247.

[101] Stevens, J., On the versal deformation of cyclic quotient singularities, *LNM*, **1462** (1991), 302–319. (Singularity theory and its applications, Warwick 1989)

[102] Stevens, J., Partial resolutions of rational quadruple points, *Int. J. of Math.*, **2** (2) (1991), 205–221.

[103] Stevens, J., Deformations of singularities, Springer LNM **1811**, 2003.

[104] Teissier, B., Cycles évanescents, sections planes et conditions de Whitney, *Asterisque*, **7–8** (1973), 285–362.

[105] Teissier, B., Déformation à type topologique constant II, *Séminaire Douady-Verdier 1972.*

[106] Teissier, B., Résolution simultanée I, II, *LNM*, **777** (1980), 71–146.

[107] Tjurina, G.-N., Locally Flat Deformations of Isolated Singularities of Complex Spaces, *Math. USSR Izvestia*, **3** (1969), 967–999.

[108] Tomari, M., A p_g–formula and elliptic singularities, Publ. R. I. M. S. Kyoto University, **21** (1985), 297–354.

[109] Ustilovsky, I., Infinitely many contact structures on S^{4m+1}, *I.M.R.N.*, **14** (1999), 781–792.

[110] Vaquié, M., Résolution simultanée de surfaces normales, *Ann. Inst. Fourier*, **35** (1985), 1–38.

[111] Wagreich, Ph., Elliptic singularities of surfaces, *Amer. J. of Math.*, **92** (1970), 419–454.

[112] Wahl, M. J., Equisingular deformations of normal surface singularities, I, *Ann. of Math.*, **104** (1976), 325–356.

[113] Wahl, M. J., Simultaneous resolution of rational singularities, *Compositio Math.*, **38** (1) (1979), 43–54.

[114] Wahl, M. J., Elliptic Deformations of Minimally Elliptic Singularities, *Math. Ann.*, **253** (1980), 241–262.

[115] Wahl, J., Smoothings of normal surface singularities, *Topology*, **20** (1981), 219–246.

[116] Yau, S. S.-T., On maximally elliptic singularities, *Transactions of the AMS*, **257**, Number 2 (1980), 269–329.

András Némethi

Rényi Institute of Mathematics
Budapest, Hungary

e-mail: nemethi@renyi.hu

BOLYAI SOCIETY
MATHEMATICAL STUDIES, 23

Deformations of
Surface Singularities
pp. 163–39.

THE VERSAL DEFORMATION OF CYCLIC QUOTIENT SINGULARITIES

JAN STEVENS

We describe the versal deformation of two-dimensional cyclic quotient singularities in terms of equations, following Arndt, Brohme and Hamm. For the reduced components the equations are determined by certain systems of dots in a triangle. The equations of the versal deformation itself are governed by a different combinatorial structure, involving rooted trees.

One of the goals of singularity theory is to understand the versal deformations of singularities. In general the base space itself is a highly singular and complicated object. Computations for a whole class of singularities are only possible in the presence of many symmetries. A natural class of surface singularities to consider consists of the affine toric singularities. These are just the cyclic quotient singularities. Their infinitesimal deformations were determined by Riemenschneider [8]. Explicit equations for the versal deformation are the result of a series of PhD-theses. Arndt [1] gave a recipe to find equations of the base space. This was further studied by Brohme [3], who proposed explicit formulas. Their correctness was finally proved by Hamm [6]. One of the objectives of this paper is to describe these equations.

Unfortunately it is difficult to find the structure of the base space from the equations. What one can do is to study the situation for low embedding dimension e. On the basis of such computations Arndt [1] conjectured that the number of irreducible components should not exceed the Catalan number $C_{e-3} = \frac{1}{e-2}\binom{2(e-3)}{e-3}$. This conjecture was proved in [11] using Kollár and Shepherd-Barron's description [7] of smoothing components as deformation spaces of certain partial resolutions. It was observed by Jan Christophersen that the components are related to special ways of writing the equations of the singularity. In terms of his continued fractions, representing zero, these equations are given in [4, §2], and in terms of subdivisions of polygons in [11, Sect. 6]. A more direct way of operating with the equations was found by Riemenschneider [2]. We use it, and the combinatorics behind it, in this paper to describe the components. A toric description of the components is given in the second part of Hamm's thesis [6].

From the toric picture one finds immediately some equations, by looking at the Newton boundary in the lattice of monomials:

$$z_{\varepsilon-1}z_{\varepsilon+1} = z_\varepsilon^{a_\varepsilon}, \qquad 2 \leq \varepsilon \leq e-1.$$

These form the bottom line of a pyramid of equations $z_{\delta-1}z_{\varepsilon+1} = p_{\delta,\varepsilon}$. In computing these higher equations choices have to be made. We derive $p_{\delta,\varepsilon}$ from $p_{\delta,\varepsilon-1}$ and $p_{\delta+1,\varepsilon}$. As $z_{\delta-1}z_{\varepsilon+1} = (z_{\delta-1}z_\varepsilon)(z_\delta z_{\varepsilon+1})/(z_\delta z_\varepsilon)$, we have two natural choices for $p_{\delta,\varepsilon}$:

$$\frac{p_{\delta,\varepsilon-1}p_{\delta+1,\varepsilon}}{p_{\delta+1,\varepsilon-1}} \quad \text{or} \quad \frac{p_{\delta,\varepsilon-1}p_{\delta+1,\varepsilon}}{z_\delta z_\varepsilon}.$$

We encode the choice by putting a white or black dot at place (δ, ε) in a triangle of dots. Only for certain systems of choices we can write down (in an easy way) enough deformations to fill a whole component. We call the corresponding triangles of dots sparse coloured triangles. We prove that the number of sparse coloured triangles of given size is the Catalan number C_{e-3}.

For the computation of the versal deformation one also starts from the bottom line of the pyramid of equations. Due to the presence of deformation parameters, divisions which previously were possible, now leave a remainder. We describe Arndt's formalism to deal with these remainders. One introduces new symbols, which in fact can be considered as new variables on the deformation space. Because they are independent of the a_ε, one obtains that the base spaces of different cyclic quotients with the same embedding dimension are isomorphic up to multiplication by a smooth factor, provided all a_ε are large enough. Also here, in writing the equations, some choices have to made. A particular system of choices was proposed by Brohme. To be able to handle the terms in the formulas, one needs a combinatorial description of them. It turns out that the number of terms grows rapidly, faster than the Catalan numbers, and a different combinatorial structure is needed. Hamm [6] discovered how rooted trees can be used. We will describe the computation of the versal deformation for embedding dimension 7 and then introduce Hamm's rooted trees, and give the equations in general in terms of these trees. We also describe the main steps in the proof that one really obtains the versal deformation.

Now that the equations are known, it is time to use them. We make a start here by showing that one recovers Arndt's equations for the versal deformation of the cones over rational normal curves (the case that all $a_\varepsilon = 2$). Furthermore, we look at the reduced base space. We start by looking at an example. We then define an ideal, using sparse coloured triangles, which has the correct reduced components. We do not touch upon the embedded components, leaving this for further research.

As one will see, notation becomes rather heavy, with many levels of indices. Although TEX allows almost anything, we have tried to restrict the indexing to a minimum. One has to admire Arndt's thesis [1], written on a typewriter. At that time, TEX was available, but Jürgen had already purchased an electronic typewriter for his Diplomarbeit. He decided to write the indices separately, diminish them with a photocopier and to glue them in the manuscript.

This paper is organised as follows. After a section introducing cyclic quotients and their infinitesimal deformations, we treat the case of embedding dimension 5 in detail. In Section 3 we define sparse coloured triangles and show how to describe the reduced components with them. In Section 4 we give the equations for the total space of the versal deformation: we describe Arndt's results, do the case of embedding dimension 6, and formulate and sketch the proof of the general result in terms of Hamm's rooted trees. In the last Section we discuss the reduced base space.

1. CYCLIC QUOTIENT SINGULARITIES

Let $G_{n,q}$ be the cyclic subgroup of $Gl(2,\mathbb{C})$, generated by $\begin{pmatrix} \zeta_n & 0 \\ 0 & \zeta_n^q \end{pmatrix}$, where ζ_n is a primitive n-th root of unity and q is coprime to n. The group acts on $\mathbb{C}^2$ and on the polynomial ring $\mathbb{C}[u,v]$. The quotient $\mathbb{C}^2/G_{n,q}$ has a singularity at the origin, which is called the cyclic quotient singularity $X_{n,q}$. The quotient map is a map of affine toric varieties, given by the inclusion of the standard lattice $\mathbb{Z}^2$ in the lattice $N = \mathbb{Z}^2 + \mathbb{Z} \cdot \frac{1}{n}(1,q)$, with as cone σ the first quadrant. The dual lattice M gives exactly the invariant monomials: $\mathbb{C}[M \cap \sigma^\vee] = \mathbb{C}[u,v]^{G_{n,q}}$. Generators of this ring are

$$z_\varepsilon = u^{i_\varepsilon} v^{j_\varepsilon}, \qquad \varepsilon = 1, \ldots, e,$$

where the numbers i_ε, j_ε are determined by the continued fraction expansion $n/(n-q) = [a_2, \ldots, a_{e-1}]$ in the following way:

$$i_e = 0, \quad i_{e-1} = 1, \quad i_{\varepsilon+1} + i_{\varepsilon-1} = a_\varepsilon i_\varepsilon$$

$$j_1 = 0, \quad j_2 = 1, \quad j_{\varepsilon-1} + j_{\varepsilon+1} = a_\varepsilon j_\varepsilon.$$

We also write $X[a]$ for $X_{n,q}$.

We exclude the case of the A_k singularities and assume that the embedding dimension e is at least 4. The equations for $X[a]$ can be given in quasi-determinantal format [9]:

$$\begin{pmatrix} z_1 & & z_2 & \cdots & z_{e-2} & & z_{e-1} \\ & z_2^{a_2-2} & & \cdots & & z_{e-1}^{a_{e-1}-2} & \\ z_2 & & z_3 & \cdots & z_{e-1} & & z_e \end{pmatrix}$$

We recall that the generalised minors of a quasi-determinant

$$\begin{pmatrix} f_1 & & f_2 & \cdots & f_{k-1} & & f_k \\ & h_{1,2} & & \cdots & & h_{k-1,k} & \\ g_1 & & g_2 & \cdots & g_{k-1} & & g_k \end{pmatrix}$$

are $f_i g_j - g_i \left(\prod_{\varepsilon=i}^{j-1} h_{\varepsilon,\varepsilon+1} \right) f_j$.

By perturbing the entries in the quasi-determinantal in the most general way one obtains the equations for the Artin component, the deformations which admit simultaneous resolution. We describe these deformations more concretely, following the notation of Brohme [3] (differing slightly from [11] in that the letters s and t are interchanged). We first remark that in general the first and the last extra term in a quasi-determinantal can be written in the matrix: just take $g_1 h_{1,2}$ as entry in the lower left corner and $f_k h_{k-1,k}$ as entry in the upper right corner. For a cyclic quotient this gives entries $z_2^{a_2-1}$ and $z_{e-1}^{a_{e-1}-1}$. We deform

(1)
$$\begin{pmatrix} z_1 & z_2 & & z_3 & \cdots & z_{e-3} & & z_{e-2} & Z_{e-1}^{(a_{e-1}-1)} \\ & & Z_3^{(a_3-2)} & & \cdots & & Z_{e-2}^{(a_{e-2}-2)} & & \\ Z_2^{(a_2-1)} & z_3+t_3 & & z_4+t_4 & \cdots & z_{e-2}+t_{e-2} & & z_{e-1} & z_e \end{pmatrix}$$

where

$$Z_\varepsilon^{(a_\varepsilon-2)} = z_\varepsilon^{a_\varepsilon-2} + s_\varepsilon^{(1)} z_\varepsilon^{a_\varepsilon-3} + \cdots + s_\varepsilon^{(a_\varepsilon-2)}$$

for $3 \leq \varepsilon \leq e - 2$ and

$$Z_\varepsilon^{(a_\varepsilon - 1)} = z_\varepsilon^{a_\varepsilon - 1} + s_\varepsilon^{(1)} z_\varepsilon^{a_\varepsilon - 2} + \cdots + s_\varepsilon^{(a_\varepsilon - 1)}$$

for $\varepsilon = 2$ and $\varepsilon = e - 1$. To obtain all infinitesimal deformations we add variables $s_\varepsilon^{(a_\varepsilon - 1)}$ for $3 \leq \varepsilon \leq e - 2$ and write the perturbation

$$Z_\varepsilon^{(a_\varepsilon - 2)} = z_\varepsilon^{a_\varepsilon - 2} + s_\varepsilon^{(1)} z_\varepsilon^{a_\varepsilon - 3} + \cdots + s_\varepsilon^{(a_\varepsilon - 2)} + s_\varepsilon^{(a_\varepsilon - 1)} z_\varepsilon^{-1},$$

which gives the coordinates $s_\varepsilon^{(a)}$, $1 \leq a \leq a_\varepsilon - 1$, $\varepsilon = 2, \ldots, e - 1$ and t_ε, $\varepsilon = 3, \ldots, e - 2$ on the vector space $T_{X[a]}^1$. We note in particular the polynomial equations

$$(2) \quad z_{\varepsilon-1}(z_{\varepsilon+1} + t_{\varepsilon+1}) = (z_\varepsilon + t_\varepsilon)\left(z_\varepsilon^{a_\varepsilon - 1} + s_\varepsilon^{(1)} z_\varepsilon^{a_\varepsilon - 2} + \cdots + s_\varepsilon^{(a_\varepsilon - 1)}\right).$$

To avoid special cases we make this formula valid for all ε by introducing variables t_2, t_{e-1} and t_e, which we set to zero.

2. EMBEDDING DIMENSION 5

The two components of the versal deformation of the cone over the rational normal curve of degree four are related to the two different ways of writing the equations. The largest, the Artin component, is obtained by deforming the 2×4 matrix

$$\begin{pmatrix} z_1 & z_2 & z_3 & z_4 \\ z_2 & z_3 & z_4 & z_5 \end{pmatrix}.$$

The equations can also be written as 2×2 minors of the symmetric 3×3 matrix

$$\begin{pmatrix} z_1 & z_2 & z_3 \\ z_2 & z_3 & z_4 \\ z_3 & z_4 & z_5 \end{pmatrix},$$

and perturbing this matrix gives as total space the cone over the Veronese embedding of $\mathbb{P}^2$. Riemenschneider observed that this generalises to all cyclic quotients of embedding dimension 5 [8]. One can even give the equations as quasi-determinantals. For the Artin component we take as described above

$$\begin{pmatrix} z_1 & z_2 & & z_3 & z_4^{a_4 - 1} \\ & & z_3^{a_3 - 2} & & \\ z_2^{a_2 - 1} & z_3 & & z_4 & z_5 \end{pmatrix}$$

and for the other component

$$\begin{pmatrix} z_1 & z_2 & z_3^{a_3-1} \\ z_2^{a_2-2} & & \\ z_2 & z_3 & z_4 \\ & z_4^{a_4-2} & \\ z_3^{a_3-1} & z_4 & z_5 \end{pmatrix}.$$

The meaning of the last symbol becomes clear if we write out the equations, which we have to do in order to generalise, as for higher embedding dimension only the Artin component has such a nice determinantal description.

We write a pyramid of equations. From the 2×4 quasi-determinantal we get

$$z_1 z_5 = z_2^{a_2-1} z_3^{a_3-2} z_4^{a_4-1}$$
$$z_1 z_4 = z_2^{a_2-1} z_3^{a_3-1} \qquad z_2 z_5 = z_3^{a_3-1} z_4^{a_4-1}$$
$$z_1 z_3 = z_2^{a_2} \qquad z_2 z_4 = z_3^{a_3} \qquad z_3 z_5 = z_4^{a_4}$$

and from the symmetric quasi-determinantal

$$z_1 z_5 = z_2^{a_2-2} \left(z_3^{a_3-1} \right)^2 z_4^{a_4-2}$$
$$z_1 z_4 = z_2^{(a_2-1)} z_3^{a_3-1} \qquad z_2 z_5 = z_3^{a_3-1} z_4^{a_4-1}$$
$$z_1 z_3 = z_2^{a_2} \qquad z_2 z_4 = z_3^{a_3} \qquad z_3 z_5 = z_4^{a_4}$$

The difference between these two systems of equations lies in the top line. Observe that $z_1 z_5 - z_2^{a_2-2} \left(z_3^{a_3-1} \right)^2 z_4^{a_4-2} = \left(z_1 z_5 - z_2^{a_2-1} z_3^{a_3-2} z_4^{a_4-1} \right) + z_2^{a_2-2} z_3^{a_3-2} z_4^{a_4-2} \left(z_2 z_4 - z_3^{a_3} \right)$.

To describe the deformation we introduce the following polynomials:

$$Z_\varepsilon^{(a_\varepsilon - k_\varepsilon)} = z_\varepsilon^{a_\varepsilon - k_\varepsilon} + s_\varepsilon^{(1)} z_\varepsilon^{a_\varepsilon - k_\varepsilon - 1} + \cdots + s_\varepsilon^{(a_\varepsilon - k_\varepsilon)}.$$

By deforming the first set of equations we obtain the Artin component:

$$z_1 z_5 = Z_2^{(a_2-1)} Z_3^{(a_3-2)} Z_4^{(a_4-1)}$$
$$z_1 z_4 = Z_2^{(a_2-1)} Z_3^{(a_3-2)} z_3 \qquad z_2 z_5 = (z_3 + t_3) Z_3^{(a_3-2)} Z_4^{(a_4-1)}$$
$$z_1 (z_3 + t_3) = z_2 Z_2^{(a_2-1)} \qquad z_2 z_4 = z_3 Z_3^{(a_3-2)} (z_3 + t_3) \qquad z_3 z_5 = z_4 Z_4^{(a_4-1)}$$

The second set of equations leads to the other component:

$$z_1 z_5 = Z_2^{(a_2-2)} \left(Z_3^{(a_3-1)} \right)^2 Z_4^{(a_4-2)}$$
$$z_1 z_4 = z_2 Z_2^{(a_2-2)} Z_3^{(a_3-1)} \qquad z_2 z_5 = Z_3^{(a_3-1)} Z_4^{(a_4-2)} z_4$$
$$z_1 z_3 = z_2 Z_2^{(a_2-2)} z_2 \qquad z_2 z_4 = z_3 Z_3^{(a_3-1)} \qquad z_3 z_5 = z_4 Z_4^{(a_4-2)} z_4$$

Together these two components constitute the versal deformation. They fit together to the deformation

$$(3) \qquad z_1 z_5 = Z_2^{(a_2-1)} Z_3^{(a_3-2)} Z_4^{(a_4-1)} + s_3^{(a_3-1)} Z_2^{(a_2-2)} Z_3^{(a_3-1)} Z_4^{(a_4-2)}$$

$$z_1 z_4 = Z_2^{(a_2-1)} Z_3^{(a_3-1)} \qquad z_2 z_5 = \widetilde{Z}_3^{(a_3-1)} Z_4^{(a_4-1)}$$

$$z_1(z_3 + t_3) = z_2 Z_2^{(a_2-1)} \qquad z_2 z_4 = Z_3^{(a_3-1)}(z_3 + t_3) \qquad z_3 z_5 = z_4 Z_4^{(a_4-1)}$$

over the base space defined by the equations

$$s_2^{(a_2-1)} s_3^{(a_3-1)} = t_3 s_3^{(a_3-1)} = s_3^{(a_3-1)} s_4^{(a_4-1)} = 0.$$

Here the factor $\widetilde{Z}_3^{(a_3-1)}$ is defined by the equation $z_3 \widetilde{Z}_3^{(a_3-1)} = (z_3 + t_3) Z_3^{(a_3-1)}$, which is possible because of the equation $t_3 s_3^{(a_3-2)} = 0$.

Remark. The equations for the versal deformation restrict (by setting the deformation variables to zero) to the equations of the singularity in the preferred form for the Artin component. A choice has to be made, and this one is sensible as the Artin component is the only component, which exists for all cyclic quotients. Observe also that the right hand side of the top equation in (3) is no longer a product. For the study of the non-Artin component, e.g., to determine adjacencies, the adapted equations are much better suited. We have therefore in general two tasks, to describe equations suited for each reduced component separately, and to give equations for the total versal deformation.

3. Equations for Components

The reduced components of the versal deformation are related to ways of writing the equations of the singularity, as shown in [4] and [11]. Here we give a description which first appeared in [2].

We have to write the equations $z_{\delta-1} z_{\varepsilon+1} = p_{\delta,\varepsilon}$. Motivated by the case of embedding dimension 5 we want the right hand side of the equations to be of the form $p_{\delta,\varepsilon} = \prod_\beta (z_\beta^{a_\beta - k_\beta})^{\alpha_\beta}$. Here the k_β and α_β depend on $\varepsilon - \delta$, but the formula should be in some sense universal, it should hold for all a_β

large enough (for $a_\beta - k_\beta$ has to be non-negative). The toric weight vectors $w_\beta \in \mathbb{Z}^2$ of the variables z_β should therefore satisfy the equations

$$w_\delta + w_\varepsilon = \sum \alpha_\beta(a_\beta - k_\beta)w_\beta,$$

the same equations as encountered by Jan Christophersen (see the Introduction of [4]).

We construct a pyramid of equations $z_{\delta-1}z_{\varepsilon+1} = p_{\delta,\varepsilon}$, where $2 \leq \delta \leq \varepsilon \leq e - 1$. We start from the base line containing the $z_{\varepsilon-1}z_{\varepsilon+1} = z_\varepsilon^{a_\varepsilon}$, and construct the next lines inductively. We have to make choices, which we encode in a subset $B(\triangle)$ of the set of pairs (δ, ε) with $1 < \delta < \varepsilon < e$. As $z_{\delta-1}z_{\varepsilon+1} = (z_{\delta-1}z_\varepsilon)(z_\delta z_{\varepsilon+1})/(z_\delta z_\varepsilon)$, we have two natural choices for $p_{\delta,\varepsilon}$: we take

$$p_{\delta,\varepsilon} = \begin{cases} \dfrac{p_{\delta,\varepsilon-1}p_{\delta+1,\varepsilon}}{p_{\delta+1,\varepsilon-1}}, & \text{if } (\delta, \varepsilon) \notin B(\triangle), \\[2ex] \dfrac{p_{\delta,\varepsilon-1}p_{\delta+1,\varepsilon}}{z_\delta z_\varepsilon}, & \text{if } (\delta, \varepsilon) \in B(\triangle). \end{cases}$$

We depict our set by a triangle $\triangle$ of the type

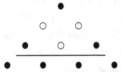

This is an example for embedding dimension $e = 7$. The dots correspond to equations $z_{\delta-1}z_{\varepsilon+1} = p_{\delta,\varepsilon}$ above the base line in the pyramid of equations. In particular, the top dot in the example corresponds to $z_1 z_7 = p_{2,6}$, and has coordinates $(2, 6)$. We colour a dot in $\triangle$ black if the corresponding point (δ, ε) is an element of $B(\triangle)$.

As the second line of equations always reads $p_{\varepsilon-1,\varepsilon} = z_{\varepsilon-1}^{a_\varepsilon-1-1} z_\varepsilon^{a_\varepsilon-1}$, $2 < \varepsilon \leq e - 1$, the lowest line of the triangle is coloured black. To characterise the coloured triangles, which give good equations, it suffices to consider only triangles $\triangle$, obtained by deleting this black line:

The original triangle will be referred to as extended triangle. We introduce some more terminology. A (broken) line l_ε, in both the triangle $\triangle$ and the extended triangle $\underline{\triangle}$, is a line connecting all dots which have ε as one of the coordinates:

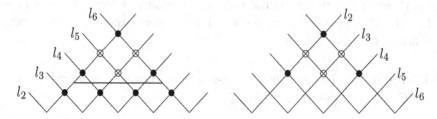

If necessary we specify a triangle by the coordinates of its highest vertex (δ, ε), as $\triangle_{\delta,\varepsilon}$. The *height* of a triangle $\triangle_{\delta,\varepsilon}$ is $\varepsilon - \delta - 1$. This is the number of horizontal lines and also the number of dots on the base line. A dot (α, β) in a triangle $\triangle_{\delta,\varepsilon}$ determines a sub-triangle $\triangle_{\alpha,\beta}$, standing on the same base line, of height $\beta - \alpha - 1$. In particular, a single dot $(\varepsilon - 1, \varepsilon + 1)$ on the base line gives a triangle $\triangle_{\varepsilon-1,\varepsilon+1}$ of height 1.

The crucial property for getting good equations is given in the following definition.

Definition 3.1. A coloured triangle $\triangle_{\delta,\varepsilon}$ is *sparse*, if for it and for every sub-triangle $\triangle_{\alpha,\beta}$ the number of black dots is at most the height of the triangle with equality if and only if its vertex is black.

Note that the example triangle above is sparse, whereas the following triangle is not sparse.

The relation $(\alpha, \gamma) \preceq (\beta, \delta)$, if $\alpha \geq \beta$ and $\gamma \leq \delta$ is a partial ordering. It means that (α, γ) lies (as black or white dot) in the triangle $\triangle_{\beta,\delta}$.

Lemma 3.2. *If two black dots lie in the region on or above a given line l_ε, then both of them lie on l_ε or they are comparable in the partial ordering $\preceq$.*

Proof. Suppose on the contrary that the black dots (α, γ) and (β, δ) on or above l_ε are not comparable in the partial ordering and that at least one of them lies strictly above l_ε. We may assume that $\gamma \leq \delta$. This implies that $\alpha < \beta$. The assumption that (α, γ) lies on or above l_ε means that $\alpha \leq \varepsilon \leq \gamma$ and likewise $\beta \leq \varepsilon \leq \delta$. Furthermore in one of these one has strict inequalities. Therefore $\beta < \gamma$. The triangle $\triangle_{\alpha,\gamma}$ contains exactly $\gamma - \alpha - 1$ black dots, the triangle $\triangle_{\beta,\delta}$ contains exactly $\delta - \beta - 1$ black dots and their intersection is the triangle $\triangle_{\beta,\gamma}$, which contains at most $\gamma - \beta - 1$ dots. So the triangle $\triangle_{\alpha,\delta}$, which has as vertex the supremum (α,δ) of (α,γ) and (β,δ) in the partial ordering, contains at least $(\gamma - \alpha - 1) + (\delta - \beta - 1) - (\gamma - \beta - 1) = \delta - \alpha - 1$ black dots other than its vertex, contradicting sparsity. ∎

Theorem 3.3. *The number of sparse coloured triangles of height $e - 4$ is the Catalan number $C_{e-3} = \frac{1}{e-2}\binom{2(e-3)}{e-3}$.*

Proof. Consider a sparse triangle $\triangle_{2,e-1}$ and let $(2,\beta)$ be the highest black dot on the line l_2. There are no black dots above the line l_β, for according to Lemma 3.2 the dot $(2,\beta)$ should lie in the triangle of such a dot, implying that it lies on l_2, but $(2,\beta)$ is the highest black dot on that line. The triangle $\triangle_{\beta,e-1}$ can be an arbitrary sparse triangle. The sparse triangle $\triangle_{3,\beta}$ determines the colour of the remaining dots on the line l_2: proceeding inductively downwards, the dot $(2,\gamma)$ has to be black if and only if there are exactly $\gamma - \beta$ black dots in the triangle $\triangle_{2,\beta}$, not lying in $\triangle_{2,\gamma}$; as $(2,\beta)$ is black, there are at least $\gamma - \beta$ black dots in this complement.

This shows that the number C_n, $n = e - 3$, of sparse coloured triangles of height $n - 1$ satisfies Segner's recursion formula for the Catalan numbers

$$C_{n+1} = C_0 C_n + C_1 C_{n-1} + \cdots + C_{n-1} C_1 + C_n C_0.$$

For more information on the Catalan numbers, see [10]. ∎

Remark. The Catalan number C_{e-3} also counts the number of subdivisions of an $(e-1)$-gon in triangles. An explicit bijection is as follows. Mark, as in [11], a distinguished vertex and number the remaining ones from 2 to $e-1$. If the vertices δ and ε are joined by a diagonal, then we colour the dot (δ,ε) black. Conversely, given a triangle $\triangle_{2,e-1}$ we join the vertices δ and ε by a diagonal, if the dot (δ,ε) is black. By Lemma 3.2 these diagonals do not intersect. We complete the subdivision with diagonals through the distinguished vertex. Sometimes it is easier to use subdivisions, but we will derive all facts we need directly from the combinatorics of sparse triangles.

To describe the equations we need the numbers $[k_2,\ldots,k_{e-1}]$ and $(\alpha_2,\ldots,\alpha_{e-1})$. These are indeed the continued fractions $[k_2,\ldots,k_{e-1}]$ representing zero [4] and the corresponding numbers satisfying $\alpha_{\varepsilon-1} + \alpha_{\varepsilon+1} = k_\varepsilon \alpha_\varepsilon$. To define these numbers out of a triangle we inductively give non-zero weights to black dots; white dots have weight zero.

Definition 3.4. Let $\triangle_{2,e-1}$ be a sparse triangle. The weight $w_{\delta,\varepsilon}$ of a black dot (δ,ε) is the sum of the weights of the dots lying in the sector above it, increased by one: $w_{\delta,\varepsilon} = 1 + \sum_{\substack{\alpha<\delta \\ \beta>\varepsilon}} w_{\alpha,\beta}$. For $2 \le \varepsilon \le e-1$ we define α_ε as the sum of the weights of dots above the line l_ε, increased by one:

$$\alpha_\varepsilon = 1 + \sum_{\substack{\alpha<\varepsilon \\ \beta>\varepsilon}} w_{\alpha,\beta}.$$

In particular, $\alpha_\varepsilon = 1$ if there are no black points above the line l_ε, so $\alpha_2 = \alpha_{e-1} = 1$.

We set k_ε to be the number of black dots on the line l_ε in the extended triangle if $\alpha_\varepsilon = 1$ and this number minus 1 otherwise.

Example.

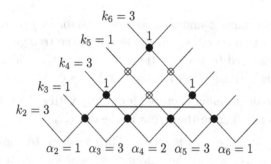

Remarks. 1. We leave it as exercise to prove that the so defined numbers α_ε and k_ε satisfy $\alpha_{\varepsilon-1} + \alpha_{\varepsilon+1} = k_\varepsilon \alpha_\varepsilon$.

2. We note the following alternative way to compute the α_ε [3, Bemerkung 1.7]. As mentioned, $\alpha_\varepsilon = 1$ if there are no black dots above the line l_ε. For every other index ε there exist unique $\beta < \varepsilon < \gamma$, such that the intersection of the lines l_β, l_ε and l_γ (in the extended triangle) consists only of black dots. Then $\alpha_\varepsilon = \alpha_\beta + \alpha_\gamma$. In fact, this is the way the numbers are determined from a subdivision of a polygon. We sketch a proof. Let (β,ε) be the highest black dot on the left half-line of l_ε, and (ε,γ) the highest black dot on the right half-line. Let (α,δ) be a black dot above the line l_ε, such that $\triangle_{\alpha,\delta}$ contains no other black dots above l_ε. Then $(\beta,\gamma) \preceq (\alpha,\delta)$, and if $(\beta,\gamma) \ne (\alpha,\delta)$, then $\triangle_{\alpha,\delta}$ does not contain enough dots. So if $\alpha_\varepsilon \ne 1$, then (β,γ) is black. Black dots above the lines l_β, l_ε and l_γ can only lie in the sector with (β,γ) as lowest point. One now computes $\alpha_\varepsilon = \alpha_\beta + \alpha_\gamma$.

We can now describe the equations belonging to a sparse triangle $\triangle_{2,e-1}$. To avoid cumbersome notation we only give the formula for the highest equation $z_1 z_e = p_{2,e-1}$, but this implies by obvious changes the formula for each equation $z_{\delta-1} z_{\varepsilon+1} = p_{\delta,\varepsilon}$, as such an equation is determined by its own sparse triangle $\triangle_{\delta,\varepsilon}$, giving its own α and k values. We will specify in the text from which triangle a specific α_ε or k_ε is computed, but we do not include this information in the notation.

Proposition 3.5. *Let the triangle $\triangle_{2,e-1}$ determine the numbers α_ε, k_ε, according to Definition 3.4. Forming the equations by taking $p_{\delta,\varepsilon} = \frac{p_{\delta,\varepsilon-1} p_{\delta+1,\varepsilon}}{z_\delta z_\varepsilon}$ if the dot (δ, ε) is black and $p_{\delta,\varepsilon} = \frac{p_{\delta,\varepsilon-1} p_{\delta+1,\varepsilon}}{p_{\delta+1,\varepsilon-1}}$ otherwise, leads to the highest equation*

$$z_1 z_e = \prod_{\beta=2}^{e-1} \left(z_\beta^{a_\beta - k_\beta} \right)^{\alpha_\beta}.$$

Proof. We fix an index ε and look at the z_ε-factor in $p_{2,e-1}$. The proof proceeds by induction on e, i.e., on the height of the triangle. The base of the induction is formed by the equations $z_{\varepsilon-1} z_{\varepsilon+1} = z_\varepsilon^{a_\varepsilon}$, which correspond to empty extended triangles, with $\alpha_\varepsilon = 1$ and $k_\varepsilon = 0$.

Suppose that the formula is proved for all $p_{\delta,\varepsilon}$ with $\varepsilon - \delta < e - 3$. There are two cases, depending on the colour of the dot $(2, e - 1)$.

Case 1: the dot $(2, e - 1)$ is white. Then we have to compare $p_{2,e-1}$, $p_{2,e-2}$, $p_{3,e-1}$ and $p_{3,e-2}$ and the values of α_ε and k_ε computed from the corresponding triangles. There are several sub-cases. In the first two we assume that $\varepsilon \neq 2, e - 1$.

1.a: Suppose both $(2, \varepsilon)$ and $(\varepsilon, e - 1)$ are black. A black dot above the line l_ε should contain both these points in its triangle, so it can only be $(2, e - 1)$, which however is assumed to be uncoloured. Therefore $\alpha_\varepsilon = 1$ in all the relevant triangles. We have that there are k_ε dots on the line l_ε in the extended triangle $\underline{\triangle}_{2,e-1}$, $k_\varepsilon - 1$ in $\underline{\triangle}_{2,e-2}$ and $\underline{\triangle}_{3,e-1}$, and $k_\varepsilon - 2$ in $\underline{\triangle}_{3,e-2}$. So indeed the z_ε factor in $p_{2,e-1}$ is equal to $z_\varepsilon^{a_\varepsilon - k_\varepsilon + 1} \cdot z_\varepsilon^{a_\varepsilon - k_\varepsilon + 1} / z_\varepsilon^{a_\varepsilon - k_\varepsilon + 2} = z_\varepsilon^{a_\varepsilon - k_\varepsilon}$.

1.b: Otherwise the segments $(2, \varepsilon) - (2, e - 1)$ and $(\varepsilon, e - 1) - (2, e - 1)$ cannot both contain black dots. Suppose the first segment, on l_2, is empty (in particular $(2, \varepsilon)$ is white). Then the number of black dots on the line l_ε is equal in both $\underline{\triangle}_{2,e-1}$ and $\underline{\triangle}_{3,e-1}$, being equal to k_ε or $k_\varepsilon + 1$ depending on the value of α_ε, and one computes also the same value for

α_ε. Also $\underline{\triangle}_{2,e-2}$ and $\underline{\triangle}_{3,e-2}$ yield the same values α'_ε and k'_ε, so we get

$$\left(z_\varepsilon^{a_\varepsilon-k_\varepsilon}\right)^{\alpha_\varepsilon} \cdot \left(z_\varepsilon^{a_\varepsilon-k'_\varepsilon}\right)^{\alpha'_\varepsilon} / \left(z_\varepsilon^{a_\varepsilon-k'_\varepsilon}\right)^{\alpha'_\varepsilon} = \left(z_\varepsilon^{a_\varepsilon-k_\varepsilon}\right)^{\alpha_\varepsilon}.$$

1.c: Suppose that $\varepsilon = 2$ or $\varepsilon = e - 1$. Consider the first case. The monomial z_2 does not occur in $p_{3,e-1}$ and $p_{3,e-2}$. Always $\alpha_2 = 1$ and the number of dots on l_2 is the same in both relevant triangles.

Case 2: the dot $(2, e - 1)$ is black. We have to compare the values of α_ε and k_ε in $p_{2,e-1}, p_{2,e-2}$ and $p_{3,e-1}$. Again we consider $\varepsilon = 2, e - 1$ separately.

2.a: Suppose both $(2, \varepsilon)$ and $(\varepsilon, e - 1)$ are black. Then $(2, e - 1)$ is the only black dot above the line l_ε, which makes $\alpha_\varepsilon = 2$, whereas $\alpha_\varepsilon = 1$ in both smaller triangles. The number of black dots on the line l_ε is $k_e + 1$ in $\underline{\triangle}_{2,e-1}$, and k_e in $\underline{\triangle}_{2,e-2}$ and $\underline{\triangle}_{3,e-1}$. So the z_ε factor in $p_{2,e-1}$ is equal to

$$z_\varepsilon^{a_\varepsilon-k_\varepsilon} \cdot z_\varepsilon^{a_\varepsilon-k_\varepsilon} = \left(z_\varepsilon^{a_\varepsilon-k_\varepsilon}\right)^2.$$

2.b: Suppose $(2, \varepsilon)$ is black and $(\varepsilon, e - 1)$ not. Then all black points above l_ε lie on l_2, all having weight 1, and there is at least one of them between $(2, \varepsilon)$ and $(2, e - 1)$, as $(\varepsilon, e - 1)$ is not black. There are $k_\varepsilon + 1$ points on l_ε in $\underline{\triangle}_{2,e-1}$ and $\underline{\triangle}_{2,e-2}$, and k_ε in $\underline{\triangle}_{3,e-1}$. The last triangle gives the value $\alpha_\varepsilon = 1$ in $p_{3,e-1}$, whereas $\alpha_\varepsilon > 1$ in the first two, and the value from $\underline{\triangle}_{2,e-1}$ is one more than that from $\underline{\triangle}_{2,e-2}$. So the z_ε factor in $p_{2,e-1}$ is equal to $\left(z_\varepsilon^{a_\varepsilon-k_\varepsilon}\right)^{\alpha_\varepsilon-1} \cdot z_\varepsilon^{a_\varepsilon-k_\varepsilon} = \left(z_\varepsilon^{a_\varepsilon-k_\varepsilon}\right)^{\alpha_\varepsilon}$.

2.c: Suppose both $(2, \varepsilon)$ and $(\varepsilon, e - 1)$ are white. The number of dots on l_ε is the same in all three relevant triangles. In the largest one $\alpha_\varepsilon > 1$, so to get the same value for k_ε we need that $\alpha_\varepsilon > 1$ also in both other triangles. This means that the triangle $\triangle_{3,e-2}$ has to have a dot above the line l_ε. Of the segments of l_2 and l_{e-1} above l_ε only one can contain black dots besides the vertex. Suppose $(j, e - 1)$ is the lowest dot of the segment on l_{e-1}. The number of dots in $\triangle_{j,e-1}$ on or under l_ε is at most $(e - 1 - \varepsilon - 2) + (\varepsilon - j - 1) = (e - 1 - j - 3)$. As the triangle contains exactly $e - 1 - j - 2$ dots other than the vertex, there has to be a dot above l_ε, which does not lie on l_{e-1} due to the choice of $(j, e - 1)$. To compute α_ε we have to look at the weights. With the convention that points above a triangle have weight 0, the inductive formula holds for points in a sub-triangle with summation over all points in the sector of the big triangle. We show by induction that for all points except the vertex $(2, e - 1)$ the weight $w_{i,j}$ computed from the big triangle, equals the sum of the weights $w'_{i,j}$ from $\triangle_{2,e-2}$ and $w''_{i,j}$ from $\triangle_{3,e-1}$. Indeed,

$$w_{i,j} = 1 + \sum w_{k,l} = 1 + 1 + \sum_{(k,l)\neq(2,e-1)} w_{k,l} = 1 + \sum w'_{k,l} + 1 + \sum w''_{k,l} =$$

$w'_{i,j} + w''_{i,j}$. The same computation shows that also the values of α_ε add. So indeed the z_ε factor in $p_{2,e-1}$ is the product of those in $p_{2,e-2}$ and $p_{3,e-1}$.

2.d: If $\varepsilon = 2$, then $\alpha_2 = 1$ and the number of points on l_2 in $\underline{\triangle}_{2,e-1}$ is k_2, whereas it is $k_2 - 1$ in $\underline{\triangle}_{2,e-2}$. The z_2 factor in $p_{2,e-1}$ is $z_2^{a_2-k_2+1}/z_2 = z_2^{a_2-k_2}$. The case $\varepsilon = e - 1$ is similar. ∎

As in the case of embedding dimension 5 we can now deform. We perturb each term $z_\varepsilon^{a_\varepsilon - k_\varepsilon}$ to

$$Z_\varepsilon^{(a_\varepsilon - k_\varepsilon)} = z_\varepsilon^{a_\varepsilon - k_\varepsilon} + \widetilde{s}_\varepsilon^{(1)} z_\varepsilon^{a_\varepsilon - k_\varepsilon - 1} + \cdots + \widetilde{s}_\varepsilon^{(a_\varepsilon - k_\varepsilon)}.$$

Here we write $\widetilde{s}_\varepsilon^{(j)}$, as these variables are not quite the same as the coordinates $s_\varepsilon^{(j)}$ on T^1, specified by the equations (2). The relation is the following: if $\alpha_\varepsilon > 1$, we set $t_\varepsilon = 0$, and $\widetilde{s}_\varepsilon^{(j)} = s_\varepsilon^{(j)}$, but if $\alpha_\varepsilon = 1$ one has

$$(z_\varepsilon + t_\varepsilon)^{\alpha_\varepsilon - 1}\left(z_\varepsilon^{a_\varepsilon - k_\varepsilon} + \widetilde{s}_\varepsilon^{(1)} z_\varepsilon^{a_\varepsilon - k_\varepsilon - 1} + \cdots + \widetilde{s}_\varepsilon^{(a_\varepsilon - k_\varepsilon)}\right) z_\varepsilon^{\alpha_\varepsilon + 1}$$

$$= (z_\varepsilon + t_\varepsilon)\left(z_\varepsilon^{a_\varepsilon - 1} + s_\varepsilon^{(1)} z_\varepsilon^{a_\varepsilon - 2} + \cdots + s_\varepsilon^{(a_\varepsilon - 1)}\right).$$

This formula makes sense, as $\alpha_\varepsilon k_\varepsilon = \alpha_{\varepsilon-1} + \alpha_{\varepsilon+1}$, so for $\alpha_\varepsilon = 1$ one has $k_\varepsilon = \alpha_{\varepsilon-1} + \alpha_{\varepsilon+1}$.

Proposition 3.6. Let $\triangle_{2,e-1}$ be a sparse triangle. Put $t_\varepsilon = 0$, if $\alpha_\varepsilon > 1$. Now form the equations $z_{\delta-1}(z_{\varepsilon+1} + t_{\varepsilon+1}) = P_{\delta,\varepsilon}$, starting from

$$P_{\varepsilon,\varepsilon} = (z_\varepsilon + t_\varepsilon)^{\alpha_\varepsilon - 1}\left(z_\varepsilon^{a_\varepsilon - k_\varepsilon} + \widetilde{s}_\varepsilon^{(1)} z_\varepsilon^{a_\varepsilon - k_\varepsilon - 1} + \cdots + \widetilde{s}_\varepsilon^{(a_\varepsilon - k_\varepsilon)}\right) z_\varepsilon^{\alpha_\varepsilon + 1},$$

if $\alpha_\varepsilon = 1$ and

$$P_{\varepsilon,\varepsilon} = \left(z_\varepsilon^{a_\varepsilon - k_\varepsilon} + s_\varepsilon^{(1)} z_\varepsilon^{a_\varepsilon - k_\varepsilon - 1} + \cdots + s_\varepsilon^{(a_\varepsilon - k_\varepsilon)}\right) z_\varepsilon^{k_\varepsilon}$$

otherwise. Take $P_{\delta,\varepsilon} = \frac{P_{\delta,\varepsilon-1} P_{\delta+1,\varepsilon}}{z_\delta(z_\varepsilon + t_\varepsilon)}$ if the dot (δ, ε) is black and $P_{\delta,\varepsilon} = \frac{P_{\delta,\varepsilon-1} P_{\delta+1,\varepsilon}}{P_{\delta+1,\varepsilon-1}}$ otherwise. This gives the highest equation

$$z_1 z_e = \prod_{\beta=2}^{e-1} \left(Z_\beta^{(a_\beta - k_\beta)}\right)^{\alpha_\beta}.$$

These equations define a flat deformation of the cyclic quotient singularity $X[a]$.

The flatness is proved explicitly in [4, 2.1.2] and [3, 2.2]. It is of course a consequence of the inductive definition of the polynomials $P_{\delta,\varepsilon}$.

In fact, one gets in this way exactly all reduced components of the versal deformation. This was proved in [11] using Kollár and Shepherd-Barron's description [7] of smoothing components as deformation spaces of certain partial resolutions. A more elementary (but not easier) approach would be to use the equations for the base space of the versal deformation, which we describe in the next section.

4. Versal Deformation

In this section we derive the equations for the versal deformation. We have to write the pyramid of equations, as in the example of embedding dimension five. The base line consist of the equations (2). These equations are lacking in symmetry: when introducing the deformation variables t_ε, say in the quasi-determinantal, there is a choice of writing them in the upper or the lower row. Arndt [1] formally symmetrises by setting $y_\varepsilon = z_\varepsilon + t_\varepsilon$. We go one step further and replace t_ε by two deformation variables. This makes that our deformation is versal, but no longer miniversal. Furthermore, there is no t_2 and t_{e-1}, but in order to avoid special cases, we allow the index ε in t_ε to take the values 2 and $e-1$.

We start from the equations $z_{\varepsilon-1}z_{\varepsilon+1} = z_\varepsilon^{a_\varepsilon}$, which we deform into

$$(4) \qquad (z_{\varepsilon-1} - l_{\varepsilon-1})(z_{\varepsilon+1} - r_{\varepsilon+1}) = z_\varepsilon^{a_\varepsilon} + \sigma_\varepsilon^{(1)} z_\varepsilon^{a_\varepsilon-1} + \cdots + \sigma_\varepsilon^{(a_\varepsilon)}.$$

We abbreviate $z_\varepsilon - r_\varepsilon = R_\varepsilon$ and $z_\varepsilon - l_\varepsilon = L_\varepsilon$. The minus sign is introduced to simplify the conditions for divisibility by R_ε or L_ε, which will be the main ingredient in our description of the base space. We write the equation (4) shortly as

$$L_{\varepsilon-1}R_{\varepsilon+1} = Z_\varepsilon^{(00)}.$$

As written, we do not even get an infinitesimal deformation: one needs $\sigma_\varepsilon^{(a_\varepsilon)} \equiv 0$ modulo the square of the maximal ideal of the parameter space. Comparison with equation (2) shows that $Z_\varepsilon^{(00)}(z_\varepsilon)$ (we use this notation to emphasise that we consider $Z_\varepsilon^{(00)}$ as polynomial in z_ε) has to be divisible by R_ε, i.e, $Z_\varepsilon^{(00)}(r_\varepsilon) = 0$. This gives an equation with non-vanishing linear part. We could as well require divisibility by L_ε. This gives an equation $Z_\varepsilon^{(00)}(l_\varepsilon) = 0$ with the same linear part.

In fact, we shall assume both conditions, $Z_\varepsilon^{(00)}{}'(r_\varepsilon) = 0$ and $Z_\varepsilon^{(00)}{}'(l_\varepsilon) = 0$. This yields then one equation with non-vanishing linear part, and one equation, not involving $\sigma_\varepsilon^{(a_\varepsilon)}$ at all, which factorises:

$$Z_\varepsilon^{(00)}(l_\varepsilon) - Z_\varepsilon^{(00)}(r_\varepsilon) = (l_\varepsilon - r_\varepsilon)\sigma_\varepsilon^{(11)} = 0.$$

This formula defines $\sigma_\varepsilon^{(11)}$, which should not be confused with one of the variables in equation (4). Those variables do not play a prominent role in the computations to come. They are important for the momodromy covering of the versal deformation, as noted by Riemenschneider and studied by Brohme [3]. There is a large covering, which induces the monodromy covering of each reduced component, obtained by considering the $\sigma_\varepsilon^{(i)}$ as elementary symmetric functions in new variables. For details we refer to [3].

We have to give the other equations. They will have the form

$$L_{\delta-1}R_{\varepsilon+1} = P_{\delta,\varepsilon}.$$

The polynomials $P_{\delta,\varepsilon}$ will be well defined modulo the ideal J, generated by the equations of the base space. To describe them we perform division with remainder.

Definition 4.1. We inductively define polynomials $Z_\varepsilon^{(ij)}$ in the variable z_ε, starting from $Z_\varepsilon^{(00)} = z_\varepsilon^{a_\varepsilon} + \sigma_\varepsilon^{(1)} z_\varepsilon^{a_\varepsilon - 1} + \cdots + \sigma_\varepsilon^{(a_\varepsilon)}$, by division by L_ε

$$(5) \qquad Z_\varepsilon^{(ij)} = L_\varepsilon Z_\varepsilon^{(i+1,j)} + \sigma_\varepsilon^{(i+1,j)},$$

and by R_ε

$$(6) \qquad Z_\varepsilon^{(ij)} = Z_\varepsilon^{(i,j+1)} R_\varepsilon + \sigma_\varepsilon^{(i,j+1)}.$$

Note that $\sigma_\varepsilon^{(i+1,j)} = Z_\varepsilon^{(ij)}(l_\varepsilon)$, and $\sigma_\varepsilon^{(i,j+1)} = Z_\varepsilon^{(ij)}(r_\varepsilon)$. From the equations (5) or (6) we obtain by substituting that

$$(7) \qquad \sigma_\varepsilon^{(i+1,j)} - \sigma_\varepsilon^{(i,j+1)} = (l_\varepsilon - r_\varepsilon)\sigma_\varepsilon^{(i+1,j+1)}.$$

The condition $Z_\varepsilon^{(00)}(l_\varepsilon) = Z_\varepsilon^{(00)}(r_\varepsilon) = 0$ translates into $\sigma_\varepsilon^{(10)} = \sigma_\varepsilon^{(01)} = 0$ and we can write

$$Z_\varepsilon^{(00)} = L_\varepsilon Z_\varepsilon^{(10)} = Z_\varepsilon^{(01)} R_\varepsilon.$$

The next line in the pyramid of equations can now be computed:

$$L_{\varepsilon-2}R_{\varepsilon+1} = \frac{(L_{\varepsilon-2}R_{\varepsilon-1})(L_{\varepsilon-1}R_{\varepsilon+1})}{L_{\varepsilon-1}R_\varepsilon} = \frac{Z_{\varepsilon-1}^{(00)} Z_\varepsilon^{(00)}}{L_{\varepsilon-1}R_\varepsilon} = Z_{\varepsilon-1}^{(10)} Z_\varepsilon^{(01)}.$$

For the higher lines we do not quite proceed as before, when describing the components. Computing with $L_{\delta-1}R_{\varepsilon+1} = (L_{\delta-1}R_\varepsilon)(L_\delta R_{\varepsilon+1})/(L_\delta R_\varepsilon)$ would be too complicated. Instead we take the asymmetric approach $L_{\delta-1}R_{\varepsilon+1} = (L_{\delta-1}R_\varepsilon)(L_{\varepsilon-1}R_{\varepsilon+1})/(L_{\varepsilon-1}R_\varepsilon) = P_{\delta,\varepsilon-1}Z_\varepsilon^{(01)}/L_{\varepsilon-1}$.

We do the next step:

$$L_{\varepsilon-3}R_{\varepsilon+1} = \frac{P_{\varepsilon-2,\varepsilon-1}Z_\varepsilon^{(01)}}{L_{\varepsilon-1}} = \frac{Z_\varepsilon^{(10)}Z_{\varepsilon-1}^{(01)}Z_\varepsilon^{(01)}}{L_{\varepsilon-1}},$$

where we now have to use the division with remainder $Z_{\varepsilon-1}^{(01)} = L_{\varepsilon-1}Z_{\varepsilon-1}^{(11)} + \sigma_{\varepsilon-1}^{(11)}$ of equation (5) to get

$$L_{\varepsilon-3}R_{\varepsilon+1} = Z_{\varepsilon-2}^{(10)}Z_{\varepsilon-1}^{(11)}Z_\varepsilon^{(01)} + \frac{Z_{\varepsilon-2}^{(10)}\sigma_{\varepsilon-1}^{(11)}Z_\varepsilon^{(01)}}{L_{\varepsilon-1}}.$$

This is not the final formula, as we can pull out a factor $L_{\varepsilon-2}$ from $Z_{\varepsilon-2}^{(10)}$ and R_ε from $Z_\varepsilon^{(01)}$ by division with remainder. Doing this successively and then using $L_{\varepsilon-2}R_\varepsilon = L_{\varepsilon-1}Z_{\varepsilon-1}^{(10)}$ gives us

$$
\begin{aligned}
(8) \qquad L_{\varepsilon-3}R_{\varepsilon+1} &= Z_{\varepsilon-2}^{(10)}Z_{\varepsilon-1}^{(11)}Z_\varepsilon^{(01)} + \frac{L_{\varepsilon-2}Z_{\varepsilon-2}^{(20)}\sigma_{\varepsilon-1}^{(11)}Z_\varepsilon^{(01)}}{L_{\varepsilon-1}} \\[1mm]
&\quad + \frac{\sigma_{\varepsilon-2}^{(20)}\sigma_{\varepsilon-1}^{(11)}Z_{\varepsilon 1}^{(01)}}{L_{\varepsilon-1}} \\[1mm]
&= Z_{\varepsilon-2}^{(10)}Z_{\varepsilon-1}^{(11)}Z_\varepsilon^{(01)} + \frac{L_{\varepsilon-2}Z_{\varepsilon-1}^{(20)}\sigma_{\varepsilon-1}^{(11)}Z_\varepsilon^{(02)}R_\varepsilon}{L_{\varepsilon-1}} \\[1mm]
&\quad + \frac{\sigma_{\varepsilon-2}^{(20)}\sigma_{\varepsilon-1}^{(11)}Z_\varepsilon^{(01)}}{L_{\varepsilon-1}} + \frac{L_{\varepsilon-2}Z_{\varepsilon-2}^{(20)}\sigma_{\varepsilon-1}^{(11)}\sigma_\varepsilon^{(02)}}{L_{\varepsilon-1}} \\[1mm]
&= Z_{\varepsilon-2}^{(10)}Z_{\varepsilon-1}^{(11)}Z_\varepsilon^{(01)} + Z_{\varepsilon-2}^{(20)}\sigma_{\varepsilon-1}^{(11)}Z_{\varepsilon-1}^{(10)}Z_\varepsilon^{(02)} \\[1mm]
&\quad + \frac{\sigma_{\varepsilon-2}^{(20)}\sigma_{\varepsilon-1}^{(11)}Z_\varepsilon^{(01)}}{L_{\varepsilon-1}} + \frac{L_{\varepsilon-2}Z_{\varepsilon-2}^{(20)}\sigma_{\varepsilon-1}^{(11)}\sigma_\varepsilon^{(02)}}{L_{\varepsilon-1}}.
\end{aligned}
$$

Further steps are not possible. For the formula to be polynomial we need that the last two summands vanish. We obtain the equations

$$
(9) \qquad \lambda_{\varepsilon-2,\varepsilon-1} := \sigma_{\varepsilon-2}^{(20)}\sigma_{\varepsilon-1}^{(11)} = 0, \qquad \rho_{\varepsilon-1,\varepsilon} := \sigma_{\varepsilon-1}^{(11)}\sigma_\varepsilon^{(02)} = 0
$$

in the deformation variables.

Example (embedding dimension 5). The computations up to now suffice. We get, modulo the ideal of the base space, the same equations as equations (3) in Section 2. To translate in the notation used there, note that there are no variables t_2 and t_4, so we set $l_2 = r_2 = l_4 = r_4 = 0$, and we take $l_3 = 0$, $r_3 = -t_3$. One gets $Z_2^{(k0)} = Z_2^{(a_2-k)}$ and $Z_4^{(0k)} = Z_4^{(a_4-k)}$, $\sigma_2^{(20)} = s_2^{(a_2-1)}$ and $\sigma_4^{(02)} = s_4^{(a_4-1)}$. For $\varepsilon = 3$ we find

$$Z_3^{(00)} = Z_3^{(01)} R_3 = Z_3^{(a_3-1)}(z_3 + t_3) = L_3 Z_3^{(10)} = z_3 \widetilde{Z}_3^{(a_3-1)}$$

and

$$Z_3^{(01)} = L_3 Z_3^{(11)} + \sigma_3^{(11)} = z_3 Z_3^{(a_3-2)} + s_3^{(a_3-1)}.$$

The formula (8) gives $\widetilde{Z}_3^{(a_3-1)}$ as factor in the second summand of the right-hand side of the equation $z_1 z_5 = P_{2,4}$, but the difference with $Z_3^{(a_3-1)}$, as given in the equations (3), lies in the ideal of the base space. Note that in general

$$Z_\varepsilon^{(i+1,j)} - Z_\varepsilon^{(i,j+1)}$$

$$= \left(Z_\varepsilon^{(i+1,j+1)} R_\varepsilon + \sigma_\varepsilon^{(i+1,j+1)} \right) - \left(L_\varepsilon Z_\varepsilon^{(i+1,j+1)} + \sigma_\varepsilon^{(i+1,j+1)} \right)$$

$$= (l_\varepsilon - r_\varepsilon) Z_\varepsilon^{(i+1,j+1)}.$$

The factor $\sigma_3^{(11)}$ in the second summand gives that we can use the equation $(l_3 - r_3)\sigma_3^{(11)} = t_3 s_3^{(a_3-1)} = 0$.

We obtained the equations (9) as necessary condition to find a polynomial $P_{\varepsilon-1,\varepsilon+1}$. We observe that they could be computed before computing $P_{\varepsilon-1,\varepsilon+1}$, as they are the result of suitable substitutions in the right hand side of the equations of the previous line: $\lambda_{\varepsilon-1,\varepsilon} = \sigma_{\varepsilon-1}^{(20)}\sigma_\varepsilon^{(11)}$ is gotten by setting $z_{\varepsilon-1} = l_{\varepsilon-1}$ and $z_\varepsilon = l_\varepsilon$ in $P_{\varepsilon-1,\varepsilon} = Z_{\varepsilon-1}^{(10)} Z_\varepsilon^{(01)}$, while $P_{\varepsilon,\varepsilon+1}$ gives $\rho_{\varepsilon,\varepsilon+1}$ by $z_\varepsilon = r_\varepsilon$ and $z_{\varepsilon+1} = r_{\varepsilon+1}$.

To find the versal deformation in general one has to proceed in the same way for the higher lines of the pyramid. Arndt has shown that this works. As the proof is only written in his thesis [1], we sketch it here.

Theorem 4.2. Let $z_{\delta-1} z_{\varepsilon+1} = p_{\delta,\varepsilon}$, $2 \leq \delta \leq \varepsilon \leq e - 1$, be the quasi-determinantal equations for a cyclic quotient singularity X of embedding dimension e. There exists a deformation $L_{\delta-1} R_{\varepsilon+1} = P_{\delta,\varepsilon}$ of these equations over a base space, whose ideal J has $\dim T_X^2 = (e-2)(e-4)$ generators, being $(l_\varepsilon - r_\varepsilon)\sigma_e^{(11)}$ for $3 \le \varepsilon \le e-2$, $\lambda_{\delta,\varepsilon}$ for $2 \le \delta < \varepsilon \le e-2$, and $\rho_{\delta,\varepsilon}$ for $3 \le \delta < \varepsilon \le e-1$. The polynomials $P_{\delta,\varepsilon}$ can be determined inductively, followed by $\lambda_{\delta,\varepsilon} = P_{\delta,\varepsilon}|_{z_\beta = l_\beta}$ and $\rho_{\delta,\varepsilon} = P_{\delta,\varepsilon}|_{z_\beta = r_\beta}$, where $\delta \le \beta \le \varepsilon$. This deformation is versal.

Sketch of proof. To find $P_{\delta,\varepsilon}$ we have to express the product $L_{\delta-1}R_{\varepsilon+1}$ in the local ring in terms of variables with indices between δ and ε. We assume that we already have the equations $L_{\beta-1}R_{\gamma+1} = P_{\beta,\gamma}$ for $\gamma - \beta < \varepsilon - \delta$, and also the base equations formed from them. Let $I_{\delta,\varepsilon}$ be the ideal of all these equation. Obviously $P_{\delta,\varepsilon}$ has to satisfy

(10) $$L_\beta R_\gamma P_{\delta,\varepsilon} \equiv P_{\beta+1,\varepsilon}P_{\delta,\gamma-1} \bmod I_{\delta,\varepsilon}$$

for all β, γ, and it can be determined from any of these equations. The other ones then follow. For the actual computation (following [3]) we use $\beta = \varepsilon - 1$ and $\gamma = \varepsilon$, but now we take $\beta = \delta$, $\gamma = \varepsilon$, so the right hand side of equation (10) becomes $P_{\delta,\varepsilon-1}P_{\delta+1,\varepsilon}$. We perform successively division with remainder by L_β and find

$$P_{\delta,\varepsilon-1} = \sum_{\beta=\delta}^{\varepsilon-1} P_{\delta,\varepsilon-1}^{(\beta)} L_\beta,$$

without remainder because of the equation $\lambda_{\delta,\varepsilon-1}$. Now we use the congruences

$$L_\beta P_{\delta+1,\varepsilon} \equiv L_\delta P_{\beta+1,\varepsilon},$$

whose validity one sees upon multiplying with $R_{\varepsilon+1}$. We conclude that

$$P_{\delta,\varepsilon-1}P_{\delta+1,\varepsilon} \equiv L_\delta \left(\sum P_{\delta,\varepsilon-1}^{(\beta)} P_{\beta+1,\varepsilon} \right).$$

Likewise, from

(11) $$P_{\delta+1,\varepsilon} = \sum Q_{\delta+1,\varepsilon}^{(\beta)} R_\beta,$$

we get, using $\rho_{\delta+1,\varepsilon}$, that

$$P_{\delta,\varepsilon-1}P_{\delta+1,\varepsilon} \equiv R_\varepsilon \left(\sum Q_{\delta+1,\varepsilon}^{(\beta)} P_{\delta,\beta-1} \right).$$

Arndt proves that, if a polynomial is divisible by L_δ and by R_ε, then it is divisible by the product $L_\delta R_\varepsilon$. To check the statement it suffices to do it for the special fibre (according to [1, 1.2.2]). One notes that the ideal $I_{\delta,\varepsilon}$ defines a flat deformation of the product of a certain cyclic quotient singularity in the variables $z_\delta, \ldots, z_\varepsilon$ with a smooth factor of the remaining coordinates, so $z_\delta = u^{n'}$, $z_\varepsilon = v^{n'}$ for a certain n'. Here indeed it holds, that if a polynomial is divisible by $u^{n'}$ and by $v^{n'}$, then it is divisible by the product $(uv)^{n'}$. Therefore there exists a polynomial $P_{\delta,\varepsilon}$ with $L_\delta R_\varepsilon P_{\delta,\varepsilon} = P_{\delta,\varepsilon-1}P_{\delta+1,\varepsilon}$.

We do not know very much about $P_{\delta,\varepsilon}$. We know that over the Artin component $Z^{(01)}_{\varepsilon-1}$ is divisible by $L_{\varepsilon-1}$, so we can define inductively $P_{\delta,\varepsilon}|_{\mathrm{AC}} = P_{\delta,\varepsilon-1}|_{\mathrm{AC}} Z^{(01)}_\varepsilon / L_{\varepsilon-1}$. Restricted to the Artin component, the difference between the so defined $P_{\delta,\varepsilon}|_{\mathrm{AC}}$ and $P_{\delta,\varepsilon}$ from above lies in the restriction of the ideal $I_{\delta,\varepsilon}$. By induction the elements of this ideal extend in the correct way, so we can use them to change $P_{\delta,\varepsilon}$, so that its restriction is equal to $P_{\delta,\varepsilon}|_{\mathrm{AC}}$.

To show flatness we lift the relations. On the Artin component we have the quasi-determinantal relations, which come in two types, depending on the use of the bottom or top line of the quasi-matrix. We give the lift for one type, the other being similar. On the Artin component one has the relation

$$L_{\gamma-1}(L_{\delta-1}R_{\varepsilon+1} - P_{\delta,\varepsilon}) = L_{\delta-1}(L_{\gamma-1}R_{\varepsilon+1} - P_{\gamma,\varepsilon}) + \frac{P_{\gamma,\varepsilon}}{R_\gamma}(L_{\delta-1}R_\gamma - P_{\delta,\gamma-1}),$$

so using an expansion like (11) for $P_{\gamma,\varepsilon}$ we find modulo the ideal $I_{\delta,\varepsilon}$ the relation

$$L_{\gamma-1}(L_{\delta-1}R_{\varepsilon+1}-P_{\delta,\varepsilon}) \equiv L_{\delta-1}(L_{\gamma-1}R_{\varepsilon+1}-P_{\gamma,\varepsilon})+\sum Q^{(\beta)}_{\gamma,\varepsilon}(L_{\delta-1}R_\beta-P_{\delta,\beta-1}).$$

For versality one needs firstly the surjectivity of the map of the Zariski tangent space of our deformation to T^1_X, and secondly the injectivity of the obstruction map $\mathrm{Ob} : (J/\mathfrak{m}J)^* \to T^2_X$, where J is the ideal of the base space. That we cover all possible infinitesimal deformations, is something we have already said and used; for a proof (which requires an explicit description of T^1_X), see [8], [1] or [11]. We neither give here an explicit description of T^2_X (see [1]). For the map Ob one starts with a map $l : J/\mathfrak{m}J \to \mathcal{O}_X$ and exhibits the following function on relations: consider a relation $\boldsymbol{r}$, i.e., $\sum f_i r_j = 0$, which lifts to $\sum F_i R_i = \sum g_j q_j$, where the g_j are the generators of the ideal J. Then $\mathrm{Ob}(l)(\boldsymbol{r}) = \sum l(g_j)q_j \in \mathcal{O}_X$. From our description of the relations we see that the equation $\rho_{\delta+1,\varepsilon}$ occurs for the first time when lifting the relation

$$L_\delta(L_{\delta-1}R_{\varepsilon+1} - P_{\delta,\varepsilon}) = L_{\delta-1}(L_\delta R_{\varepsilon+1} - P_{\delta+1,\varepsilon}) + \frac{P_{\delta+1,\varepsilon}}{R_{\delta+1}}(L_{\delta-1}R_{\delta+1} - P_{\delta,\delta}).$$

This more or less shows that one really needs all equations for the base space. ∎

Note that this result indeed determines the ideal of the base space, but does not give explicit formulas for specific generators. Looking at the equations, say for the total space, one might recognise the numbers $[1, 2, 1]$ and $[2, 1, 2]$, suggesting that an explicit formula can be somehow given in terms of Catalan combinatorics. To show that the situation is more complicated, we will derive the equations of the next line.

We compute the polynomial $P_{\varepsilon-3,\varepsilon}$. It will be more complicated than formula (8), so better notation is desirable to increase readability. Following Brohme [3] we use a position system. In stead of the complicated symbols $Z_{\varepsilon-\gamma}^{(ij)}$ and $\sigma_{\varepsilon-\gamma}^{(ij)}$ we write only the upper index ij; the lower index is not needed, if we write factors with the same $\varepsilon - \gamma$ below each other. To distinguish between $Z_{\varepsilon-\gamma}^{(ij)}$ and $\sigma_{\varepsilon-\gamma}^{(ij)}$ we write the ij representing $Z_{\varepsilon-\gamma}^{(ij)}$ in bold face.

Example. The symbol

$$
\begin{array}{cccc}
\mathbf{30} & \mathbf{20} & \mathbf{10} & \mathbf{03} \\
& 20 & 11 & \\
& 11 & 12 &
\end{array}
$$

represents the monomial $Z_{\varepsilon-3}^{(30)}\sigma_{\varepsilon-2}^{(11)}\sigma_{\varepsilon-2}^{(20)}Z_{\varepsilon-2}^{(20)}\sigma_{\varepsilon-1}^{(12)}\sigma_{\varepsilon-1}^{(11)}Z_{\varepsilon-1}^{(10)}Z_{\varepsilon}^{(03)}$.

A factor $L_{\varepsilon-\gamma}$ in the denominator will be represented by $\overline{L}$, whereas an extra factor $L_{\varepsilon-\gamma}$ in the numerator will be written in bold face. We start from $P_{\varepsilon-3,\varepsilon-1}Z_{\varepsilon}^{(01)}/L_{\varepsilon-1}$, being the sum of two terms. These will be transformed using the division with remainder (5) and (6) and previous equations. One gets some terms, which occur in the final answer, and some terms, which will be transformed again. Terms that will be transformed, are written in italics, and should be considered erased, when transformed. So $P_{\varepsilon-3,\varepsilon-1}Z_{\varepsilon}^{(01)}/L_{\varepsilon-1}$ is, modulo previous equations, equal to the sum of the not italicised terms up to a line in italics, plus the terms in that line (disregarding all text in between). The final result consists of a polynomial of 8 terms, (P.1)–(P.8), and two types of terms with $L_{\varepsilon-1}$ in the denominator, called (L.1)–(L.4) and (R.1)–(R.4). Now we start:

$$(\text{I}.1)\quad \begin{array}{cccc} 10 & 11 & 01 & 01 \\ & & \overline{L} & \end{array} \qquad + \qquad (\text{I}.2)\quad \begin{array}{cccc} 20 & 10 & 02 & 01 \\ & 11 & \overline{L} & \end{array}$$

$$(\text{P}.1)\quad \begin{array}{cccc} 10 & 11 & 11 & 01 \end{array} \qquad + \qquad (\text{P}.2)\quad \begin{array}{cccc} 20 & 10 & 12 & 01 \\ & 11 & & \end{array}$$

$$(\text{I}.3)\quad \begin{array}{ccc} 10 & 11 & 01 \\ & 11 & \\ & \overline{L} & \end{array} \qquad + \qquad (\text{I}.4)\quad \begin{array}{ccc} 20 & 10 & 01 \\ & 11 & 12 \\ & & \overline{L} \end{array}$$

Now take out factors $L_{\varepsilon-2}$ and R_{ε}, giving $L_{\varepsilon-1}Z^{(10)}_{\varepsilon-1}$ and two remainders:

$$(\text{P}.3)\quad \begin{array}{cccc} 10 & 21 & 10 & 02 \\ & & 11 & \end{array} \qquad + \qquad (\text{P}.4)\quad \begin{array}{cccc} 20 & 20 & 10 & 02 \\ & & 11 & 12 \end{array}$$

$$(\text{I}.5)\quad \begin{array}{cccc} 10 & & & 01 \\ & 21 & 11 & \\ & & \overline{L} & \end{array} \qquad + \qquad (\text{I}.6)\quad \begin{array}{ccc} 20 & & 01 \\ & 20 & 12 \\ & 11 & \overline{L} \end{array}$$

$$(\text{R}.1)\quad \begin{array}{cccc} & L & & \\ 10 & 21 & & \\ & & 11 & 02 \\ & & \overline{L} & \end{array} \qquad + \qquad (\text{R}.2)\quad \begin{array}{ccc} & L & \\ 20 & 20 & \\ & 11 & 12 & 02 \\ & & \overline{L} \end{array}$$

Taking out $L_{\varepsilon-3}$ and R_{ε} from (I.5) and (I.6) gives $Z^{(10)}_{\varepsilon-2}Z^{(01)}_{\varepsilon-1}$ and two remainders:

$$(\text{I}.7)\quad \begin{array}{cccc} 20 & 10 & 01 & 02 \\ & 21 & 11 & \\ & & \overline{L} & \end{array} \qquad + \qquad (\text{I}.8)\quad \begin{array}{cccc} 30 & 10 & 01 & 02 \\ & 20 & 12 & \\ & 11 & \overline{L} & \end{array}$$

$$(\text{L}.1)\quad \begin{array}{cccc} & & 01 & \\ 20 & 21 & 11 & \\ & & \overline{L} & \end{array} \qquad + \qquad (\text{L}.2)\quad \begin{array}{cccc} & & & 01 \\ 30 & 20 & 12 & \\ & 11 & \overline{L} & \end{array}$$

$$(\text{R}.3)\quad \begin{array}{cccc} & L & & \\ 20 & & & \\ & 21 & 11 & 02 \\ & & \overline{L} & \end{array} \qquad + \qquad (\text{R}.4)\quad \begin{array}{cccc} & L & & \\ 30 & & & \\ & 20 & 12 & 02 \\ & 11 & \overline{L} & \end{array}$$

Now we follow the steps of the computation of $P_{\varepsilon-2,\varepsilon}$.

$$
(\text{P.5}) \quad
\begin{matrix}
20 & 10 & 11 & 02 \\
 & 21 & 11 &
\end{matrix}
\qquad + \qquad
(\text{P.6}) \quad
\begin{matrix}
30 & 10 & 11 & 02 \\
 & 20 & 12 & \\
 & 11 & &
\end{matrix}
$$

$$
(\text{I.9}) \quad
\begin{matrix}
20 & 10 & & 02 \\
 & 21 & 11 & \\
 & & 11 & \\
 & & L &
\end{matrix}
\qquad + \qquad
(\text{I.10}) \quad
\begin{matrix}
30 & 10 & & 02 \\
 & 20 & 11 & \\
 & 11 & 12 & \\
 & & L &
\end{matrix}
$$

$$
(\text{P.7}) \quad
\begin{matrix}
20 & 20 & 10 & 03 \\
 & 21 & 11 & \\
 & & 11 &
\end{matrix}
\qquad + \qquad
(\text{P.8}) \quad
\begin{matrix}
30 & 20 & 10 & 03 \\
 & 20 & 11 & \\
 & 11 & 12 &
\end{matrix}
$$

$$
(\text{L.3}) \quad
\begin{matrix}
20 & & & 02 \\
 & 20 & 11 & \\
 & 21 & 11 & \\
 & & L &
\end{matrix}
\qquad + \qquad
(\text{L.4}) \quad
\begin{matrix}
30 & & & 01 \\
 & 20 & 11 & \\
 & 20 & 12 & \\
 & 11 & L &
\end{matrix}
$$

$$
(\text{R.5}) \quad
\begin{matrix}
 & L & & \\
20 & 20 & & \\
 & 21 & 11 & 03 \\
 & & 11 & \\
 & & L &
\end{matrix}
\qquad + \qquad
(\text{R.6}) \quad
\begin{matrix}
 & L & & \\
30 & 20 & & \\
 & 20 & 11 & 03 \\
 & 11 & 12 & \\
 & & L &
\end{matrix}
$$

Remarks. 1. It is easy to see that the terms (L.i) vanish modulo the ideal J: the terms (L.1) and (L.2) vanish because of the equation $\lambda_{\varepsilon-3,\varepsilon-1}$, and the terms (L.3) and (L.4) both vanish by equation $\lambda_{\varepsilon-2,\varepsilon-1}$.

The terms (R.i) are more difficult. Taken together, (R.2) and (R.5) vanish by equation $\rho_{\varepsilon-2,\varepsilon}$. The term (R.1) vanishes by $\rho_{\varepsilon-1,\varepsilon}$, as does less evidently (R.3). For (R.6) one uses $\lambda_{\varepsilon-2,\varepsilon-1}$. The term (R.4) is the most complicated. We multiply $\rho_{\varepsilon-2,\varepsilon}$ with $\sigma^{(20)}_{\varepsilon-2}$ to get ${}^{20\ 12\ 02}_{\ \ 11} + {}^{20\ 11\ 03}_{21\ 11}$, in which the second summand vanishes by equation $\lambda_{\varepsilon-2,\varepsilon-1}$.

2. The term (P.8) lies in the ideal, so one could leave it out. However, to have a general formula it is better to keep it.

3. Arndt [1] gives a slightly different, more symmetric result (without showing his computation). He has also 8 terms, (P.8) is missing and (P.4) is replaced by two terms, which together are equivalent to it, modulo the base ideal:

$$
\begin{matrix}
20 & 20 & 01 & 02 \\
 & 11 & 12 &
\end{matrix}
\qquad + \qquad
\begin{matrix}
20 & 20 & 02 & 02 \\
 & 11 & 11 &
\end{matrix}
$$

In our computation we work systematically from the right to the left. Once we take out a factor $L_\gamma R_\varepsilon$ and replace it by $P_{\gamma+1,\varepsilon-1}$, we basically repeat an earlier computation. Brohme [3] has given an inductive formula for the resulting terms (P.i) and the remainder terms (R.i). The problem lies in showing that the remainder terms lie in the ideal of the base space. This problem was solved by Martin Hamm [6] on the basis of a more direct, combinatorial description of the occurring terms. Each term is represented by a rooted tree, which we draw horizontally (Hamm puts as usually the root at the top). Accordingly we call for length of a tree, what is usually called its height.

We consider as example the term (P.7). We first draw the tree such that each vertex directly correspond to a position in the symbol for (P.7) above, but then we transform it so that the bottom line is straight. This will be the way we draw all trees in the sequel.

We explain how to compute the numbers ij in the symbol from the tree, with the highest node at distance γ from the root giving $Z_{\varepsilon-\gamma}^{(ij)}$, and the other nodes $\sigma_{\varepsilon-\gamma}^{(ij)}$. Given a tree T, the resulting polynomial in these variables will be denoted by $P(T)$. We write $\lambda(T)$ for the corresponding term in $\lambda_{\delta,\varepsilon}$, obtained by putting $z_\beta = l_\beta$, and $\rho(T)$ for the term obtained with $z_\beta = r_\beta$.

Definition 4.3. Let T be a rooted tree. To each node $a \in T$ we associate two numbers, i and j, as follows. The second number j is the number of child nodes of a. Let $p(a)$ be the parent of a; if there exists a node b lying directly above a, let $p(b)$ be its parent. Then the number i is the number of nodes between $p(a)$ and $p(b)$ (with $p(a)$ and $p(b)$ included), or the number of nodes above $p(a)$ (with $p(a)$ included), if there is no node lying above a.

We also represent the remainder terms (R.i) by a tree. Such a term has the form $L_\gamma R_{\delta,\varepsilon}^{(i)}/L_{\varepsilon-1}$. The tree will give $R_{\delta,\varepsilon}^{(i)}$. If a tree T is given, we write $R(T)$ for this polynomial. As example we consider (R.5). We observe that its symbol contains the same pairs of numbers as (P.7), except that some on the top are missing, and that the root represents $\sigma_\varepsilon^{(ij)}$ instead of $Z_\varepsilon^{(ij)}$.

We represent it by the following tree.

To the open nodes (except the root) we do not attach numbers ij, but these nodes do contribute to the numbers ij for the other nodes.

Example. For $P_{\varepsilon-3,\varepsilon}$ we find the following trees

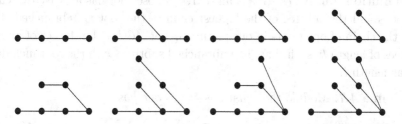

and we have the following trees for the remainder.

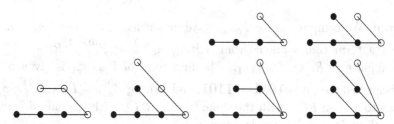

We now characterise the tree representing terms $P(T)$ in the polynomial $P_{\delta,\varepsilon}$, which is obtained as above from $P_{\delta,\varepsilon-1}Z_\varepsilon^{(01)}/L_{\varepsilon-1}$, by working systematically from the right to the left. Let $a \in T$ be a node in a rooted tree, then $T(a)$ is the (maximal) subtree with a as root.

Definition 4.4. An α-tree is a rooted tree satisfying the following property: if two nodes a and b have the same parent, and if b lies above a, then the subtree $T(b)$ is shorter than $T(a)$. By $\mathcal{A}(k)$ we denote the set of all α-trees of length k.

Theorem 4.5. *The polynomial $P_{\delta,\varepsilon}$ is given by*

$$P_{\delta,\varepsilon} = \sum_{T \in \mathcal{A}(\varepsilon-\delta+1)} P(T).$$

This Theorem claims two things: that the α-trees give exactly the polynomial terms in the computation, and secondly that this is really the polynomial $P_{\delta,\varepsilon}$ we are after. To show the second part we have to prove that the remainder terms lie in the ideal generated by the equations of the base space. As we have seen with (R.4) above, the use of the equations leads to terms, which do not occur in the computation itself. We have to characterise the corresponding trees. This leads to the concept of γ-tree (Hamm has also β-trees [6]). We postpone the definition, and first consider a sub-class of the α-trees.

Definition 4.6. An $\alpha\gamma$-tree is an α-tree, whose root has at least two child nodes and the subtree of the highest child of the root is unbranched (this is the chain of open dots in our pictures). Let $\mathcal{AC}(k,l)$ be the set of all $\alpha\gamma$-trees of length k, such that the unbranched subtree (with the root included) has length l.

Lemma 4.7. *Modulo the equations $\lambda_{\gamma,\varepsilon-1}$ one has*

$$P_{\delta,\varepsilon-1}Z_\varepsilon^{(01)}/L_{\varepsilon-1} = \sum_{T\in\mathcal{A}(\varepsilon-\delta)} P(T) + \sum_\gamma \sum_{T\in\mathcal{AC}(\varepsilon-\delta,\varepsilon-\gamma)} L_\gamma R(T)/L_{\varepsilon-1}.$$

Proof. We compute as for $P_{\varepsilon-3,\varepsilon}$. We first consider the rest terms $R(T)$. Such a term comes about from writing $Z_\varepsilon^{(0,k)} = Z_\varepsilon^{(0,k+1)}R_\varepsilon + \sigma_\varepsilon^{(0,k+1)}$. We replace $L_\gamma R_\varepsilon$ by $P_{\gamma+1,\varepsilon-1}$. The first term of $P_{\gamma+1,\varepsilon-1}$ is given by an unbranched tree, it is $\mathbf{1011\cdots1101}$, and writing $Z_{\varepsilon-1}^{(01)} = L_{\varepsilon-1}Z_{\varepsilon-1}^{(11)} + \sigma_{\varepsilon-1}^{(11)}$ leads to a term $P(T)$ with the same tree as $R(T)$ (with the only difference that all nodes are denoted by black dots).

We are left to show that the polynomial terms are exactly those represented by α-trees. This is done by induction. We can construct an α-tree of length $\varepsilon - \delta$ by taking a root, an α-tree of length $\varepsilon - \delta - 1$ as lowest subtree, and as its complement an arbitrary α-tree of length at most $\varepsilon - \delta - 1$ (conversely, given an α-tree of length $\varepsilon - \delta$, the lowest subtree starting from the root, but not including it, is α-tree of length $\varepsilon - \delta - 1$, while its complement has length at most $\varepsilon - \delta - 1$). Doing this in all possible ways gives all α-trees of length $\varepsilon - \delta$. All these trees occur by our construction: in all monomials of $P_{\delta,\varepsilon-1}Z_\varepsilon^{(01)}/L_{\varepsilon-1}$ we simultaneously take out factors L_γ, until we finally are left with $\lambda_{\delta,\varepsilon-1}Z_\varepsilon^{(01)}/L_{\varepsilon-1}$. Each $L_\gamma R_\varepsilon$ is replaced by $P_{\gamma+1,\varepsilon-1}$, and here we repeat the same computation as for $P_{\gamma+1,\varepsilon}$, except that the upper index of $Z_\varepsilon^{(0k)}$ is different. This means that we place all possible trees of length at most $\varepsilon - \gamma - 1$ above the given tree. $\blacksquare$

Remark. The above proof gives an inductive formula for the number of α-trees of length k:

$$\#\mathcal{A}(k) = \#\mathcal{A}(k-1) \cdot \sum_{i=0}^{k-1} \#\mathcal{A}(i).$$

One has $\#\mathcal{A}(0) = 1$, $\#\mathcal{A}(1) = 1$, $\#\mathcal{A}(2) = 2$, $\#\mathcal{A}(3) = 8$ and $\#\mathcal{A}(4) = 96$. As we have seen in the example, some of the terms lie in the ideal J, generated by the equations of the base space. For $k = 4$ already 55 of the 96 terms lie in J, leaving "only" 41 terms. Still this number is considerably larger than the relevant Catalan number (14 in this case).

As already mentioned, the use of ρ-equations brings us outside the realm of α-trees. We retain some properties, which are automatically satisfied for α-trees. The definition becomes rather involved.

Definition 4.8. A γ-tree of length k is a rooted tree satisfying the following properties:

   (i) there is only one node at distance k from the root, and it lies on the bottom line,

  (ii) the number of child nodes of a node at distance d from the root is at most $k - d$,

 (iii) a node a has a child node, if there exists a node b lying above a with the same parent,

 (iv) the root has at least two child nodes and the subtree of the highest child of the root is unbranched.

By $\mathcal{G}(k, l)$ we denote the set of all γ-trees of length k, such that the unbranched subtree (with the root included) has length l.

Example. We consider the term (R.4) above. The sum of the following two terms

is a multiple of the equation $\rho_{\varepsilon-2,\varepsilon}$, and the second graph is not an α-tree.

We have to show that the sum of all remainder terms (i.e., the sum of the $R(T)$ over all $\alpha\gamma$-trees) lies in the ideal J generated by the base equations. We do this by showing that the sum of $R(T)$ over all γ-trees lies in the ideal, as does the sum over all γ-trees which are not α-trees.

Lemma 4.9. *The sum* $\sum_{T\in\mathcal{G}(\varepsilon-\delta,\varepsilon-\gamma)\setminus\mathcal{AG}(\varepsilon-\delta,\varepsilon-\gamma)} R(T)$ *lies in the ideal generated by the* λ-*equations.*

Proof. Let T be a γ-tree, which is not an α-tree. Then there exists a node a, such that the subtree $T(a)$ is an α-tree, but directly above a lies a node b with the same parent, such that the bottom line of the subtree $T(b)$ is at least as long as $T(a)$. Denote by $R\bigl(T|T(a)\bigr)$ the monomial obtained by only multiplying the factors of $R(T)$ corresponding to the nodes lying in $T(a)$. We claim that $R\bigl(T|T(a)\bigr) = \lambda\bigl(T(a)\bigr)$. We have to compute the numbers ij. The second number, of child nodes, is determined by $T(a)$ only. The number i also coincides in $R(T)$ and $P\bigl(T(a)\bigr)$, except when c is a node without nodes above it in $T(a)$. Then its value in $R(T)$ is one more than in $P\bigl(T(a)\bigr)$, so the same as in $\lambda\bigl(P(a)\bigr)$, proving the claim. Replacing $T(a)$ in T by an another α-tree of the same length gives a another γ-tree, which is not an α-tree. So the sum of $R(T)$ over all γ-trees, differing only in the α-tree with root a, is a multiple of a λ-equation. If T has several such subtrees, we consider all possible replacements, and get the product of λ-equations. ∎

The next task is to find terms of ρ-equations in a given tree. For this we introduce the operation of taking away one child node at each highest node. This can be done for any γ-tree.

Definition 4.10. Let T be a γ-tree. We determine inductively a subtree $G(T)$ with the same root as T by the following condition: if $a_1,\ldots,a_p$ are the nodes in $G(T)$ at distance d from the root, then they have the same child nodes in $G(T)$ as in T, except for the highest node a_p, where we take away the highest child node.

Example.

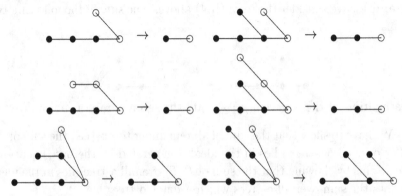

Lemma 4.11. *Suppose $G(T)$ is an α-tree. Then $R\big(T|G(T)\big) = \rho\big(G(T)\big)$ if and only if the number of child nodes in T is at least 1 for every highest lying node in $G(T)$.*

The sum $\sum R(T)$ over all trees satisfying the conditions of the lemma lies in the ideal generated by the ρ-equations. If the number of child nodes in T is at least 1 for every highest lying node in $G(T)$, but $G(T)$ is not an α-tree, then one can find as before a term of a λ-equation.

The most difficult case is when the condition on the number of child nodes is not satisfied. An example is the remainder term (R.6), which is represented by the last two pictures in the previous example. The term contains a factor, which is a term in a λ-equation, but the corresponding dots are not connected by an edge. There is a way to connect the edges differently, bringing a λ-term into evidence. For this we refer to [6, pp. 30–40]. We conclude:

Lemma 4.12. *The sum $\sum_{T \in \mathcal{G}(\varepsilon-\delta,\varepsilon-\gamma)} R(T)$ lies in the ideal generated by the λ and ρ-equations.*

Together with Lemma 4.9 this shows that the remainder

$$\sum_{T \in \mathcal{AG}(\varepsilon-\delta,\varepsilon-\gamma)} R(T)$$

lies in the ideal J, thereby concluding the proof of Theorem 4.5.

Example (The base space for $e = 6$, see [1, 3]). There are 8 equations, which read as

$$(l_3 - r_3)\sigma_3^{(11)}, \qquad (l_4 - r_4)\sigma_4^{(11)},$$

$$\sigma_2^{(20)}\sigma_3^{(11)}, \qquad \sigma_3^{(20)}\sigma_4^{(11)}, \qquad \sigma_3^{(11)}\sigma_4^{(02)}, \qquad \sigma_4^{(11)}\sigma_5^{(02)},$$

$$\sigma_2^{(20)}\sigma_3^{(21)}\sigma_4^{(11)} + \sigma_2^{(30)}\sigma_3^{(20)}\sigma_3^{(11)}\sigma_4^{(12)}, \qquad \sigma_3^{(11)}\sigma_4^{(12)}\sigma_5^{(02)} + \sigma_3^{(21)}\big(\sigma_4^{(11)}\big)^2\sigma_5^{(03)}.$$

We note the relations

$$\sigma_3^{(20)} - \sigma_3^{(11)} = (l_3 - r_3)\sigma_3^{(21)}, \qquad \sigma_4^{(02)} - \sigma_4^{(11)} = (r_4 - l_4)\sigma_4^{(12)}.$$

We can take $\sigma_2^{(20)}$, $\sigma_2^{(30)}$, $l_3 - r_3$, $\sigma_3^{(11)}$, $\sigma_3^{(21)}$, $r_4 - l_4$, $\sigma_4^{(11)}$, $\sigma_4^{(12)}$, $\sigma_5^{(02)}$ and $\sigma_5^{(03)}$ as independent coordinates. The relation with the coordinates in Section 1, see formula (2), is the following: $l_3 - r_3 = t_3$, $\sigma_3^{(11)} = s_3^{(a_3-1)}$, $\sigma_3^{(12)} = s_3^{(a_3-2)}$. Also for $\varepsilon = 2$ and $\varepsilon = 5$ it is simple: $\sigma_\varepsilon^{(ij)} = s_\varepsilon^{(a_\varepsilon-i-j)}$, but for $\varepsilon = 4$ it is more complicated: $r_4 - l_4 = -t_4$, $\sigma_4^{(11)} = s_4^{(a_4-1)}$, while $\sigma_4^{(12)} = \tilde{s}_4^{(a_4-2)}$, where $\tilde{s}_\varepsilon^{(\nu)}$ is defined as following [1, 5.1.1], see also [3, p. 38], by the equality

$$Z_\varepsilon^{(00)} = (z_\varepsilon + t_\varepsilon) \sum_{\nu=0}^{a_\varepsilon-1} s_\varepsilon^{(\nu)} z_\varepsilon^{a_\varepsilon-1-\nu} = z_\varepsilon \sum_{\nu=0}^{a_\varepsilon-1} \tilde{s}_\varepsilon^{(\nu)} (z_\varepsilon + t_\varepsilon)^{a_\varepsilon-1-\nu},$$

where we put $s_\varepsilon^{(0)} = 1$. This implies that

$$\tilde{s}_\varepsilon^{(\nu)} = \sum_{\mu=0}^{\nu} \binom{a_\varepsilon - 2 - \mu}{a_\varepsilon - 2 - \nu} (-t_\varepsilon)^{\mu-\nu} s_\varepsilon^{(a_\varepsilon-1-\mu)}.$$

The primary decomposition gives five reduced components and one embedded component. The five components are parametrised by the five sparse coloured triangles of height 2.

$$\sigma_3^{(11)} = \sigma_4^{(11)} = 0$$

$$\sigma_2^{(20)} = l_3 - r_3 = \sigma_4^{(11)} = \sigma_4^{(12)} = 0$$

$$\sigma_3^{(11)} = \sigma_3^{(21)} = r_4 - l_4 = \sigma_5^{(02)} = 0$$

$$\sigma_2^{(20)} = \sigma_2^{(30)} = l_3 - r_3 = r_4 - l_4 = \sigma_4^{(11)} = \sigma_5^{(02)} = 0$$

$$\sigma_2^{(20)} = l_3 - r_3 = \sigma_3^{(11)} = r_4 - l_4 = \sigma_5^{(02)} = \sigma_5^{(03)} = 0$$

The embedded component is supported at $\sigma_2^{(20)} = l_3 - r_3 = \sigma_3^{(11)} = r_4 - l_4 = \sigma_4^{(11)} = \sigma_5^{(02)} = 0$, which is the locus of singularities of embedding dimension 6.

The primary decomposition, given in the example above, holds if all a_ε are large enough, meaning that $a_\varepsilon \geq \max(k_\varepsilon)$, where the k_ε depend on the possible triangles. By openness of versality one deduces the structure for all cyclic quotient singularities of the given embedding dimension. The formulas for the base space and the total space of the deformation hold in all cases, with suitable interpretations. We note that $Z_\varepsilon^{(ij)}$ is a monic polynomial in z_ε of degree $a_\varepsilon - i - j$, obtained by division with remainder. Therefore $Z_\varepsilon^{(ij)} = 1$ if $i + j = a_\varepsilon$ and $Z_\varepsilon^{(ij)} = 0$ if $i + j > a_\varepsilon$. For the remainder terms $\sigma_\varepsilon^{(ij)}$ we find therefore that $\sigma_\varepsilon^{(ij)} = 1$ if $i + j = a_\varepsilon + 1$ and $\sigma_\varepsilon^{(ij)} = 0$ if $i + j > a_\varepsilon + 1$. By using these values the formulas hold. Also the description of the reduced components holds in general, if one takes an equation $1 = 0$ to mean that the component is absent.

Example (The cone over the rational normal curve $[1, 3]$).

Proposition 4.13. *For the cone over the rational normal curve of degree* $e - 1$ *the versal deformation is given by the equations* $L_{\delta-1}R_{\varepsilon+1} = P_{\delta,\varepsilon}$ *with* $P_{\varepsilon,\varepsilon} = Z_\varepsilon^{(00)}$ *and*

$$P_{\delta,\varepsilon} = Z_\delta^{(10)} Z_\varepsilon^{(01)} + \sum_{\gamma=1}^{\varepsilon-\delta-1} \sigma_{\varepsilon-\gamma}^{(11)} Z_{\delta+\gamma}^{(10)}$$

for $\varepsilon - \delta > 0$.

Proof. We derive the formula from Hamm's description of the equations (i.e., Theorem 4.5). All terms in the equations containing $Z_\varepsilon^{(ij)}$ with $i + j > 2$ and $\sigma_\varepsilon^{(ij)}$ with $i + j > 3$ are absent. We characterise the remaining α-trees. We have of course a simple chain, giving rise to the first term in the formula. Suppose we have a factor $\sigma_{\varepsilon-\gamma}^{(11)}$, coming from a node a on the bottom line. Then there is a node lying directly above it, having the same parent. The unique child node of a has $i = 2$. If it has itself a child node, then necessarily $ij = 21$, and there is a node lying above it. This process continues until we come to an end node with $ij = 20$. If the parent of a is not the root, then necessarily $ij = 12$ for it, so there is a node lying above it. For the node above a we find then that $i = 2$, and there lies a whole chain above the chain starting with this node. In this way we proceed to the root, which has $ij = 02$. We find that the tree has the following shape: from each node on the right of the node a originates a chain of maximal length. An

example is

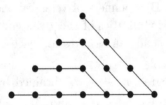

Finally we observe that the lowest lying child node of the root or of a node with $ij = 12$ cannot have $i = 2$, and that the next to last node on the left cannot have $ij = 12$, so there has to be a node a with $ij = 11$.

Consider now $P(T)$, if T is not a chain. The only node with $ij = 10$ lies as end-node on the highest chain, so indeed $P(T) = \sigma_{\varepsilon-\gamma}^{(11)} Z_{\delta+\gamma}^{(10)}$. ∎

It follows that

$$\lambda_{\delta,\varepsilon} = \sum_{\gamma=0}^{\varepsilon-\delta-1} \sigma_{\varepsilon-\gamma}^{(11)} \sigma_{\delta+\gamma}^{(20)}$$

and

$$\rho_{\delta,\varepsilon} = \sigma_{\delta}^{(11)} \sigma_{\varepsilon}^{(02)} + \sum_{\gamma=1}^{\varepsilon-\delta-1} \sigma_{\varepsilon-\gamma}^{(11)} \sigma_{\delta+\gamma}^{(11)}.$$

With $l_\varepsilon - r_\varepsilon = t_\varepsilon$, $\sigma_\varepsilon^{(11)} = s_\varepsilon$, $\sigma_\varepsilon^{(20)} = s_\varepsilon + t_\varepsilon$ and $\sigma_\varepsilon^{(02)} = s_\varepsilon - t_\varepsilon$ we get the same formulas as Arndt gives [1, 5.1.4].

Note that $\left(\sigma_\varepsilon^{(11)}\right)^3 = \sigma_\varepsilon^{(11)} \rho_{\varepsilon-1,\varepsilon+1} - \sigma_{\varepsilon-1}^{(11)} \rho_{\varepsilon-1,\varepsilon}$, so $\sigma_\varepsilon^{(11)}$ lies in the radical of the ideal for $3 < \varepsilon < e - 1$; for $2 < \varepsilon < e - 2$ one has a formula with λ-equations. So indeed the Artin component is the only component, if $e > 5$.

Other applications of the explicit equations include

- the discriminant of the components and adjacencies, studied by Christophersen [4] and Brohme [3],
- embedded components. For low embedding dimension Brohme found all components. He made a general conjecture [3, 4.4].

5. Reduced Base Space

The ideal J of the base space is described explicitly by Theorems 4.2 and 4.5. We have to determine the radical $\sqrt{J}$ of this ideal. We are able to do this explicitly for low embedding dimension, and formulate a conjecture in general. We prove that the proposed ideal describes the reduced components. The combinatorics involved resembles that described by Jan Christophersen in his thesis [5]. To prove the conjectural part one has to show that the monomials we give below, really lie in $\sqrt{J}$, something we do not do here.

Example ($e = 6$ continued). We multiply $\rho_{3,5}$, the last one of the 8 equations for the base space, by $\sigma_4^{(11)}$. Then the first summand contains the factors $\sigma_4^{(11)}\sigma_5^{(02)}$ so lies in the ideal J. Therefore also the second term $\sigma_3^{(21)}\big(\sigma_4^{(11)}\big)^3\sigma_5^{(03)}$ lies in J, and $\sigma_3^{(21)}\sigma_4^{(11)}\sigma_5^{(03)}$ lies in the radical $\sqrt{J}$. Then also the first summand of $\rho_{3,5}$ lies in $\sqrt{J}$. If we multiply $\lambda_{2,4}$ with $\sigma_4^{(11)}$, then the second summand lies in J. We find that the first summand of $\lambda_{2,4}$ lies in the radical, so also the second summand. One has $\sigma_3^{(11)}\big(\sigma_3^{(20)} - \sigma_3^{(11)}\big) = \sigma_3^{(11)}(l_3 - r_3)\sigma_3^{(21)}$, which lies in the ideal, so not only the second summand $\sigma_2^{(30)}\sigma_3^{(20)}\sigma_3^{(11)}\sigma_4^{(12)}$, but also $\sigma_2^{(30)}\big(\sigma_3^{(11)}\big)^2\sigma_4^{(12)}$ and therefore $\sigma_2^{(30)}\sigma_3^{(11)}\sigma_4^{(12)}$ lie in $\sqrt{J}$. We find the following equations

$$(l_3 - r_3)\sigma_3^{(11)}, \qquad (l_4 - r_4)\sigma_4^{(11)},$$

$$\sigma_2^{(20)}\sigma_3^{(11)}, \qquad \sigma_3^{(20)}\sigma_4^{(11)}, \qquad \sigma_3^{(11)}\sigma_4^{(02)}, \qquad \sigma_4^{(11)}\sigma_5^{(02)},$$

$$\sigma_2^{(20)}\sigma_3^{(21)}\sigma_4^{(11)}, \qquad \sigma_2^{(30)}\sigma_3^{(11)}\sigma_4^{(12)}, \qquad \sigma_3^{(11)}\sigma_4^{(12)}\sigma_5^{(02)}, \qquad \sigma_3^{(21)}\sigma_4^{(11)}\sigma_5^{(03)}.$$

This ideal is not reduced, as it contains $\big(\sigma_3^{(11)}\big)^2\sigma_4^{(11)} = \sigma_3^{(11)}\sigma_3^{(20)}\sigma_4^{(11)} - \sigma_3^{(11)}(l_3 - r_3)\sigma_3^{(21)}\sigma_4^{(11)}$, but not $\sigma_3^{(11)}\sigma_4^{(11)}$. But it is easy to find the reduced components from the given equations.

Our first, rough conjecture is that each summand of the equations $\lambda_{\delta,\varepsilon}$, $\rho_{\delta,\varepsilon}$ lies in the radical $\sqrt{J}$. Let us look at $\rho_{\varepsilon-3,\varepsilon}$. We note that (P.4) and (P.5) yield the same term, being $21\,\genfrac{}{}{0pt}{}{21}{11}\,\genfrac{}{}{0pt}{}{11}{12}\,03$ and $21\,\genfrac{}{}{0pt}{}{11}{21}\,\genfrac{}{}{0pt}{}{12}{11}\,03$ respectively. As we have the equation $21\,11\,03$ in the radical, these terms do not contribute new equations. As (P.8) itself already lies in the ideal, we are left with

5 terms (a Catalan number!). One computes that indeed each summand
lies in the radical. We look at the term in $\rho_{\varepsilon-3,\varepsilon}$, coming from (P.6):

$$
\begin{array}{cccc}
31 & 11 & 12 & 03 \\
& 20 & 12 & \\
& 11 & &
\end{array}
$$

We claim that it is associated to the extended triangle

The easiest way to see this is via the numbers k_ε and α_ε, being $[k] =$
$[3,1,2,2]$ and $(\alpha) = (1,3,2,1)$ in this case. One sees that there are α_ε
factors $\sigma_\varepsilon^{(ij)}$, and they all have $i+j = k_\varepsilon + 1$. The other terms can be
parametrised in the same way by the other extended triangles. The same
picture parametrises the term in $\lambda_{\varepsilon-3,\varepsilon}$, coming from (P.6):

$$
\begin{array}{cccc}
40 & 20 & 21 & 12 \\
& 20 & 12 & \\
& 11 & &
\end{array}
$$

In the radical we find $31\ 11\ 12\ 03$ and $40\ 11\ 12\ 12$. For the last term we
compute as follows: $\sigma_{\varepsilon-1}^{(12)}\big(\sigma_{\varepsilon-1}^{(21)} - \sigma_{\varepsilon-1}^{(12)}\big) = \sigma_{\varepsilon-1}^{(12)}(l_{\varepsilon-1} - r_{\varepsilon-1})\sigma_{\varepsilon-1}^{(22)} =$
$\big(\sigma_{\varepsilon-1}^{(11)} - \sigma_{\varepsilon-1}^{(02)}\big)\sigma_{\varepsilon-1}^{(22)}$, and we observe that the term contains the factors $\sigma_{\varepsilon-2}^{(20)}$
and $\sigma_{\varepsilon-2}^{(11)}$.

We can now make our conjecture more precise. As remarked before, we
do not quite get the radical $\sqrt{J}$ of the ideal of the base space, but an
intermediate ideal, obtained from the summands in the generators of J. As
variables we use $l_\varepsilon - r_\varepsilon$, and the $\sigma_\varepsilon^{(ij)}$, which are connected by the relations

$$
\sigma_\varepsilon^{(i+1,j)} - \sigma_\varepsilon^{(i,j+1)} = \big(l_\varepsilon - r_\varepsilon\big)\sigma_\varepsilon^{(i+1,j+1)}.
$$

Conjecture. *For the ideal J of the base space of the versal deformation of
a cyclic quotient singularity of embedding dimension e and its radical $\sqrt{J}$
holds that $\sqrt{J} = \sqrt{J'} \supset J' \supset J$, where J' is the ideal generated by
$(l_\varepsilon - r_\varepsilon)\sigma_\varepsilon^{(11)}$, for $2 < \varepsilon < e-1$ and monomials $\lambda(\triangle_{\delta,\varepsilon})$, $2 \le \delta < \varepsilon < e-1$, and
$\rho(\triangle_{\delta,\varepsilon})$, $2 < \delta < \varepsilon \le e-1$, parametrised by sparse coloured triangles $\triangle_{\delta,\varepsilon}$,*

of the form $\prod_{\beta=\delta}^{\varepsilon} \sigma_{\beta}^{(i_\beta j_\beta)}$. The numbers i_β, j_β are determined as follows: if $\alpha_\beta > 1$, then in both $\lambda(\triangle_{\delta,\varepsilon})$ and $\rho(\triangle_{\delta,\varepsilon})$

$$i_\beta = \#\{\text{black dots on right half-line } l_\varepsilon\}$$

$$j_\beta = \#\{\text{black dots on left half-line } l_\varepsilon\}$$

but if $\alpha_\beta = 1$, then in $\lambda(\triangle_{\delta,\varepsilon})$

$$i_\beta = \#\{\text{black dots on right half-line } l_\varepsilon\} + 1$$

$$j_\beta = \#\{\text{black dots on left half-line } l_\varepsilon\}$$

and in $\rho(\triangle_{\delta,\varepsilon})$

$$i_\beta = \#\{\text{black dots on right half-line } l_\varepsilon\}$$

$$j_\beta = \#\{\text{black dots on left half-line } l_\varepsilon\} + 1$$

Example.

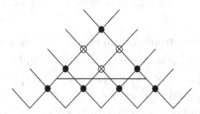

One has $\lambda(\triangle_{\delta,\varepsilon}) = {}_{40}\,{}_{11}\,{}_{22}\,{}_{11}\,{}_{13}$ and $\rho(\triangle_{\delta,\varepsilon}) = {}_{31}\,{}_{11}\,{}_{22}\,{}_{11}\,{}_{04}$.

Remark. The generators of J' correspond to certain terms in generators of J, so there is a special subclass of α-trees, counted by the Catalan numbers. It would be interesting to characterise them. The five trees of length 3 can be seen from the previous pictures. We now list all 14 trees of length 4.

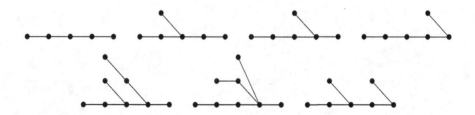

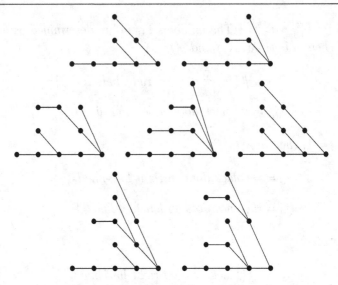

Inductive proofs about the reduced components often use the procedure of blowing up and blowing down [11, 1.1]. The term comes from the analogy with chains of rational curves on a smooth surface, which can be described by continued fractions. For sparse coloured triangles it means the following [3, Lemma 1.8].

Blowing up is a way to obtain an extended triangle of height $e - 2$ from an extended triangle of height $e - 3$. Choose an index $2 \leq \varepsilon \leq e$. We define a shift function $s : \{2, \ldots, e-1\} \to \{2, \ldots, \varepsilon-1\} \cup \{\varepsilon+1, \ldots, e\}$ by $s(\beta) = \beta$ if $\beta \in \{2, \ldots, \varepsilon - 1\}$ and $s(\beta) = \beta + 1$ if $\beta \in \{\varepsilon, \ldots, e - 1\}$. The blow-up $\mathrm{Bl}_\varepsilon(\underline{\triangle})$ of $\underline{\triangle}$ at the index ε is the triangle with $\big(s(\beta), s(\gamma)\big) \in B\big(\mathrm{Bl}_\varepsilon(\underline{\triangle})\big)$ if and only if $(\beta, \gamma) \in B(\underline{\triangle})$, and from the points on the line l_ε only $(\varepsilon - 1, \varepsilon)$ and $(\varepsilon, \varepsilon+1)$ are black. If $\varepsilon = 2$, then only $(2, 3)$ is black, while only $(e - 1, e)$ is black if $\varepsilon = e$. By deleting the base line we get the blow-up $\mathrm{Bl}_\varepsilon(\triangle)$. In terms of pictures this means that one moves the sector, bounded by l_ε and $l_{\varepsilon-1}$ with lowest point $(\varepsilon - 1, \varepsilon)$, one position up, and moves the arising two triangles sideways, to make room for a new line l_ε, which has no black dots in $\mathrm{Bl}_\varepsilon(\triangle)$. If $\varepsilon = 2$ or $\varepsilon = e$ one just adds an extra line without black dots to the triangle.

Example.

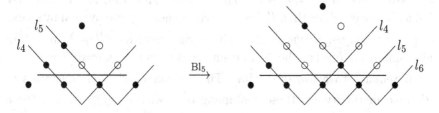

The inverse process is called blowing down at ε. This is possible at ε, for $2 < \varepsilon < e$, if the dot $(\varepsilon - 1, \varepsilon + 1)$ is black; by Lemma 3.2 the line l_ε does not contain black dots. Actually, if l_ε is empty, but $\alpha_\varepsilon > 1$, i.e., there are black dots above it, then it follows that $(\varepsilon - 1, \varepsilon + 1)$ is black: otherwise there cannot be enough black dots in a triangle with black vertex on the lowest level.

Proposition 5.1. *The ideal J' has $C_{e-3} = \frac{1}{e-2}\binom{2(e-3)}{e-3}$ reduced components.*

Proof. If $\sigma_\varepsilon^{(11)} = 0$ for all $2 < \varepsilon < e - 1$, the equations are satisfied: if a triangle $\triangle$ contains black dots, there has to be at least one on the base line, at $(\varepsilon - 1, \varepsilon + 1)$, so $\sigma_\varepsilon^{(11)} = 0$ for that ε; an equation $\lambda(\triangle)$ for an empty triangle ends with a $\sigma_\varepsilon^{(11)}$ for some $\varepsilon < e - 1$, and $\rho(\triangle)$ starts with a $\sigma_\varepsilon^{(11)}$ for some $\varepsilon > 2$. So the Artin component is a component.

Suppose now that there exists an ε with $\sigma_\varepsilon^{(11)} \neq 0$. Let J'_ε be the saturation of J' by $\sigma_\varepsilon^{(11)}$, i.e., $J'_\varepsilon = \cup\left(J' : \left(\sigma_\varepsilon^{(11)}\right)^i\right)$. It yields the equation $l_\varepsilon - r_\varepsilon = 0$, so $\sigma_\varepsilon^{(20)} = \sigma_\varepsilon^{(11)} = \sigma_\varepsilon^{(02)}$. We conclude that $\sigma_{\varepsilon-1}^{(20)} = \sigma_{\varepsilon-1}^{(11)} = 0$ and $\sigma_{\varepsilon+1}^{(11)} = \sigma_{\varepsilon+1}^{(02)} = 0$ (if $\varepsilon = 2$ or $\varepsilon = e - 1$ the statements have to be modified somewhat). As $\sigma_{\varepsilon-1}^{(20)} - \sigma_{\varepsilon-1}^{(11)} = \sigma_{\varepsilon-1}^{(21)}(l_{\varepsilon-1} - r_{\varepsilon-1})$, one has $\sigma_{\varepsilon-1}^{(21)}(l_{\varepsilon-1} - r_{\varepsilon-1}) \in J'_\varepsilon$, and likewise $\sigma_{\varepsilon+1}^{(12)}(l_{\varepsilon+1} - r_{\varepsilon+1}) \in J'_\varepsilon$.

Consider a monomial $\lambda(\triangle_{\beta,\gamma})$ or $\rho(\triangle_{\beta,\gamma})$, containing $\sigma_\varepsilon^{(11)}$ (or $\sigma_\varepsilon^{(20)}$ if $\beta = \varepsilon$, or $\sigma_\varepsilon^{(02)}$ if $\varepsilon = \gamma$). Then $\triangle_{\beta,\gamma} = \mathrm{Bl}_\varepsilon(\triangle_{\beta,\gamma-1})$, and the monomial in question is obtained from $\lambda(\triangle_{\beta,\gamma-1})$ or $\rho(\triangle_{\beta,\gamma-1})$ by leaving the $\sigma_\delta^{(ij)}$ unchanged for $\delta < \varepsilon - 1$, replacing $\sigma_{\varepsilon-1}^{(ij)}$ by $\sigma_{\varepsilon-1}^{(i+1,j)}$, inserting $\sigma_\varepsilon^{(11)}$, replacing $\sigma_\varepsilon^{(ij)}$ by $\sigma_{\varepsilon+1}^{(i,j+1)}$ and $\sigma_\delta^{(ij)}$ by $\sigma_{\delta+1}^{(ij)}$ for $\delta > \varepsilon$. We claim that the polynomials considered so far, together with the monomials, not involving $\sigma_{\varepsilon-1}^{(ij)}$, $\sigma_\varepsilon^{(ij)}$ and $\sigma_{\varepsilon+1}^{(ij)}$ at all, generate the ideal J'_ε. It follows then that this ideal, up to renaming the coordinates as above, and up to some linear equations, is an ideal of the same type as J', but one embedding dimension lower. By induction we conclude that J'_ε describes components, parametrised by sparse coloured triangles, blown up at ε. By varying ε, with $\sigma_\varepsilon^{(11)} \neq 0$, we obtain all components (except the Artin component, which we already have).

It remains to prove the claim. The not yet considered generators of J' come in two types, those containing $\sigma_\varepsilon^{(ij)}$ with $i + j > 2$, and those not containing a $\sigma_\varepsilon^{(ij)}$ at all, but ending with $\sigma_{\varepsilon-1}^{(ij)}$ or starting with $\sigma_{\varepsilon+1}^{(ij)}$. Regarding the first type, we prove that such a monomial is a multiple of one of the claimed generators, by induction on the length of the monomial. For this we note that the claim holds for the monomial if and only if it holds for the monomial, obtained by blowing down the triangle at δ with $\delta \neq \varepsilon-1, \varepsilon+1$. The base of the induction is the case of monomials containing $\sigma_{\varepsilon-1}^{(20)}, \sigma_{\varepsilon-1}^{(11)}, \sigma_{\varepsilon+1}^{(11)}$ or $\sigma_{\varepsilon+1}^{(02)}$. As to the second type, we consider those starting with $\sigma_{\varepsilon+1}^{(ij)}$. If the term is of the form $\lambda(\triangle\varepsilon + 1, \gamma)$, then it starts with $\sigma_{\varepsilon+1}^{(i0)}$. One has $\sigma_{\varepsilon+1}^{(i0)} = \sigma_{\varepsilon+1}^{(i-1,1)} + \sigma_{\varepsilon+1}^{(i1)}(l_{\varepsilon+1} - r_{\varepsilon+1})$. The term obtained by replacing $\sigma_{\varepsilon+1}^{(i0)}$ by $\sigma_{\varepsilon+1}^{(i1)}$, is one of our generators. We are left with monomials, starting with $\sigma_{\varepsilon+1}^{(i-i,1)}$; such monomials also come from $\rho(\triangle\varepsilon + 1, \gamma)$. For those the claim is again shown by induction, using blowing down. ∎

REFERENCES

[1] Arndt, Jürgen, *Verselle Deformationen zyklischer Quotientensingularitäten,* Diss. Hamburg, 1988.

[2] Behnke, Kurt and Riemenschneider, Oswald, *Quotient Surface Singularities and Their Deformations,* in: Singularity theory (Trieste, 1991), 1–54, World Sci. Publ., River Edge, NJ, 1995.

[3] Brohme, Stephan, Monodromieüberlagerung der versellen Deformation zyklischer Quotientensingularitäten Diss. Hamburg 2002. URN: urn:nbn:de:gbv:18-6733.

[4] Christophersen, Jan, *On the components and discriminant of the versal base space of cyclic quotient singularities,* in: Singularity theory and its applications, Part I (Coventry, 1988/1989), 81–92, Lecture Notes in Math., **1462**, Springer, Berlin, 1991.

[5] Christophersen, Jan, *The Combinatorics of the Versal Base Space of Cyclic Quotient Singularities,* in: Obstruction spaces for rational singularities and deformations of cyclic quotients. Thesis, Oslo s.a.

[6] Hamm, Martin, Die verselle Deformation zyklischer Quotientensingularitäten: Gleichungen und torische Struktur. Diss. Hamburg 2008. URN: urn:nbn:de:gbv:18-37828.

[7] Kollár, J. and N. I. Shepherd-Barron, *Threefolds and deformations of surface singularities.* Invent. math. **91** (1986), 299–338.

[8] Riemenschneider, Oswald, Deformationen von Quotientensingularitäten (nach zyklischen Gruppen), *Math. Ann.,* **209** (1974), 211–248.

[9] Riemenschneider, Oswald, Zweidimensionale Quotientensingularitäten: Gleichungen und Syzygien, *Arch. Math.,* **37** (1981), 406–417.

[10] Stanley, Richard P., *Enumerative combinatorics. Vol. 2.,* Cambridge Studies in Advanced Mathematics, **62**. Cambridge University Press, Cambridge, 1999.

[11] Stevens, Jan, *On the versal deformation of cyclic quotient singularities,* in: Singularity theory and its applications, Part I (Coventry, 1988/1989), 302–319, Lecture Notes in Math., **1462**, Springer, Berlin, 1991.

Jan Stevens

Matematiska Vetenskaper
Göteborgs universitet
SE 412 96 Göteborg
Sweden

e-mail: stevens@chalmers.se

BOLYAI SOCIETY
MATHEMATICAL STUDIES, 23

Deformations of
Surface Singularities
pp. 203–228.

COMPUTING VERSAL DEFORMATIONS OF SINGULARITIES WITH HAUSER'S ALGORITHM

JAN STEVENS

Hauser's algorithm provides an alternative approach to the computation of versal deformations, not based on step by step extending infinitesimal deformations. We use this method to compute nontrivial examples.

INTRODUCTION

In describing complex analytic spaces locally by equations one needs in general more equations than the codimension. Although linearly independent, these equations will not be algebraically independent, but are related by so called syzygies. Therefore the perturbations of the equations cannot be chosen independently. This makes the description of all possible perturbations into a nontrivial problem.

Normally one extends infinitesimal deformations step for step. This can lead to a never ending computation, which has to be cut off after a finite number of steps. Most calculations on record avoid this problem by only considering deformations of negative degree of quasi-homogeneous singularities, where the base space also has polynomial quasi-homogeneous equations. An alternative method, based on an idea of Teissier's (see [20]), was developed by Herwig Hauser in his thesis [10], see also [12, 11]. Given a singularity $(X_0, 0) \subset (\mathbb{C}^n, 0)$ defined by a system of equations $f = (f_1, \ldots, f_k)$, one first determines the versal unfolding of the map $f : (\mathbb{C}^n, 0) \to (\mathbb{C}^k, 0)$ without bothering about syzygies and then computes the stratum over which one has a deformation of the original singularity. In general the map f is not of finite singularity type, so its versal unfolding is infinite dimensional and one is forced to work in a suitable category of infinite dimensional spaces, e.g., Banach analytic spaces. The result is an infinite system of equations in an infinite number of variables, most of which can be eliminated.

The examples given by Hauser himself mostly concern codimension two determinantal singularities, which can be treated much easier with other methods [15]. Here we study more complicated examples. The original problem we were interested in, was to find the versal deformation of the cone over a hypercanonical curve (such a curve is the intersection of the projective cone over a rational normal curve with a quadric hypersurface). The cone over the rational normal curve of degree 4 itself is first serious test case. The computation is rather involved. It can be short-cut by using the extra information that the versal deformation is only of negative degree. By using similar tricks results for the hypercanonical cone of degree 8 are obtained. For some other singularities we discuss how one in principle can get the versal deformation.

The method embeds the deformation problem in a larger unobstructed one, which is easy to solve. The price one has to pay is the introduction of many new variables. The equations tend to become unmanageable. This is a intrinsic difficulty in deformation computations. The practical limitations of Hauser's algorithm are approximately the same as for the standard method. Without quasi-homogeneity one does not come very long. All this suggests that one should refrain from computing versal deformations, except in very special cases. Only very symmetric equations are suitable for computation – however one may hope that they are representative for the general case.

After a short introduction to versal deformations and a quick description of the algorithm we go into details for some of the steps. The examples we study in detail are the cone over the rational normal curve of degree 4 and the cone over a hypercanonical curve of degree 8. We add some remarks on computations in the multiplicity five case. Finally we show how Hauser's ideas can be used to find deformations of the Stanley–Reisner ring of the icosahedron. We conclude with a discussion of the results.

1. VERSAL DEFORMATIONS

Let $(X_0, 0) \subset (\mathbb{C}^n, 0)$ be the germ of a complex analytic space, given by equations $f_1 = \cdots = f_k = 0$. A *deformation* of X_0 is determined by a deformation of its local ring $\mathcal{O}_{X_0} = \mathbb{C}\{x_1, \ldots, x_n\}/(f_1, \ldots, f_k)$ as $\mathbb{C}$-

algebra. This means that the underlying $\mathbb{C}$-module structure is the same, but the multiplication is perturbed. The $\mathbb{C}$-module structure is particularly easy if $\mathcal{O}_{X_0}$ is Cohen–Macaulay of dimension d: for a suitable choice of coordinates (sufficiently general will do) it is a free $\mathbb{C}\{x_1, \ldots, x_d\}$-module (in fact, this property can be taken as the definition of Cohen–Macaulay). Such an additive realisation can be found by computing a standard basis. Associativity gives a nontrivial constraint on the possible perturbations. However, one mostly uses perturbations of generators of the ideal. The connection with the previous point of view is given by the concept of flatness. Recall that by definition a *deformation* $\pi : X_S \to S$ of X_0 over a base germ $(S, 0)$ is a flat map π such that X_0 is isomorphic to $\pi^{(-1)}(0)$. Flatness of the map π means that $\mathcal{O}_{X_S}$ is a flat $\mathcal{O}_S$-module. In particular, if the map is finite, so X_0 is a fat point, this means that $\mathcal{O}_{X_S}$ is a free $\mathcal{O}_S$-module. Flatness captures the idea that the underlying $\mathbb{C}$-module structure is constant in a deformation over the base space S.

To give a workable definition of flatness we start from a finite free resolution of $\mathcal{O}_{X_0}$

$$(1) \qquad 0 \longleftarrow \mathcal{O}_{X_0} \longleftarrow \mathcal{O}_{\mathbb{C}^n} \overset{f}{\longleftarrow} \mathcal{O}_{\mathbb{C}^n}^k \overset{r}{\longleftarrow} \mathcal{O}_{\mathbb{C}^n}^l,$$

where f is the row vector $(f_1, \ldots, f_k)$ containing the generators of the ideal of X_0 and r is a matrix whose columns generates the module of relations. Each column gives rise to a relation, or *syzygy*, $\sum_i f_i r_{ij} = 0$. We may realise X_S as embedded in $\mathbb{C}^n \times S$. The condition for flatness is now that the resolution (1) of $\mathcal{O}_{X_0}$ lifts to a resolution of $\mathcal{O}_{X_S}$:

$$(2) \qquad 0 \longleftarrow \mathcal{O}_{X_S} \longleftarrow \mathcal{O}_{\mathbb{C}^n \times S} \overset{F}{\longleftarrow} \mathcal{O}_{\mathbb{C}^n \times S}^k \overset{R}{\longleftarrow} \mathcal{O}_{\mathbb{C}^n \times S}^l,$$

or alternatively, that every syzygy $\sum_i f_i r_{ij} = 0$ between the generators of the ideal of X_0 lifts to a syzygy $\sum_i F_i R_{ij} = 0$ between the generators of the ideal of X_S in $\mathbb{C}^n \times S$. A proof of this characterisation of flatness can be found in [9] and [12].

The lifting of relations imposes no extra conditions for complete intersections. In that case the resolution is given by the Koszul complex. The syzygy module is generated by the trivial syzygies $f_i \cdot f_j - f_j \cdot f_i = 0$ and they can be lifted to $F_i \cdot F_j - F_j \cdot F_i = 0$. In particular, every perturbation of the equations f_i defines a deformation. Less obvious is the case of codimension two Cohen–Macaulay singularities (see [15]), which by the Hilbert–Burch theorem have a resolution of the form

$$0 \longleftarrow \mathcal{O}_{X_0} \longleftarrow \mathcal{O}_{\mathbb{C}^n} \overset{\Delta}{\longleftarrow} \mathcal{O}_{\mathbb{C}^n}^{k+1} \overset{r}{\longleftarrow} \mathcal{O}_{\mathbb{C}^n}^k \longleftarrow 0,$$

where Δ may be taken to be the vector of maximal minors of the matrix r. Starting from a set of generators $(f_0, \ldots, f_k)$ one computes a syzygy matrix r whose minors are up to a unit the f_i. Deformations are obtained by perturbing the syzygy matrix.

A *versal* deformation is the most general one, in the sense that all other deformations can be obtained from it. A deformation $X_S \to S$ of X_0 is *versal* if for every deformation $X_T \to T$ there exists a map $\psi : T \to S$ such that X_T is isomorphic to the pull-back $\psi^* X_S$; moreover, if a map $\psi' : T' \to S$ is already given on a subspace $T' \subset T$, then ψ can be chosen as an extension of ψ'. Such a map is not unique, but if it is unique on the level of tangent spaces, the deformation is called semi-universal or *miniversal*. By a theorem of Grauert versal deformations of isolated singularities exist. For an extended discussion see [19].

The isomorphism classes of first order infinitesimal deformations form a vector space, usually called T^1, which (by definition of versality) is also the Zariski tangent space to the base space of the miniversal deformation. In the computer algebra system SINGULAR [8] the computation of T^1 is implemented. The obstructions to extend infinitesimal deformations land in the vector space T^2 (also computable with SINGULAR). Its dimension gives an upper bound for the number of equations defining the base space. In fact, one usually constructs the base space as fibre of a non-linear obstruction map ob : $T^1 \to T^2$.

2. HAUSER'S ALGORITHM

The basic idea is to first describe a minimal perturbation of the equations f_i and only after that be concerned about the relations. To compute the versal unfolding of a map $f : (\mathbb{C}^n, 0) \to (\mathbb{C}^k, 0)$ one determines the tangent space

$$T_f = (f_1, \ldots, f_k) \cdot \mathcal{O}_{\mathbb{C}^n,0}^k + \left(\frac{\partial f}{\partial x_1}, \ldots, \frac{\partial f}{\partial x_n} \right) \cdot \mathcal{O}_{\mathbb{C}^n,0}$$

to the $\mathcal{K}$-orbit of f. Let (g_i) be a monomial basis for the quotient $K_f = \mathcal{O}_{\mathbb{C}^n,0}^k / T_f$. The versal unfolding is then $F : (\mathbb{C}^n \times S, 0) \to (\mathbb{C}^k, 0)$, given by $F(x, s) = f(x) + \sum s_i g_i$, with S isomorphic to K_f. If one starts with a map f, defining a singularity, which is not a complete intersection, then the $\mathbb{C}$-dimension of K_f is in general not finite and one has to take the base space $(S, 0)$ in a suitable category of infinite dimensional spaces, like the category of inductive limits of objects and morphisms of the category of germs of Banach-analytic spaces. In fact, one has to extend the concept of versality to such a category. We do not enter into details, which can be found in Hauser's papers [11, 12].

Let $(X, 0)$ be the fibre of the map F and $\pi : (X, 0) \to (S, 0)$ be the induced morphism. In the present infinite-dimensional situation it still holds that π is flat if and only if the relations $fr = 0$ between the f_i can be lifted to relations $FR = 0$ between the F_i. The maximal subspace $(Z, 0) \subset (S, 0)$ over which π is flat is called the *flattener* of π. It can be determined using standard basis techniques. Given the unfolding F and the relation matrix r one considers Fr and reduces it to normal form, in which only specific monomials in x occur. For flatness the $\mathcal{O}_{S,0}$-coefficients of these monomials have to vanish, which gives us the required equations for $(Z, 0)$. One uses the generalised Weierstraß division theorem, which is not algorithmic. For important special cases an algorithm exists.

Before describing the algorithm in more detail, we give a simple example.

Example 1. Let X_0 be the curve consisting of the three coordinate axes in $\mathbb{C}^3$. We take $f : \mathbb{C}^3 \to \mathbb{C}^3$ given by $f(x, y, z) = (yz, xz, xy)$. By hand or with a computer algebra system (using Gröbner bases with a global ordering, as the singularity is homogeneous) one computes easily the versal unfolding

$$yz + ay + bz + \sum_{i=0} d_i x^i,$$

$$xz + cz + \sum_{i=0} e_i y^i,$$

$$xy + \sum_{i=0} g_i z^i.$$

To write only a finite number of deformation variables we collect the unfolding parameters in power series, writing $D = \sum d_i x^i$, $E = \sum e_i y^i$ and $G = \sum g_i z^i$. Then $F = (yz + ay + bz + D, xz + cz + E, xy + G)$. We lift relations:

$$(xy + G)z - (xz + cz + E)y + (yz + ay + bz + D)c$$

$$= -y(E - ac) + z(G + bc) + Dc$$

$$(xy + G)(z + a) + (xz + cz + E)b - (yz + ay + bz + D)x$$

$$= -xD + z(G + bc) + Ga + Eb$$

The right hand side cannot be reduced further because only pure power of the variables (x, y, z) occur. We take coefficients of x, y and z and conclude that $D = 0$, $E = ac$, so $e_0 = ac$, $e_i = 0$ for $i > 0$, and that $G = -bc$, so $g_0 = -bc$, $g_i = 0$ for $i > 0$. The versal deformation of X_0 is $(yz + ay + bz, xz + cz + ac, xy - bc)$. As the singularity is Cohen–Macaulay of codimension two, it and its versal deformation are determinantal [15] and the defining matrix is given by the lift of the relations:

$$\begin{pmatrix} z & -y & c \\ z+a & b & -x \end{pmatrix}.$$

Hauser's algorithm. *The versal deformation can by computed by the following steps.*

(1) *Write the singularity $(X_0, 0)$ as fibre $(f^{-1}(0), 0)$ of a suitable chosen map $f : (\mathbb{C}^n, 0) \to (\mathbb{C}^k, 0)$.*

(2) *Compute a monomial basis for the quotient $K_f = \mathcal{O}^k_{(\mathbb{C}^n, 0)}/T_f$ and construct the versal unfolding $F : (\mathbb{C}^n \times S, 0) \to (\mathbb{C}^k, 0)$ of f.*

(3) *Compute a basis r of the module of relations between the components of f.*

(4) *Lift by division of the components of Fr the relation matrix r to a matrix R such that each component of FR is in reduced normal form.*

(5) *Solve the infinite system of equations given by the vanishing of the $\mathcal{O}_{(S,0)}$-coefficients of the series FR, to find the base space $(Z, 0) \subset (S, 0)$.*

The vector space T^1 is obtained by solving the linearised equations.

Choosing generators of the ideal of $X_0 \subset \mathbb{C}^n$ realises X_0 as fibre $f^{-1}(0)$ of a map germ $f : \mathbb{C}^n \to \mathbb{C}^k$. A good choice of f simplifies the computations to come. A natural requirement is that the components $(f_1, \ldots, f_k)$ form a standard basis of the ideal. Furthermore, the ring of d-dimensional Cohen–Macaulay singularity should be realised as free $\mathbb{C}\{x_1, \ldots, x_d\}$-module. In the simple example above, of the coordinate axes in $\mathbb{C}^3$, this was not necessary, but in general we will have problems lifting the relations without this assumption. We postpone the description of the normal form to the next section, where we go into details on standard bases.

The infinite system of equations will come out as a finite system of equations between power series with variable coefficients. We give a simple example, of how such a system can look like.

Example 2. Let $C = \sum c_i y^i$ be a power series and consider an equation of the form $(y - b)C = a$, where a and b do not depend on y. We get the infinite system of equations

$$-bc_0 = a$$
$$c_0 - bc_1 = 0$$
$$c_1 - bc_2 = 0$$
$$\vdots$$

We conclude that $c_0 = bc_1 = b^2 c_2 = \cdots$, and as we are looking for a solution in the neighbourhood of the origin we may assume that $|b| < 1$ and $|c_i| < 1$ for all i. We conclude that $c_0 = 0$ and similarly that $c_i = 0$ for all i, and that $a = 0$.

3. Standard Bases

Let $\mathcal{O}$ denote the ring $\mathbb{C}\{x_1, \ldots, x_n\}$. We are interested in submodules M of the free module $\mathcal{O}^q$. Each element $f \in \mathcal{O}^q$ can be written $f = \sum f_{\alpha,i} x^\alpha e_i$ with $f_{\alpha,i} \in \mathbb{C}$, and the e_i a basis of $\mathcal{O}^q$. Choose a multiplicative well-ordering $<$ on the monomials $x^\alpha e_i$, e.g. with help of an injective linear form $L : \mathbb{N}^n \times \{1, \ldots, q\} \to \mathbb{R}_+$. For a non-zero $f \in \mathcal{O}^q$ we define the exponent

$$\exp(f) = \min\{(\alpha, i) \mid f_{\alpha,i} \neq 0\}$$

and its initial form as in $(f) = f_{\alpha,i} x^\alpha e_i$ with $(\alpha, i) = \exp(f)$. One sets $\exp(0) = \infty$ and in $(0) = 0$. For a submodule $M \subset \mathcal{O}^q$ one has semigroup of exponents $\exp(M) = \{\exp(f) \mid f \in M \setminus 0\}$ and the module of initial forms in $(M) = \{\text{in}(f) \mid f \in M\}$. We define

$$\Delta(M) = \{g \in \mathcal{O}^q \mid g_{\alpha,i} = 0 \text{ if } (\alpha, i) \in \exp(M)\}.$$

Definition. A *standard basis* of the module M is a finite collection $(f_1, \ldots, f_k)$ of elements of M, such that their initials forms $(\text{in}(f_1), \ldots, \text{in}(f_k))$ generate the module in (M).

On the polynomial ring $\mathbb{C}[x_1, \ldots, x_n]$ we can consider the highest power and look at the ideal of leading forms. A standard basis in that context is

usually called Gröbner basis. Standard bases exist and can be computed
with the Mora tangent cone algorithm, implemented for example in SINGU-
LAR [8]. This enables us to do step 2 and 3 in the algorithm. To be more
precise, the input has to consist of polynomials and one computes not with
$\mathcal{O}$ but with the localisation $K[x_1, \ldots, x_n]_{\mathfrak{m}}$ of the polynomial ring at the
maximal ideal $\mathfrak{m} = (x_1, \ldots, x_n)$, where $K = \mathbb{Q}$ or a finite field.

In the power series case more can be shown, using the generalised
Weierstraß division theorem, see [7, 12]. For algebraic power series a division
algorithm exists, for the hypersurface case see [1].

Theorem 1.

   a) *Let $(f_1, \ldots, f_k)$ generate M. Every $g \in \mathcal{O}^q$ can be written as*

$$g = \sum m_i f_i + h$$

   with $m_i \in \mathcal{O}$ and a unique rest $h \in \Delta(M)$.

   b) *Equivalently, the natural projection $\Delta(M) \to \mathcal{O}^q/M$ is an isomorphism of vector spaces.*

One may choose generators f_i of M such that $f_i - \mathrm{in}\,(f_i) \in \Delta(M)$.
Indeed, let g_j be any set of generators, whose initial terms generate $\mathrm{in}\,(M)$.
By division $g_j - \mathrm{in}\,(g_j) = \sum m_{ji} g_i + r_j$. Put $f_j = \mathrm{in}\,(g_j) + r_j$ and let
$N \subset M$ be the module, generated by the f_j. Then $\Delta(N) = \Delta(M)$, so by
the Division Theorem $N = M$. In particular the f_i form a standard basis.

There is also a parametric version of the Division Theorem, which even
holds in the infinite dimensional case [12, A.II.4]. Let $(S, 0)$ be a parameter
space. Let M be a submodule of finite type of $\mathcal{O}^q_{S \times \mathbb{C}^n, 0}$ and M_0 its image
in $\mathcal{O}^q_{\mathbb{C}^n, 0} = \mathcal{O}^q$. We now set

$$\Delta(M) = \left\{ g \in \mathcal{O}^q_{S \times \mathbb{C}^n, 0} \mid g_{\alpha, i}(s) = 0 \text{ if } (\alpha, i) \in \exp(M_0) \right\}.$$

Proposition 2.

   a) *The natural projection $\Delta(M) \to \mathcal{O}^q_{S \times \mathbb{C}^n, 0}/M$ is surjective.*

   b) *Equivalently, let $(F_1, \ldots, F_k)$ generate M. Every $G \in \mathcal{O}^q_{S \times \mathbb{C}^n, 0}$ can be
   written as*

$$G = \sum M_i F_i + H$$

   with $H \in \Delta(M)$ and the image h of H in $\Delta(M_0)$ is unique.

To do step 4 of the algorithm we lift a relation fr by applying division with parameters to Fr: we write $\sum F_i r_i = \sum Q_i F_i + H$ with $H \in \Delta(I)$, where I is the ideal generated by the F_i. As $\sum f_i r_i = 0$ we may assume that $Q_i \in \mathfrak{m}_S$. We then define R by $R_i = r_i - Q_i$. In general the division is only "theoretically" constructive, as Hauser puts it [11]. But under special assumptions there is a finite division algorithm.

It is important to choose variables and equations for the singularity appropriately. For a d-dimensional Cohen–Macaulay singularity of multiplicity m one takes a Noether normalisation, i.e., one realises X as finite cover of degree m of a smooth d-dimensional germ. This makes $\mathcal{O}_X$ into a free $\mathbb{C}\{y_1, \ldots, y_d\}$-module with a finite basis consisting of monomials in the variables x_i. We take a compatible ordering, such that the initial forms of the generators of the ideal depend only on the x-variables. For a general singularity this is not possible, but it is very useful to choose coordinates such that the initial forms depend on as few coordinates as possible (although this was not necessary in the simple Example 1).

For zero-dimensional singularities and more generally for Cohen–Macaulay singularities, a finite division algorithm is possible if one increases the number of elements in the standard basis [12, A.II.3]. Suppose that $f_i - \mathrm{in}\,(f_i) \in \Delta(I)$. In general, reducing a monomial $x^m \notin \Delta(I)$ with an f_i introduces higher order terms: if $m = \alpha + \exp(f_i)$ and $x^\beta \in \Delta(I)$ is a monomial occurring in f_i, then we need to reduce $x^{\alpha+\beta}$. The condition, which Hauser gives (and which already occurs with Janet), is now that there exists a f_j with $\alpha + \beta = \alpha' + \exp(f_j)$ such that $\alpha' < \alpha$ in the given ordering. If this condition holds for all $x^\beta \in \Delta(I)$, then we can choose our reductions with smaller and smaller multipliers and the process is finite. A sufficient condition is that all exponents on the staircase of I are exponents of standard basis elements, see also the example below. Each new standard basis element gives rise to many new unfolding parameters.

It is desirable to keep the number of variables as small as possible. In practice the most important condition allowing a finite division algorithm, is that the standard basis is at the same time a Gröbner basis for a suitable global ordering. This means that the initial forms of the equations are also the maximal terms for a different ordering on the monomials. This condition is satisfied for quasi-homogeneous singularities.

Example 3. Consider the Tyurina algebra of the singularity $f(x,y) = x^3 y^2 + xy^5 + x^7$. Our ring is $A = \mathbb{C}\{x,y\}/I$, where $I = \left(f, \frac{\partial f}{\partial x}, \frac{\partial f}{\partial y}\right)$. The ring A defines a fat point of multiplicity 19. A SINGULAR computation gives that $\dim T_A^1 = 26$. As the singularity is Cohen–Macaulay of codimension two, all deformations are determinantal. It is therefore easy to describe the versal deformation; we do not do that here.

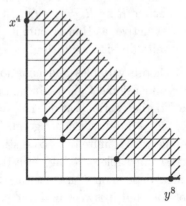

A standard basis $(g_1, \ldots, g_5)$ for I is:

$$\left(2x^3y + 5xy^4, 3x^2y^2 + y^5 + 7x^6, xy^5, x^7, y^8 \right)$$

To compute the versal unfolding of $g : \mathbb{C}^2 \to \mathbb{C}^5$ we determine the initial forms of the ideal $\left(\frac{\partial g}{\partial x}, \frac{\partial g}{\partial y} \right) \subset A^5$:

$$\begin{pmatrix} 2x^3 \\ 0 \\ 0 \\ 0 \\ 0 \end{pmatrix}, \begin{pmatrix} 6x^2y \\ 0 \\ 0 \\ 0 \\ 0 \end{pmatrix}, \begin{pmatrix} 0 \\ 6xy^3 \\ 0 \\ 0 \\ 0 \end{pmatrix}, \begin{pmatrix} 5xy^4 \\ 0 \\ 0 \\ 0 \\ 0 \end{pmatrix}, \begin{pmatrix} 0 \\ x^6 \\ 0 \\ 0 \\ 0 \end{pmatrix}, \begin{pmatrix} 0 \\ y^6 \\ 0 \\ 0 \\ 0 \end{pmatrix}, \begin{pmatrix} y^7 \\ 0 \\ 0 \\ 0 \\ 0 \end{pmatrix}.$$

We need $14 + 12 + 19 + 19 + 19 = 83$ unfolding parameters. To ensure a finite division algorithm one has to complete the standard basis of the ideal I with the exponents on the sides of the staircase (but not in the corners), giving 7 new elements, which increases the number of unfolding parameters with 7×19.

The staircase gets a better form by applying a general linear transformation, for which we can take $y \mapsto x + y$. We compute the new standard basis:

$$\left(3x^4 - 3x^2y^2 + 11x^2y^3 + 10xy^4 - y^5 - 6xy^5 - 13y^6 - 45y^7, \right.$$

$$12x^3y + 12x^2y^2 - 14x^2y^3 - 10xy^4 + 4y^5 - 6xy^5 + 22y^6 + 105y^7,$$

$$\left. 3x^2y^4 + 3xy^5 - 4y^7, \quad xy^6, \quad y^8 \right)$$

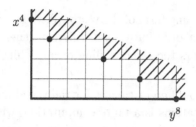

For the global ordering with weights $(2,1)$ these polynomials form a Gröbner basis with the same leading terms. A SINGULAR computation gives

$$\big(12x^3y - 14x^2y^3 - 6xy^5 + 105y^7 + 12x^2y^2 - 10xy^4 + 22y^6 + 4y^5,$$

$$y^8, \quad xy^6, \quad 3x^2y^4 + 3xy^5 - 4y^7,$$

$$3x^4 + 11x^2y^3 - 6xy^5 - 45y^7 - 3x^2y^2 + 10xy^4 - 13y^6 - y^5 \big).$$

The unfolding is by terms of lower weight, so the normal form given by the Gröbner basis leads to the desired lifting.

Example 4. One of the simplest examples of a singularity which is not Cohen–Macaulay is the one-point union of two planes in $\mathbb{C}^4$. We choose coordinates (w, z, y, x) such that the last two are coordinate functions on both planes. We take as generators of the ideal

$$\big(z^2 - xz, zw - xw, w^2 - yw, yz - xw \big),$$

just as in [12, III.2, Ex. 3]. As the singularity is homogeneous we compute with a global ordering, which is the product ordering with the degree reverse lexicographical ordering on the first three coordinates. This gives the following additive representation of the ring:

$$\mathbb{C}\{x, y\} + \mathbb{C}\{x, y\}w + \mathbb{C}\{x\}z.$$

The image of $\big(\frac{\partial f}{\partial w}, \frac{\partial f}{\partial z}, \frac{\partial f}{\partial y}, \frac{\partial f}{\partial x} \big)$ in $(\mathcal{O}/I)^4$ is for a suitable ordering on the generators of the free module generated by

$$\begin{pmatrix} 0 \\ z - x \\ -y \\ 2z - x \end{pmatrix}, \begin{pmatrix} -z + x \\ 0 \\ 0 \\ w - y \end{pmatrix}, \begin{pmatrix} 2z - x \\ w \\ 0 \\ y \end{pmatrix}, \begin{pmatrix} 0 \\ 0 \\ w \\ -z \end{pmatrix}, \begin{pmatrix} x^2 \\ yx \\ y^2 \\ 0 \end{pmatrix}.$$

In particular, the leading term of the vector $(0, z-x, -y, 2z-x)$ is $(0, z, 0, 0)$. By taking a different ordering we get different generators, but we can get from the other vectors the same leading terms, while $(0, z - x, -y, 2z - x)$ gives $(0, 0, 0, 2z)$.

This gives us two different choices for the versal unfolding. For the first one lifting of the relations is a finite computation, whereas the second one is Hauser's example where such an algorithm is impossible. To be more precise, let capital deformation variables stand for power series in x and y, while lower case ones depend only on x. Then the first choice gives the unfolding

$$z^2 - xz + A + bz + (Cy + c)w,$$

$$zw - xw + Dy + d + iz,$$

$$w^2 - yw + e + fz + gy,$$

$$yz - xw + Hy + h.$$

We have split the constant term in the second and fourth equation because the lifting requires to divide them by y. We do not give the results of the lifting. It is not difficult to conclude that the singularity is rigid. We find e.g. two equations $yH - xi + h = 0$ and $xi + h = 0$. As $h - xi$ is independent of y we get by equating coefficients that $H = 0$ and from there $h = i = 0$. Continuing we obtain that all variables have to vanish.

The second choice of the unfolding is almost the same, but the difference is that the term iz occurs in the fourth equation in stead of the second one: $yz - xw + Hy + h + iz$. This is an equation of the form $f = yz - az - b$. There is no finite algorithm to divide a power series $\sum c_i y^i z$ by f. The first reduction gives $c_0 z + b \sum c_i y^{i-1} + a \sum c_i y^{i-1} z$, containing again infinitely many terms divisible by the initial form yz. To obtain the division one has to rewrite $\sum c_i y^i z$ as a power series in $(y - a)$.

4. COMPUTATIONS

4.1. Test case: Pinkham's example

We compute the well-known versal deformation of the cone over the rational normal curve of degree four [14]. We start from the determinantal equations

$$\begin{pmatrix} y_0 & x_1 & x_2 & x_3 \\ x_1 & x_2 & x_3 & y_4 \end{pmatrix}.$$

Projection onto the (y_0, y_4)-plane realises the singularity as finite covering of degree 4. This gives the additive realisation of the ring as

$$\mathbb{C}\{y_0, y_4\} + x_1\mathbb{C}\{y_0, y_4\} + x_2\mathbb{C}\{y_0, y_4\} + x_3\mathbb{C}\{y_0, y_4\}.$$

We compute standard bases with the global product ordering for the coordinates $(x_2, x_1, x_3; y_0, y_4)$, where we take x_2 first to get more symmetric formulas. We write the ideal as

$$x_1^2 - x_2y_0, \qquad x_2^2 - y_0y_4, \qquad x_3^2 - x_2y_4,$$

$$x_1x_2 - x_3y_0, \qquad x_2x_3 - x_1y_4, \qquad x_1x_3 - y_0y_4.$$

We compute the versal unfolding. This can be even done by hand. We use the same indexing system for the unfolding parameters as for the variables. The torus action on the singularity gives a bigrading, for which the variables x_i have weight $(4 - i, i)$, and their index is the second component of the weight. The variables a_i are constants, the b_i are power series in y_0 and c_i, d_i depend on y_0 and y_4. The variables of power series type are indexed using the weight of their constant term. To distinguish between variables of the same weight we also use names like a'. So both a_2 and a_2' are constants, c_1' depends on y_0 and y_4. With this notation the versal unfolding becomes

$$x_1^2 - x_2y_0 + a_1x_1 + c_{-1}x_3 + d_2,$$

$$x_2^2 - y_0y_4 + a_3x_1 + c_1x_3 + d_4,$$

$$x_3^2 - x_2y_4 + b_5x_1 + c_3x_3 + d_6,$$

$$x_1x_2 - x_3y_0 + a_2x_1 + c_0x_3 + d_3,$$

$$x_2x_3 - x_1y_4 + b_4x_1 + c_2x_3 + d_5,$$

$$x_1 x_3 - y_0 y_4 + b_3 x_1 + a_2' x_2 + c_1' x_3 + d_4'.$$

The initial forms all belong to $\mathbb{C}\{x_1, x_2, x_3\}$, so it is easy to lift the relations. The normal form of the lifted relations FR does not contain terms quadratic in the x_i. The equations for the base space are the coefficients of the x_1, x_2, x_3 and 1. As there are 8 relations, we get 32 equations. The equations, which are the constant terms of the lifted relations, are of degree three in the unfolding variables, but they are consequences of the other ones. The other equations are not linearly independent. We end up with 21 equations. Six of them express the d_i in the other variables:

$$d_6 + (y_4 - b_4)(a_2 - a_2') - c_2 b_4 + c_1 b_5 + b_3 a_3 = 0,$$

$$d_5 - y_4 c_1 + a_3 a_2' = 0,$$

$$d_4 - 2y_0 b_4 + c_2^2 - c_1 c_3 + c_0 b_4 - c_1' a_3 + (c_2 - a_2)a_2' = 0,$$

$$d_4' - y_0 b_4 + c_2 a_2' = 0,$$

$$d_3 + (y_0 - c_0)(c_3 - b_3) + (c_2 - 2a_2)c_1' + c_1 a_2' = 0,$$

$$d_2 - y_0 a_2 - y_0 a_2' + c_0 a_2' = 0.$$

We have six equations in which y_4 explicitly occurs:

(3) $\qquad (y_4 - b_4)(c_3 - b_3 + a_3) + c_2 b_5 - b_5 a_2 + b_4 a_3 = 0,$

(4) $\qquad (y_4 - b_4)(c_2 + a_2) + c_1 b_5 + b_3 a_3 = 0,$

(5) $\qquad (y_4 - b_4)(c_1 - c_1' + a_1) - c_2 b_3 + b_4 c_1 + b_3 a_2 = 0,$

(6) $\qquad (y_4 - b_4)c_1' - (y_0 - c_0)b_5 - a_3 a_2' = 0,$

(7) $\qquad y_4 c_0 - y_0 b_4 + c_2 a_2' - a_2 a_2' = 0,$

(8) $\qquad y_4 c_{-1} - y_0 b_3 + c_1' a_2' - a_1 a_2' = 0,$

three equations in which only y_0 explicitly occurs:

$$(y_0 - c_0)(c_3 - b_3 + a_3) + c_2 c_1' - c_1' a_2 + c_0 a_3 = 0,$$

$$(y_0 - c_0)(c_2 + a_2) + c_1 c_1' + c_{-1} a_3 = 0,$$

$$(y_0 - c_0)(c_1 - c_1' + a_1) - c_{-1} c_2 + c_0 c_1 + c_{-1} a_2 = 0,$$

and finally six equations where no deformation variable occurs linearly. These last ones turn out to be implied by the other equations.

We first determine T^1. For this we solve the linearised equations. By taking the coefficients of the monomials in y_4 and y_0 we get infinitely many equations. They can however be written in our notation. From equations (3), (4) and (5) we find $c_3 - b_3 + a_3 = c_2 + a_2 = c_1 - c_1' + a_1 = 0$. The linearisation of equation (6) is $y_4 c_1' - y_0 b_5 = 0$. As b_5 only depends on y_0 we conclude that $c_1' = 0$ and thereafter $b_5 = 0$. Equations (7) and (8) give $c_0 = c_{-1} = b_4 = b_3 = 0$. Altogether we find that all b and c variables vanish except the constant terms of c_1, c_2 and c_3, for which we have $c_i + a_i = 0$. The dimension of T^1 is four: variables are a_1, a_2, a_3 and a_2'.

Now we solve the equations themselves. We use equation (8) to eliminate c_{-1}, then equation (3) to eliminate c_3 and equation (6) for c_1'. We write the remaining three equations (4), (5) (with c_1' eliminated) and (7) in matrix form:

$$
\begin{pmatrix} y_4 - b_4 & b_5 & 0 \\ -b_3 & y_4 & b_5 \\ a_2' & 0 & y_4 \end{pmatrix} \begin{pmatrix} c_2 + a_2 \\ c_1 + a_1 \\ c_0 \end{pmatrix} = \begin{pmatrix} b_5 a_1 - b_3 a_3 \\ y_0 b_5 + b_4 a_1 - 2b_3 a_2 + a_3 a_2' \\ y_0 b_4 + 2a_2 a_2' \end{pmatrix}.
$$

The right-hand side does not depend on y_4. As the coefficient matrix is of the form $y_4 I + A$, the argument of example 2 applies to show that $c_2 + a_2 = c_1 + a_1 = c_0 = 0$. From equations (8), (3) and (6) we then find with the same argument that $c_1' = c_3 + b_3 + a_3 = c_{-1} = 0$. Equation (6) now reduces to $-(y_0 - c_0)b_5 - a_3 a_2' = 0$, from which we conclude that $b_5 = 0$. From equations (7), and (8) we get that $b_4 = b_3 = 0$. The equations now reduce to the expected three quadratic equations $a_3 a_2' = a_2 a_2' = a_1 a_2' = 0$ for the base space. The other equations are then also satisfied. We also determine the d_i.

We have finally obtained the versal deformation. It is

$$x_1^2 - x_2 y_0 + x_1 a_1 + y_0 a_2 + y_0 a_2',$$

$$x_2^2 + x_1 a_3 - x_3 a_1 - y_0 y_4 - a_2^2 + a_1 a_3,$$

$$x_3^2 - x_2 y_4 - x_3 a_3 - y_4 a_2 + y_4 a_2',$$

$$x_2 x_1 + x_1 a_2 - x_3 y_0 + y_0 a_3,$$

$$x_2 x_3 - x_1 y_4 - x_3 a_2 - y_4 a_1,$$

$$x_1 x_3 + x_2 a_2' - y_0 y_4,$$

over the base space given by $a_3 a_2' = a_2 a_2' = a_1 a_2' = 0$.

Remark 5. The Artin component is given by $a_2' = 0$ and the total family over it can be written as determinantal

$$\begin{pmatrix} y_0 & x_1 + a_1 & x_2 + a_2 & x_3 \\ x_1 & x_2 - a_2 & x_3 - a_3 & y_4 \end{pmatrix}.$$

Usually one writes all deformation variables on either the top or bottom line, but there is no reasonable choice of term ordering which gives an unfolding inducing such a deformation.

Remark 6. The computation of T^1 shows that there are only deformations of negative weight. As the used unfolding is quasi-homogeneous, we can restrict to the subspace of negative weight to find the versal deformation. This means that the b_i and c_i are independent of y_4 and y_0, so we have only finitely many variables and we have only to solve the equations obtained from equations (3) to (8) by taking coefficients of y_4, y_0 and 1. This is very easy.

4.2. Hypercanonical cones

The next computation concerns again a homogeneous singularity, but this time there are also deformations of positive weight. We look at cones over non-rational curves. To have the simplest possible equations we consider hyperelliptic curves, embedded with a multiple of the hyperelliptic involution.

So let C be a hyperelliptic curve and let K_0 be the line bundle corresponding to the hyperelliptic pencil. Then $(g + l)K_0$ is ample for $l \geq 1$ and gives the lth transcanonical embedding [6]. In particular for $l = 1$ one gets the hypercanonical embedding. The embedded curve lies on a scroll of type $(g + l, l - 1)$; the hypercanonical curve is the intersection of the projective cone over the rational normal curve of degree $g + 1$ with a quadric hypersurface. Let X be the affine cone over a hypercanonical curve. For the actual computation we will restrict ourselves to the case $g = 3$.

With the same notation as in the previous section we write the cone over the normal curve as determinantal

$$(9) \qquad \begin{pmatrix} y_0 & x_1 & \cdots & x_{g-1} & x_g \\ x_1 & x_2 & \cdots & x_g & y_{g+1} \end{pmatrix}$$

and take as quadric

(10) $Q = z^2 + p_0 y_0^2 + p_1 x_1 y_0 + \cdots + p_g x_g y_0 + p_{g+1} y_0 y_{g+1}$

$$+ p_{g+2} x_1 y_{g+1} + \cdots + p_{2g+1} x_g y_{g+1} + p_{2g+2} y_{g+1}^2.$$

The ring $\mathcal{O}_X$ has the following additive realisation:

$$\mathbb{C}\{y_0, y_{g+1}\}\langle 1, z, x_1, x_1 z, \ldots, x_g, x_g z\rangle.$$

The infinitesimal deformations and the obstruction space for these singularities are known. The dimension of T_X^1 was computed by Drewes [5]:

$$\dim T_X^1(l) = \begin{cases} 1, & l = -2, \\ 3g, & l = -1, \\ 4g - 3, & l = 0, \\ g - 2, & l = 1. \end{cases}$$

Of these are $2(2g - 2)$ deformations of the scroll, while the remaining ones are obtained by perturbing Q. Each deformation (of degree -1) of the scroll gives rise to two deformations of the hypercanonical cone, one of degree -1 given by exactly the same formula, and one of degree 0, obtained by multiplying with z. The perturbations of Q do not involve z. There is one of degree -2, $g + 2$ of degree -1 (perturbing with an arbitrary linear term), while the ones in degree 0 and 1 depend on the equation. For a generic equation one can perturb with $x_2 y_0, \ldots, x_{g-1} y_{g+1}$ and $x_2 y_0 y_{g+1}, \ldots, x_{g-1} y_0 y_{g+1}$ (this is the result of a computation for the special case $z^2 + y_0^2 + y_{g+1}^2$).

The dimension of T^2 is easily determined with the methods of [4]; in fact the version of that paper contained in Jan Christophersen's thesis contains a discussion of these singularities. One has $\dim T_X^2(-2) = \dim T_X^2(-1) = g(g - 2)$ and the isomorphism between these two groups is multiplication with z. There is a deformation to two singularities, both isomorphic to the cone over the rational normal curve of degree $g + 1$ (simply take $Q = t$) and in this deformation the dimension of T^2 is constant. This shows that the number of equations for the base space is equal to the dimension of T^2.

We now apply Hauser's algorithm. The versal unfolding of the system of equations (9) and (10) can be obtained from the versal unfolding of the equations for the cone over the rational normal curve, together with perturbations of the quadric. In particular, for $g = 3$ we use the results of the previous subsection. Each unfolding parameter $a_i, \ldots, d_i$ there doubles into $A_i + a_i z, \ldots, D_i + d_i z$. The quadric can be deformed independently of the cone, so the unfolding is the same as the deformation given above. In particular, there are only finitely many parameters in the unfolding of the quadric.

The quadric is only involved in Koszul relations, so it suffices to lift the determinantal relations. This is not so easy, as there are many variables. We can use the fact that we have written our ring as a finite $\mathbb{C}\{y_0, y_{g+1}\}$-module: we consider y_0 and y_{g+1} also as unfolding parameters, and make a coordinate transformation on the parameter space. We write the perturbation of Q, given by equation (10), as

$$z^2 + s_1 x_1 + \cdots + s_{g+1} x_{g+1} + S.$$

The term $p_0 y_0^2$ in Q is subsumed in S, whereas $p_1 x_1 y_0$ contributes to $s_1 x_1$. We can regard this as deformation of the non-reduced singularity given by the determinantal (9) and $Q = z^2$. By introducing the weights $\text{wt} \, x_i = \text{wt} \, y_0 = \text{wt} \, y_{g+1} = 3$, $\text{wt} \, z = 2$, $\text{wt} \, s_i = 1$, $\text{wt} \, S = 4$ and weights 3 and 1 for the unfolding parameters occurring in the determinantal equations we get quasi-homogeneous equations with all perturbations of negative weight. The lift of the relations is now easy. The equations for the base space are obtained from the coefficients of $1, z, x_1, x_1 z, \ldots, x_g, x_g z$.

We have done the computation in the case $g = 3$. One has to take all unfolding parameters into account. It is not difficult to solve the linearised equations. This gives indeed T_X^1 as described above. But the equations themselves are difficult to handle. To get a finite number of variables one can restrict to the subspace of deformations of weight at most 0. In this case this means forgetting the 1-dimensional $T_X^1(1)$, which only corresponds to an equisingular deformation. But this reduction is not sufficient. One could expect that the versal deformation does not really depend on the moduli of the curve. Therefore one can try to compute without the deformation variables, which only change the p_i in the quadric (10), and even use a particular simple form for Q, namely $z^2 + y_0^2 + y_4^2$. It turns out that many variables can be eliminated, and one comes almost down to the dimension of the relevant subspace of T_X^1, For one or two variables one finds a polynomial equation with nontrivial linear part. The remaining equations contain very many monomials. We have not succeeded in completing the computation.

Although the obtained equations were very complicated, we expected a very simple result. Not only does the singularity deform into two cones over the rational normal curve of degree 4, it also deforms into the 2-star singularity [13] (perturb Q with a general linear form); this singularity is the double cover of the cone over the rational normal curve of degree four, branched in a general hyperplane section. By the results of De Jong and Van Straten [13] the base space of a rational quadruple point depends up to a smooth factor only on one discrete invariant of the singularity. So the expected result is that the hypercanonical cone with $g = 3$ has the same base space as the 2-star singularity. The equations can be obtained by a simple procedure from the three base space equations for the cone over the rational normal curve. Recall that each variable a_i doubled into $A_i + a_i z$. We put $z^2 + S = 0$, and look at the coefficients of 1 and z (this is like taking the real and imaginary parts of the product of two complex numbers). The expected ideal of the base space is then generated by

$$A_2' a_3 + A_3 a_2', \quad A_2' a_2 + A_2 a_2, \quad A_2' a_1 + A_1 a_2,$$
$$A_3 A_2' - a_3 a_2' S, \quad A_2 A_2' - a_2 a_2' S, \quad A_1 A_2' - a_1 a_2' S.$$

These formulas are independent of the moduli p_i. In fact, this should also be the equations for the infinite dimensional base space of the nonreduced singularity with $Q = z^2$ (i.e., all $p_i = 0$). As the equations do not involve the s_i we can let the s_i be power series in the y-variables. But S has to be independent of the space variables, so we also introduce (power series) variables s_0 and s_4. We compute with the deformation of Q given by

$$z^2 + x_2 s_2 + x_1 s_1 + x_3 s_3 + y_0 s_0 + y_4 s_4 + S.$$

With the weights used above for the lifting (wt x_i = wt y_0 = wt y_4 = 3, wt $z = 2$, wt $s_i = 1$, wt $S = 4$ and weights 3 and 1 for the other variables) we have now quasi-homogeneous equations with all variables of positive weight. In contrast to the previous computation, where variables of weight 0 did not allow to control the degree of the polynomials involved, the degrees are now fixed and we have less variables. There are now only 44 deformation parameters.

We do not get the expected base equations, but rather complicated expressions. As deterring example we write the last one:

$$2A_1 A_2' - 2a_1 a_2' S + 2A_1 (a_2')^2 s_2 + a_1^2 (a_2')^2 s_1 s_2 - a_1 a_2^2 a_2' s_2^2 + a_1^2 a_3 a_2' s_2^2$$

$$- 2a_1 (a_2')^3 s_2^2 + 4A_2 (a_2')^2 s_3 + 2a_1 (a_2')^3 s_1 s_3 + 2a_2^2 (a_2')^2 s_2 s_3$$

$$- 3a_1 a_3 (a_2')^2 s_2 s_3 - 4a_2 (a_2')^3 s_2 s_3 - 2a_3 (a_2')^3 s_3^2 + 2A_3 (a_2')^2 s_4$$

$$- 2a_2^2(a_2')^2 s_1 s_4 + 2a_1 a_3(a_2')^2 s_1 s_4 - 2a_3(a_2')^3 s_2 s_4 = 0.$$

But one sees that it starts with the expected terms. The other equation involving A_1 is

$$A_2' a_1 + A_1 a_2' - (a')^2(a_1 s_2 + 2a_2 s_3 - a_3 s_4) = 0$$

and an easy coordinate transformation brings it in the desired form. After a similar transformation for A_2 and A_3 one only needs to transform S to get all equations right. Now we apply the transformation to the equations itself. The change in the quadric can be undone with a coordinate transformation in the space variables. The scrollar equations remain quite complicated. We give only the quadric

$$z^2 + x_2 s_2 + x_1 s_1 + x_3 s_3 + S + y_0 s_0 + y_4 s_4$$

and the first of the scrollar equations:

$$x_1\big(x_1 + za_1 + A_1 + a_2'(a_1 s_2 + 2a_2 s_3 - a_3 s_4)\big) + x_3 a_1 a_2' s_4$$

$$- (x_2 - za_2 - za_2' - A_2 - A_2' - a_2 a_2' s_2)y_0 + 2\big(za_2 - A_2 + a_2^2 s_2\big)(a_2')^2 s_4.$$

In these formulas we still have to replace the s_i to reintroduce the higher degree terms.

4.3. Multiplicity five

We did some computations for quasi-homogeneous singularities of multiplicity five. By restricting to negative degrees one ensures that there are only finitely many unfolding parameters. As their number tends nevertheless to be large, computations have to be done with a computer algebra system, like SINGULAR [8]. The calculation itself requires only a few lines of SINGULAR. In fact, creating the necessary ring is the most involved part. Given a system of generators f of the ideal, one first computes a standard basis for the jacobian module in the local ring. From this one determines the unfolding F. One then computes a basis r for the relation module. One then reduces the product Fr to normal form with the ideal f, and takes coefficients of the monomials. We did this in two cases, where the result was already known by other means.

- For the cone over the rational curve of degree 5 the algorithm easily gives the versal deformation in negative degrees. To understand the result it is important that the unfolding parameters have names reflecting their degree under the torus action. A general formula, valid for all multiplicities, was already given by Jürgen Arndt in his thesis [3] using different methods. He also first solves an unobstructed problem and takes care of flatness afterwards.

- The monomial curve $\left(t^5, t^6, t^7, t^8, t^9\right)$ of minimal multiplicity 5 occurs as hyperplane section of rational singularities. With a different method the base space was determined in [17]. The computation with Hauser's algorithm is similar to that for the cone over the rational normal curve. It leads to 68 equations of degree at most two for the base space, in 67 variables. The dimension of T^1 is 19. One can eliminate 48 variables with 48 equations with non-zero linear part to get 20 equations in 19 variables, but this introduces terms of higher degree, and the 20 equations involve many monomials more than the original 67 equations. It is important to choose monomial orders and the form of the unfolding in such a way that the perturbed equations of lowest degree are as simple as possible. Then also the base equations of lowest degree will be manageable.

We did not try the hypercanonical cone of genus 4. We note that it deforms into a 2-star singularity of multiplicity 5. Such a rational singularity has many smoothing components, cf. [19, Ch. 14], so the base equations are complicated, and involve probably the perturbations of degree -1 of the quadric. Therefore the trick of the case $g = 3$ does not work, and finding the versal deformation with any method seems out of reach.

5. THE STANLEY–REISNER RING OF THE ICOSAHEDRON

One way to study algebraic surfaces is to look at the boundary of their moduli space. Of particular interest are highly singular surfaces, which consist of a union of planes. They have a very easy local structure, but can can be quite complicated globally. The combinatorics is captured in a simplicial complex. In particular, the icosahedron occurs, if one considers semistable degenerations of $K3$-surfaces. Looking for degenerations with icosahedral symmetry [18] one can easily describe several possible special fibres, but it is difficult to describe the total family. One of the simplest candidates is given by the Stanley–Reisner ring of the icosahedron. Here we describe how one gets many equations for the deformation, again using many additional variables.

We first recall how one can associate a projective scheme to a simplicial complex [16]. So let Δ be a simplicial complex with set of vertices $V = \{v_1, \ldots, v_n\}$. A monomial on V is an element of $\mathbb{N}^V$. Each subset of V determines a monomial on V by its characteristic function. The support of a monomial $M : V \to \mathbb{N}$ is the set $\operatorname{supp} M = \{v \in V \mid M(v) \neq 0\}$. The set Σ_Δ of monomials whose support is not a face is an ideal, generated by the monomials corresponding to minimal non-simplices. Consider now the polynomial ring $K[V]$ over an algebraically closed field K. The monomials in the Σ_Δ generate the Stanley–Reisner ideal $I_\Delta \subset K[V]$. The *Stanley–Reisner ring* is $A_\Delta = K[V]/I_\Delta$.

Consider now the icosahedron as simplicial complex. The associated projective scheme X consists of 20 projective planes in $\mathbb{P}^{11}$ intersecting each other just as the faces of the icosahedron. We use use variables $x_0, \ldots, x_{11}$. Unfortunately there does not seem to be a numbering system for the vertices of the icosahedron, which makes the symmetries evident. We refer to the figure for our numbering. The antipodal involution is $x_i \mapsto x_{11-i}$.

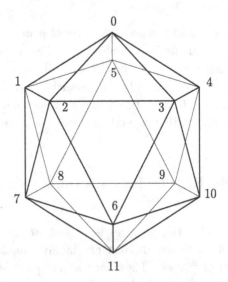

We look at infinitesimal deformations of the affine cone $C(X)$ over our projective scheme X, see also [18, § 5.9]. To find them one can apply the results of [2] or compute directly. All deformations are unobstructed ($T^2_{C(X)} = 0$). We have $T^1_{C(X)}(\nu) = 0$ for $\nu < 0$ and $\dim T^1_{C(X)}(0) = 30$.

The deformations of degree zero give embedded deformations of X. The dimension of $H^0(\Theta_X)$ equals 11, which fits with the fact that X deforms to smooth $K3$-surfaces $(30 - 11 = 19)$ of degree 20 and genus 11.

The Stanley–Reisner ideal is generated by 36 monomials, six corresponding to pairs of opposite vertices, like x_0x_{11}, and 30 monomials like x_0x_6. The infinitesimal deformations can be given by

$$(11) \qquad x_0x_6 + \delta_{0,6}x_2x_3, \qquad x_0x_{11},$$

while the other equations can be obtained by using the group action. By taking all deformation variables equal we get a icosahedral invariant deformation. When adding higher order terms the equations become complicated. Using the fact that the ideal is invariant under the full symmetry group of the icosahedron the first equation can be written as

$$
\begin{aligned}
x_0x_6 + &\delta_1 x_2 x_3 \\
&+ \delta_2\left(x_0^2 + x_6^2\right) + \delta_3\left(x_2^2 + x_3^2\right) + \delta_4\left(x_5^2 + x_{11}^2\right) + \delta_5\left(x_8^2 + x_9^2\right) \\
&+ \delta_6\left(x_1^2 + x_4^2 + x_7^2 + x_{10}^2\right) + \delta_7 x_8 x_9 + \delta_8(x_1 x_7 + x_4 x_{10}) \\
&+ \delta_9(x_0 x_2 + x_0 x_3 + x_2 x_6 + x_3 x_6) + \delta_{10}(x_0 x_5 + x_6 x_{11}) \\
&+ \delta_{11}(x_0 x_1 + x_0 x_4 + x_6 x_7 + x_6 x_{10}) \\
&+ \delta_{12}(x_1 x_5 + x_4 x_5 + x_7 x_{11} + x_{10} x_{11}) \\
&+ \delta_{13}(x_1 x_2 + x_2 x_7 + x_3 x_4 + x_3 x_{10}) \\
&+ \delta_{14}(x_5 x_8 + x_5 x_9 + x_{11} x_8 + x_{11} x_9) \\
&+ \delta_{15}(x_1 x_8 + x_7 x_8 + x_4 x_9 + x_{10} x_9),
\end{aligned}
$$

where all δ_i are power series in δ_1. With a coordinate transformation one can achieve that $\delta_9 = 0$. For the other type of equation we find three more deformation variables.

To apply Hauser's method we have to lift relations. This process is not finite, but we can make it finite by adding extra equations, namely all 272 monomials of degree 3 lying in the ideal. The icosahedral action divides them into six types. Each equation can be perturbed with 92 monomials. Due to the symmetry 246 extra deformation variables suffice. An example of a new equation and the corresponding relation to be lifted is

$$\left(x_0x_6^2\right) \cdot 1 - (x_0x_6) \cdot x_6 = 0.$$

The lift can be done in one step. We end up with many quadratic equations in $246 + 17$ variables. By solving the linear part of these equations we find indeed the infinitesimal deformation (11) with all δ_{ij} equal.

To do the computation we first made a list of all group elements, acting by permutations of the coordinates. We then generated a sample perturbed equation for each type. In process of lifting we let the group elements act on these sample equations. The equations are the coefficients of the 92 monomials in the lifts of the different types of relations; again by symmetry it suffices to look at a few samples.

The 92 monomials come in three orbits of the group: there are 12 third powers like x_0^2, 20 face monomials like $x_0 x_1 x_2$ and 60 monomials supported on the edges, like $x_0^2 x_1$. The equations split correspondingly into three sets, and each of the additional 246 variables occurs only in one set. We already find a 1-dimensional infinitesimal deformation by only computing the linearisation of the equations coming from the edge monomials. Further simplification comes from the induced action of the antipodal involution on the deformation variables. It suffices to look at the invariant equations. One has also the action of the whole group. The invariant equations do not give sufficiently many equations, but the complementary equations do. After eliminating some variables, which occur only linearly, we are left with 102 inhomogeneous quadratic equations in $63 + 17$ variables, each consisting of circa 40 monomials. Although an enormous simplification from the original system of equations, it is still too difficult to eliminate the additional variables. We have the equations of the curve of icosahedral invariant $K3$-surfaces, but we are unable to find its properties (what is its genus?).

6. Discussion

Compared with the usual procedure, Hauser's algorithm requires less computations, e.g., there is no need to compute the module T^2. All computations were easily set up. One quickly gets equations for the base space, unfortunately in general in much more variables (infinitely many). The example of the cone over the rational normal curve of degree four shows that already in simple cases it is difficult to solve the infinite system of equations. One obtains a finite number of equations in a finite number of variables by restricting to deformations of negative weight of quasi-homogeneous singularities. This is a severe restriction, but it is the same restriction one runs into, if one wants to ensure a finite computation in the usual approach. Actually, in Hauser's method one can allow deformations of weight zero and still have polynomials. The risk is that the ensuing system of equations is too complicated to be solved.

Only in a few cases we found explicit results. This is not a defect of the algorithm, but reflects the fundamental difficulty in computing versal deformations, that one ends up with a messy system of long equations, which is too complicated to be of any use. In general one should refrain from computing versal deformations. Only special equations, preferable with much symmetry, are suitable for computation. One has to hope that they are representative for the general case.

REFERENCES

[1] Alonso, María Emilia, Mora, Teo and Raimondo, Mario, A computational model for algebraic power series, *J. Pure Appl. Algebra,* **77** (1992), 1–38.

[2] Altmann, Klaus and Christophersen, Jan Arthur, Cotangent cohomology of Stanley–Reisner rings, *Manuscripta Math.,* **115** (2004), 361–378.

[3] Arndt, Jürgen, *Verselle Deformationen zyklischer Quotientensingularitäten,* PhD thesis, Hamburg. 1988.

[4] Behnke, Kurt and Christophersen, Jan Arthur, Hypersurface sections and obstructions (rational surface singularities), *Compositio Math.,* **77** (1991), 233–268.

[5] Drewes, R., Infinitesimal Deformationen von Kegeln über transkanonisch eingebettenten hyperelliptischen Kurven, *Abh. Math. Sem. Univ. Hamburg,* **59** (1989), 269–280.

[6] Eisenbud, David, Transcanonical embeddings of hyperelliptic curves, *J. Pure Appl. Algebra,* **19** (1980), 77–83.

[7] Galligo, André, Théorème de division et stabilité en géométrie analytique locale, *Ann. Inst. Fourier,* **29**(2) (1979), 107–184.

[8] Greuel, G.-M., Pfister, G. and Schönemann, H., SINGULAR 3.0. *A Computer Algebra System for Polynomial Computations*, Centre for Computer Algebra, University of Kaiserslautern, 2005. http://www.singular.uni-kl.de.

[9] Hironaka, Heisuke, *Stratification and flatness,* in: Real and complex singularities, pp. 199–265. Sijthoff and Noordhoff, Alphen aan den Rijn, 1977.

[10] Hauser, Herwig, *Sur la construction de la déformation semi-universelle d'une singularité isolée,* PhD thesis, Paris Orsay, 1980.

[11] Hauser, Herwig, *An algorithm of construction of the semiuniversal deformation of an isolated singularity,* in: *Singularities, Part 1 (Arcata, Calif., 1981),* pp. 567–573, Proc. Sympos. Pure Math. **40**. Amer. Math. Soc., Providence, R.I., 1983.

[12]   Hauser, Herwig, La construction de la déformation semi-universelle d'un germe de variété analytique complexe, *Ann. Sci. École Norm. Sup.*, **18** (1985), 1–56.

[13]   de Jong, Theo and van Straten, Duco, On the base space of a semi-universal deformation of rational quadruple points, *Ann. of Math.*, **134** (1991), 653–678.

[14]   Pinkham, Henry C., *Deformations of algebraic varieties with G_m action*, Astérisque, No. **20**. Société Mathématique de France, Paris, 1974.

[15]   Schaps, Malka, *Déformations non singulières de courbes gauches,* in: Singularités à Cargèse, pp. 121–128. Astérisque, Nos. **7** et **8**. Soc. Math. France, Paris, 1973.

[16]   Stanley, Richard P., *Combinatorics and commutative algebra,* Second edition. Progress in Mathematics **41**. Birkhuser Boston, Inc., Boston, MA, 1996.

[17]   Stevens, J., The versal deformation of universal curve singularities, *Abh. Math. Sem. Univ. Hamburg,* **63** (1993), 197–213.

[18]   Stevens, Jan, Semistable $K3$-surfaces with icosahedral symmetry, *Enseign. Math.*, **48** (2002), 91–126.

[19]   Stevens, Jan, *Deformations of singularities,* Lecture Notes in Mathematics. **1811**. Springer-Verlag, Berlin, 2003.

[20]   Teissier, Bernard, *The hunting of invariants in the geometry of discriminants,* in: Real and complex singularities, pp. 565–678. Sijthoff and Noordhoff, Alphen aan den Rijn, 1977.

Jan Stevens

Matematiska Vetenskaper
Göteborgs universitet
SE 412 96 Göteborg
Sweden

e-mail: stevens@chalmers.se

BOLYAI SOCIETY
MATHEMATICAL STUDIES, 23

Deformations of
Surface Singularities
pp. 229–287.

TREE SINGULARITIES: LIMITS, SERIES AND STABILITY

DUCO VAN STRATEN

A tree singularity is a surface singularity that consists of smooth components, glued along smooth curves in the pattern of a tree. Such singularities naturally occur as degenerations of certain rational surface singularities. To be more precise, they can be considered as *limits* of certain *series* of rational surface singularities with reduced fundamental cycle. We introduce a general class of limits, construct series deformations for them and prove a stability theorem stating that under the condition of *finite dimensionality* of T^2 the base space of a semi-universal deformation for members high in the series coincides up to smooth factor with the "base space of the limit". The simplest tree singularities turn out to have already a very rich deformation theory, that is related to problems in plane geometry. From this relation, a very clear topological picture of the Milnor fibre over the different components can be obtained.

INTRODUCTION

The phenomenon of *series of isolated singularities* has attracted the attention of many authors. It is obligatory to quote Arnol'd ([3], Vol. I, p. 243):

> *"Although series undoubtedly exist, it is not altogether clear what it means."*

The very formulation is intended to be vague, and should maybe remind us that mathematics is an experimental science, and only forms concepts and definitions in the course of exploration and discovery. In any case, the word *series* is used to denote a collection of singularities $\{X_i\}_{i\in I}$, where I is some partially ordered set, which "belong together in some sense". The archetypical examples are the A_k and D_k series of surface singularities:

There are several ad hoc ways of saying that these singularities "belong together", but to quote Arnol'd again (p. 244):

"However a general definition of series of singularities is not known. It is only clear that the series are associated with singularities of infinite multiplicity (for example $D \sim x^2 y, T \sim xyz$), so that the hierarchy of series reflects the hierarchy of non-isolated singularities."

Most attempts have been to formalize certain aspects of the series phenomenon. D. Siersma and R. Pellikaan started studying hypersurface singularities with one-dimensional singular locus ([51], [52], [36], [38], [39]). These objects can be thought of as the *limits* of the simplest types of series of isolated singularities. A precision of this limit idea can be found in the notion of *stem*, due to D. Mond ([40]). In the thesis of R. Schrauwen [49] the notion of series is developed for plane curves from a topological point of view. It would be interesting to extend these ideas to isolated hypersurface singularities of arbitrary dimension.

The series phenomenon was observed by Arnol'd for hypersurfaces, but for non-hypersurfaces series also undoubtedly exist; for this one just has to take a look at the tables of rational triple points obtained by M. Artin ([4]) or of the minimally elliptic singularities as compiled by H. Laufer ([30]). A series here is characterized in terms of resolution graphs: the effect of increasing the index of the series is that of the introduction of an extra (-2)-curve in a chain of the resolution. In my thesis [58] the appropriate limits for series of normal surface singularities were identified as the class of *weakly normal Cohen–Macaulay* surface singularities. So a one-index series of normal surface singularities is associated with a Cohen–Macaulay surface germ X with an irreducible curve Σ as a singular locus, transverse to which X has ordinary crossings (A_1).

Intimately related to the notions of a limit and its associated series are the ideas of *regularity* and *stability:* certain properties of the series members X_i do in fact not depend on i, at least for i big enough, and the limit X has a corresponding property. Many examples of these phenomena are known. For example, the Milnor number will grow linearly with the index [68], [39], multiplicity and geometric genus will stay constant, and the monodromy varies in a regular, predictable manner [53].

In the deformation theory of rational surface singularities one also encounters these phenomena. From the work of J. Arndt [2] on the base space of the semi-universal deformation of cyclic quotient singularities, and the work of T. de Jong and the author on rational quadruple points [23], the idea emerged of *stability of base spaces.* This is intended to mean that in a *good series* $\{X_i\}_{i \in I}$ something like the following should happen:

1. $T^1_{X_i}$ grows linearly with the index i in a series.

2. $T^2_{X_i}$ is constant (or stabilizes at a certain point).

3. "The" obstruction map $Ob : T^1_{X_i} \longrightarrow T^2_{X_i}$ becomes *independent* of the *series deformations*, and consequently

4. the base
$$\mathcal{B}_i = Ob^{-1}(0), \quad Ob : T^1_{X_i} \longrightarrow T^2_{X_i}$$
retains the same overall structure, in the sense that it gets multiplied by a smooth factor.

Of course, this is rather inprecise, but maybe the following series of singularities gives some feeling of what we are after.

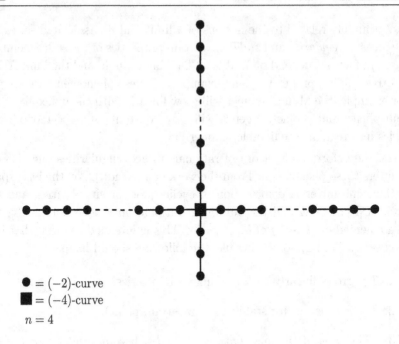

$\bullet = (-2)$-curve
$\blacksquare = (-4)$-curve
$n = 4$

Resolution graph of a rational quadruple point

Apparently, these singularities form a four-index series $X_{a,b,c,d}$ of rational quadruple points. From [21] it follows that one has for the base space:

$$\mathcal{B}_{a,b,c,d} = B(n) \times S,$$

where $n = \min(a,b,c,d)$, $B(n)$ is a very specific space with n irreducible components and S is some smooth space germ. So, up to a smooth factor, the base space of $X_{a,b,c,d}$ only depends on the *shortest arm length* n. This n determines the *core* $B(n)$ of the deformation space, but every time we increase the shortest length, we pick up a new component! From [25], we know that T^2 is also determined by the smallest arm length. So although $X_{a,b,c,d}$ clearly "is" a four-index series, stability of the base space can only be seen by considering it as a (one-index series of a) three-index series of singularities. The *limit* obtained by sending the three shortest arms to infinity is a first example of what we call a *tree singularity*: a union of smooth planes intersecting in smooth curves in the pattern of a tree. In this case, there is a central plane with three smooth curves in it. The curves all have mutual contact of order n, and to these curves three other smooth planes are glued.

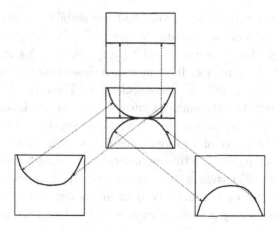

Glueing of planes

The usefulness of the stability phenomenon is obvious: as a degeneration, these limits have a simpler structure, and as a rule their deformation theory will be easier to understand than that of a series member. But one has to pay a price: as these limits no longer have isolated singularities, they do not have a semi-universal deformation in the usual sense. Their base spaces are in any sense *infinite dimensional*. This causes some inconveniences, but the work of Hauser [19] shows that a good theory can be developed in the framework of Banach-analytic spaces. We take here another approach: we will work consistently with the deformation functors and smooth transformations between them.

The purpose of this paper is twofold. In the first place it is intended as a heuristic guide to the understanding of [25]. By introducing the concept of a tree singularity we hope to clarify some of the ideas behind [25], where sometimes technicalities obscure simple and strong geometrical ideas.

In the second place we have a few theorems about series and the stability phenomenon that deserve formulation and exposition. There are many open ends here, and maybe the paper can interest others to prove more general results in this direction.

The organization of the paper is as follows. in §1 we review the basic theory of weakly normal Cohen–Macaulay surface germs. We will call such object simply *limits*. In §2 we show that such limit deforms in a *series* of normal surface singularities, whose resolution graphs can be described explicitly. Most of these ideas can be found in [58]. We will use these concepts as a sort of working definition, and in no way as the last word on these. In §3 we formulate and prove the basic stability results: the *theorem of the core* (3.5), and the *stability theorem* (3.7). The projection method of [22] is used, but clearly here is something very general going on, and a better understanding is wanted for. In §4 we take a closer look at a particular class of limits, the afore mentioned *tree singularities*. These tree singularities are the limits of series of the simplest surface singularities imaginable: those which are rational and have reduced fundamental cycle. A key notion in [25] was that of a *limit tree* of a rational singularity with reduced fundamental cycle. This is an abstraction to systematically distinguish between *long* and *short* chains of (-2)-curves in the resolution graph. Another way of thinking about a limit tree of a singularity is as an *assignment of the singularity as a series member of a limit*. Things are not always straightforward, as a singularity might very well be member of more than one series, with very different limits, unlike the situation with A_k and D_k. For these tree singularities, the deformation theory has a rather simple description. We will give an interpretation of the module generators for T^1 and T^2 as found in [25] in the case of tree singularities, and review the equations for the bases spaces. The base spaces for even the simplest tree singularities and their series members turn out to be extremely interesting, and can be interpreted in terms of elementary plane geometry. In particular, the Milnor fibre of the series members over the different components has a simple description in terms of certain configurations of curves and points. In the paper [26] with T. de Jong we have given a more systematic account of this *picture method*. With this method one now gets some insight in the dazzling complexity of deformation theory of rational singularities, and hopefully the reader will be convinced after reading this paper that the answer to the question:

"How many smoothing components does this singularity have?"

probably in most cases will be:

"Many!"

(Unless you are somewhere at the beginning of the series ...)

1. Limits and Tree Singularities

In this section we introduce a certain class of non-isolated surface singularities, called *limits*. We review some basic properties and notions of these singularities, and we will see in the next section how limits give rise to series of isolated singularities. Most of this can be found in [58]. Furthermore, we introduce a particularly simple class of limits that we call *tree singularities*.

We will consider germs X of analytic spaces, or small contractible Stein representatives thereof. Σ usually will denote the singular locus of X, and $p \in \Sigma$ the base point of the germ.

Notation 1.1. Let X be a reduced germ of an analytic space, Σ its singular locus, $\mathcal{O}_X$ its local ring, and $\mathcal{K}_X$ its total quotient ring.

The *normalization* of X is denoted by

$$n : \widetilde{X} \longrightarrow X.$$

The *weak normalization* of X is denoted by

$$w : \widehat{X} \longrightarrow X.$$

Recall that the semi-local rings of $\widetilde{X}$ and $\widehat{X}$ are given by:

$$\mathcal{O}_{\widetilde{X}} = \{f \in \mathcal{K}_X \mid f_{|X-\Sigma} \in \mathcal{O}_{X-\Sigma} \text{and } f \text{ is bounded}\}$$

$$\mathcal{O}_{\widehat{X}} = \{f \in \mathcal{O}_{\widetilde{X}} \mid f \text{ extends continuously to } X\}$$

so one has the inclusions $\mathcal{O}_X \subset \mathcal{O}_{\widehat{X}} \subset \mathcal{O}_{\widetilde{X}} \subset \mathcal{K}_X$.

A space is called *normal* if n is an isomorphism, *weakly normal* if w is an isomorphism. Normalization and weak normalization have obvious universal properties. Furthermore, the weak normalization has the property that if $h : Y \longrightarrow X$ is an holomorphic homeomorphism, then w can be factorized as $w = h \circ \overline{h}$ for some $\overline{h} : \widehat{X} \longrightarrow Y$. This explains the usefulness of the weak normalization and its alternative name *maximalization*. For more details we refer to the standard text books like [18], [14].

Example 1.2. For each m there is exactly one weakly normal curve singularity $Y(m)$ of multiplicity m, to know the union of the m coordinate lines L_p, $p = 1, \ldots, m$ in $\mathbb{C}^m$:

$$Y(m) = \{(y_1, \ldots, y_m) \mid y_i \cdot y_j = 0, \ i \neq j\}$$
$$= \cup_{p=1}^m \{(y_1, \ldots, y_m) \mid y_i = 0, \ i \neq p\}$$
$$= \vee_{p=1}^m L_p$$

Weakly normal surface singularities have a more complicated and interesting structure. If we assume X also to be *Cohen–Macaulay* (note that in dimension two normality implies Cohen–Macaulay, but weak normality does not), then there is a simple geometrical description of weak normality.

Definition 1.3. A Cohen–Macaulay surface germ is called a *limit* if it satisfies one of the following three equivalent conditions:

1. X is weakly normal.

2. $X - \{p\}$ is weakly normal.

3. For points $q \in X - \{p\}$ we have the following analytic local normal forms:

$$q \in X - \Sigma; \quad \mathcal{O}_{(X,q)} \approx \mathbb{C}\{x_1, x_2\}$$

$$q \in \Sigma - \{p\}; \quad \mathcal{O}_{(X,q)} \approx \mathbb{C}\{x, y_1, \ldots, y_m\}/(y_i \cdot y_j; \; i \neq j)$$

Here, of course, m can depend on the choice of q.

Proof. The equivalence of 2. and 3. is clear in view of example (1.2). Obviously 1. $\Rightarrow$ 2. and 2. $\Rightarrow$ 1. follows from the fact that Cohen–Macaulay implies that all holomorphic functions on $X - \{p\}$ extend to X. ∎

The following gluing construction is very useful:

Proposition 1.4. *Let be given maps of analytic spaces* $\pi : \widetilde{\Sigma} \longrightarrow \Sigma$ *and* $\iota : \widetilde{\Sigma} \longrightarrow \widetilde{X}$. *If* π *is finite and* ι *is a closed embedding, then the push-out* X *in the category of analytic spaces exists, i.e. there is a diagram*

$$
\begin{array}{ccc}
\widetilde{\Sigma} & \overset{\iota}{\longrightarrow} & \widetilde{X} \\
\pi \downarrow & & \downarrow \\
\Sigma & \longrightarrow & X
\end{array}
$$

with the obvious universal property. The map $\widetilde{X} \longrightarrow X$ *is also finite, and the map* $\Sigma \longrightarrow X$ *is also a closed embedding. Furthermore,* $\widetilde{X} - \widetilde{\Sigma} \approx X - \Sigma$.

We say that X is obtained from $\widetilde{X}$ by *gluing* the subspace $\widetilde{\Sigma}$ to Σ. For a proof, see [27] or [58], where in the local case explicit algebra generators of $\mathcal{O}_X$ are given.

The above construction is also "universal" in the sense that any finite, generically 1-1 map $\widetilde{X} \longrightarrow X$ between reduced spaces can be obtained that way. To formulate this more precisely, we will fix the following notation associated to such a map:

Notation 1.5. Given a finite, generically 1-1 map $\widetilde{X} \longrightarrow X$ between reduced spaces, we define the *conductor* to be:

$$\mathcal{C} = \mathrm{Hom}_X(\mathcal{O}_{\widetilde{X}}, \mathcal{O}_X) \subset \mathcal{O}_X.$$

Put $\mathcal{O}_\Sigma = \mathcal{O}_X/\mathcal{C}$ and $\mathcal{O}_{\widetilde{\Sigma}} = \mathcal{O}_{\widetilde{X}}/\mathcal{C}$ for the structure sheaves of the corresponding sub spaces $\widetilde{\Sigma}$ and Σ. It is now a tautology that we have a diagram

$$
\begin{array}{ccccccccc}
0 & \longrightarrow & \mathcal{C} & \longrightarrow & \mathcal{O}_X & \longrightarrow & \mathcal{O}_\Sigma & \longrightarrow & 0 \\
 & & {\scriptstyle\approx}\downarrow & & \downarrow & & \downarrow & & \\
0 & \longrightarrow & \mathcal{C} & \longrightarrow & \mathcal{O}_{\widetilde{X}} & \longrightarrow & \mathcal{O}_{\widetilde{\Sigma}} & \longrightarrow & 0
\end{array}
$$

so X can be seen as obtained from $\widetilde{X}$ by gluing $\widetilde{\Sigma}$ to Σ.

This now leads to a characterization of limits in terms of the normalization:

Proposition 1.6. *Let X be a surface germ, $n : \widetilde{X} \longrightarrow X$ the normalization. Then X is a limit if and only if:*

1. *$\widetilde{X}$ is purely two-dimensional.*

2. *Σ and $\widetilde{\Sigma}$ are reduced curve germs (with structure as in 1.5).*

3. *$H^0_{\{0\}}(\mathcal{O}_{\widetilde{\Sigma}}/\mathcal{O}_\Sigma) = 0$, i.e. $\mathcal{O}_{\widetilde{\Sigma}}/\mathcal{O}_\Sigma$ is $\mathcal{O}_\Sigma$-torsion free.*

Proof. (See [58], (1.2.20)) If X is a limit, then $\widetilde{X}$ will be a normal surface (multi-) germ, and the curves Σ and $\widetilde{\Sigma}$ will be reduced, by the local normal forms 1.3 and the fact that $\mathcal{C}$ is defined as a Hom. Via a local cohomology computation using the push-out diagram, the Cohen–Macaulayness of X comes down to condition 3. ∎

EXAMPLES OF LIMITS

Partition singularities 1.7.

Suppose that we have a germ X that sits in a push-out diagram as in (1.4):

$$\begin{array}{ccc} \widetilde{\Sigma} & \hookrightarrow & \widetilde{X} \\ \downarrow & & \downarrow \\ \Sigma & \hookrightarrow & X \end{array}$$

Suppose furthermore that $\widetilde{X}$, $\widetilde{\Sigma}$, and Σ are all *smooth* (multi-) germs. Hence, Σ is a single smooth branch, and $\widetilde{\Sigma}$, $\widetilde{X}$ both consist of r smooth pieces, where r is the number of irreducible components of X. The map $\widetilde{\Sigma} \hookrightarrow \widetilde{X}$ is the standard inclusion, and in appropriate coordinates the map $\widetilde{\Sigma} \longrightarrow \Sigma$ is given by $t_i \mapsto t_i^{m_1}$. Hence, X is completely described by the partition of $m = \Sigma_{i=1}^r m_i$ into r numbers. We call X a *partition singularity,* and write $X = X(\pi)$, where $\pi = (m_1, m_2, \ldots, m_r)$. This space has the following more or less obvious properties:

$$\text{mult}\big(X(\pi)\big) = m, \quad \text{embdim}\big(X(\pi)\big) = m + 1, \quad \text{type}\big(X(\pi)\big) = m - 1.$$

The singular locus of $X(\pi)$ is the line Σ, and the generic transverse singularity is the curve $Y(m)$ of (1.2). The general hyperplane section of $X(\pi)$ is the partition curve of type π as defined in [11]. These partition singularities are in some sense the building blocks from which all other limits are constructed, see (2.3). Note also that $X(1,1) = A_\infty$ and $X(2) = D_\infty$.

Projections 1.8. Consider a normal surface singularity $\widetilde{X} \subset \mathbb{C}^N$, and consider a general linear projection $L : \mathbb{C}^N \longrightarrow \mathbb{C}^3$. Let X be the image of $\widetilde{X}$ in $\mathbb{C}^3$. Then X will have an ordinary double curve outside the special point. As a hypersurface X is Cohen–Macaulay, hence X is a limit, and moreover, the map $l : \widetilde{X} \longrightarrow X$ can be identified with the normalization map.

In the proof of theorem (3.5) we will use a slightly more general situation in which $\widetilde{X}$ is assumed to be a limit rather than a normal space. The corresponding X will be a limit if and only if $\widetilde{X}$ has only transverse A_1 outside the special point.

Tree singularities 1.9. A *tree singularity* is a singularity X that satisfies one of the following equivalent conditions:

1. X is the total space of a δ-constant deformation of the curve $Y(m)$ of (1.2) to a curve with only nodes. Note that the curve $Y(m)$ has δ-invariant equal to $m - 1$, and only deforms into singularities $Y(k)$, with $k \leq m$ ([12]). Note that the general fibre will have m components and $m - 1$ nodes, so the components have to intersect in the pattern of a tree.

2. X has the curve $Y(m)$ of (1.2) as a general hyperplane section and is the union of m smooth irreducible components X_p, $p = 1, \ldots, m$ and $L_p \subset X_p$. Two such components X_p and X_q intersect in the point 0, or in a smooth curve $\Sigma_{\{p,q\}}$. The graph with vertices corresponding to the components X_p and edges corresponding to the curves $\Sigma_{\{p,q\}}$ is a tree T. So a tree singularity is obtained by gluing smooth planes along smooth curves in the pattern of a tree.

Example 1.10. We illustrate these two different ways of looking at a tree singularity with two pictures.

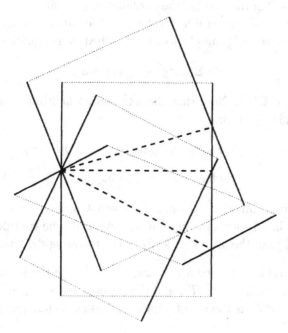

A δ-constant deformation of $Y(4)$

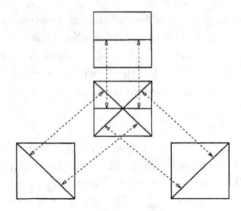

The same tree singularity as a glueing

To describe a tree singularity completely, we not only need T, but also a description of the curves Σ_{pq} in the planes X_p and X_q. This can be done as follows. We choose coordinates $x, y_1, y_2, \ldots, y_m$, such that the hyperplane section $x = 0$ describes $Y(m)$ in the coordinates of (1.2). As the plane X_p intersects $x = 0$ in the line L_p, the variables x, y_p form a coordinate system on X_p. As $x = 0$ is a general hyperplane section, all curves Σ_{qp} in the plane X_p are transverse to L_p, and hence are described by an equation of the form:

$$\Sigma_{qp} : y_p + a_{qp}(x) = 0$$

for some $a_{qp} \in \mathbb{C}\{x\}$. Note that the intersection multiplicity of the curves Σ_{rp} and Σ_{qp} is equal to:

$$i\left(\Sigma_{rp}, \Sigma_{qp}\right) = \mathrm{ord}_x\left(\phi(r, q; p)\right) =: \rho(r, q; p)$$

$$\text{where } \phi(r, q; p) := a_{rp} - a_{qp}.$$

These difference functions ϕ (and the contact orders ρ) play a very important role in all sorts of computations and are to be considered as more fundamental than the a_{qp}. This leads to the following definition:

Definition 1.11. Let T be a tree and let us denote the set of vertices by $v(T)$, the set of edges by $e(T)$, and the set of oriented edges by $o(T)$. The set of corners $c(T)$ is the set of triples $(r, q; p)$ such that $\{r, p\} \in e(T)$ and $\{q, p\} \in e(T)$.

A *decorated tree* $\boldsymbol{T} = (T, \phi)$ is a tree T, together with a system ϕ of functions,

$$\phi(r, q; p) \in x\mathbb{C}\{x\} \quad (r, q; p) \in c(T)$$

anti-symmetric in the first two indices and satisfying the cocycle condition:

$$\phi(s,r;p) + \phi(r,q;p) + \phi(q,s;p) = 0.$$

Furthermore, it is assumed that none of the ϕ's is identically zero.

So every tree singularity X with a function $x \in \mathcal{O}_X$ defining the general hyperplane section, gives us a decorated tree $\boldsymbol{T}$. Conversely, one has the following:

Proposition 1.12. *Let* $\boldsymbol{T} = (T, \phi)$ *be a decorated tree. Consider the power series ring R with variables x, z_{qp}, where $(p,q) \in o(T)$. Denote by $\boldsymbol{C}_{pq}$ the unique chain in T from p to q. Let $X(\boldsymbol{T})$ be defined by the following system of equations:*

$$z_{pq}z_{rs} = 0 \qquad \text{for all } (p,q), (r,s) \in o(T) \text{ such that } p,r \in \boldsymbol{C}_{qs}$$
$$z_{rp} - z_{qp} = \phi(r,q;p) \text{ for all corners } (r,q;p) \in c(T).$$

Then $X(\boldsymbol{T})$ is the tree singularity with decorated tree $\boldsymbol{T}$. The irreducible components of $X(\boldsymbol{T})$ are

$$X_t, \quad t \in v(T) \text{ defined by } z_{sr} = 0, \ s \in \boldsymbol{C}_{tr}.$$

Proof. The hyperplane section of $X(\boldsymbol{T})$ is readily seen to be $Y(m)$: modulo x one has $z_{rp} = z_{qp}$, so the quadratic equations reduce to those of $Y(m)$ given in (1.2). From this it also follows that $X(\boldsymbol{T})$ is of dimension ≤ 2. Now choose a splitting of the cocycle ϕ; i.e. we write

$$z_{qp} = y_p + a_{qp} \ ; \ a_{qp} \in \mathbb{C}\{x\}.$$

Define X_t as the set where $z_{sr} = 0$, $s \in \boldsymbol{C}_{tr}$. Then on X_t one has coordinates x, y_t and the other y_r are expressed via:

$$y_r + a_{sr} = 0$$

where $s \in \boldsymbol{C}_{tr}$ is such that and $\{s,r\} \in e(T)$, so indeed X_t is a smooth surface. Furthermore, for any $t \in v(T)$ and any given (p,q) and $(r,s) \in o(T)$ such that p and $r \in \boldsymbol{C}_{qs}$ we have that $r \in \boldsymbol{C}_{ts}$ or $p \in \boldsymbol{C}_{tq}$, because T is a tree.

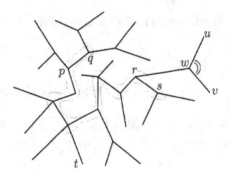

The tree equations $z_{pq}z_{rs} = 0$, and a corner $(u, v; w) \in c(T)$

This means that for each t the quadratic equations $z_{pq}z_{rs}$ is zero on X_t. Hence X_t is a component of $X(\boldsymbol{T})$. If $\{p, q\} \in e(T)$ then $X_p \cap X_q$ is described by the equation

$$y_p + a_{qp} = 0$$

in the plane X_p. As none of the ϕ's is identically zero, all these curves are distinct. So in each plane we find precisely the right curves to give as incidence diagram the tree T. As the hyperplane section of $X(\boldsymbol{T})$ was the reduced curve $Y(m)$, this indeed proves that $X(\boldsymbol{T})$ is the total space of a δ-constant deformation of $Y(m)$. ■

Remark 1.13. There is another, in some sense simpler, but more redundant form to write the equations for $X(\boldsymbol{T})$. We introduce for each pair $p \neq q$ a variable z_{pq} and consider the equations:

$$z_{pq}z_{qp} = 0$$

$$z_{rp} - z_{qp} = \phi(r, q; p).$$

Here we extend ϕ to all triples of distinct elements by putting: $\phi(r, q; p) := \phi(s, t; p)$ if $p \in \boldsymbol{C}_{rq}$ and where s and t are determined by the rule that $s \in \boldsymbol{C}_{rp}$ and $\{s, p\} \in e(T)$, etc. One puts $\phi(r, q; p) = 0$ in case that $p \notin \boldsymbol{C}_{qr}$. This has the effect that $z_{qp} = z_{rp}$ for such triples. This second form of the equations correspond exactly to the form used in [25], where these were called the canonical equations. The canonical equations for a rational surface singularity with reduced fundamental cycle read

$$z_{pq}z_{qp} = f_{pq}$$

$$z_{rp} - z_{qp} = \phi(r, q; p).$$

The system of functions f_{pq}, $\phi(r, q; p)$ has to satisfy a certain system of compatibility equations (the "Rim-equations"), see [25]. A fundamental fact is that the f_{pq} for $\{p, q\} \in e(T)$ and $\phi(r, q; p)$ for $(r, q; p) \in c(T)$ uniquely determine the others, so a rational surface singularity with reduced fundamental cycle is determined by data T, f, where T is a decorated tree, and $f = \{f_{pq} \in \mathbb{C}\{x\}, \{p, q\} \in e(T)\}$ a system of functions. In this way, one can see the tree singularity as a degeneration by putting all f_{pq} for $\{p, q\} \in e(T)$ equal to zero.

2. SERIES OF SINGULARITIES

In this section we will see how to associate with each limit X a certain (multi-) series of singularities. Such a series is constructed by deforming the singularities of an *improvement* $\pi : Y \longrightarrow X$. We will describe how the resolution graphs of the series members can be understood as *root graphs of the improvements*.

As our approach to series is based on deformation theory, it might be profitable for the reader to have a look at the appendix as well. As we want to construct series by deforming X, we first take a look at the overall structure of $\mathrm{Def}(X)$ on the infinitesimal level:

Proposition 2.1. *Let X be a limit, and Σ its singular locus and $q \in \Sigma - \{0\}$ a point of multiplicity m. If $m \neq 2$ then*

(1) $T^1_{(X,q)}$ *is a free $\mathcal{O}_{(\Sigma,q)}$-module of rank $m.(m-1)$.*

(2) $T^2_{(X,q)}$ *is a free $\mathcal{O}_{(\Sigma,q)}$-module of rank $(1/2).m.(m-1).(m-3)$.*

(*If $m = 2$ these ranks are 1 and 0, respectively.*) In particular, T^1_X is finite dimensional if and only if X is normal, and T^2_X is finite dimensional if and only if the multiplicity of the transverse singularities does not exceed 3.

Proof. By the local normal form (1.3), this is really a statement about the curve $Y(m)$ of (1.2). For this the calculation of T^1 and T^2 is an easy exercise. See also [17], and [11]. For more information about the semi-universal deformation of this curve, see [57]. ∎

Formally, the base space $\mathcal{B}_X$ of a limit X is the fibre over 0 of a map $Ob : T_X^1 \longrightarrow T_X^2$. If T^2 is finite dimensional, then there always will be deformations, because T_X^1 has infinite dimension. To get some control over these deformations it is useful to study deformations of an improvement Y of X. A normal surface singularity X can be studied effectively using a resolution, i.e. a proper map $\pi : Y \longrightarrow X$ where Y is smooth, and $\pi_*(\mathcal{O}_Y) = \mathcal{O}_X$. If X is a limit and has a curve Σ as singular locus, then one can first normalize X to get $\tilde{X}$, and then resolve $\tilde{X}$ to Y. The resulting map $\pi : \tilde{Y} \longrightarrow X$ is still proper, but because we removed the singular curve, we no longer have $\pi_*(\mathcal{O}_{\tilde{Y}}) = \mathcal{O}_X$. In order to preserve this property, we have to "glue back" the identification of points that was lost during normalization. The prize one has to pay is that the resulting space Y now has become singular. These singularities however can be controlled. Improvements were first considered by N. Shepherd-Barron [50]. For improvements of surfaces with more general transverse singular loci, see [54].

Definition 2.2. Let X be a limit. $\pi : Y \longrightarrow X$ is called an *improvement* of X if it satisfies the following properties:

(1) π is proper.

(2) $\pi : Y - E \approx X - \{p\}$, where $E = \pi^{-1}(p)$, the exceptional locus.

(3) Y has only partition singularities.

Proposition 2.3. *Improvements exist.*

Proof. Let $n : \tilde{X} \longrightarrow X$ be the normalization, $\Sigma \in X$ and $\tilde{\Sigma} \in \tilde{X}$ the locus of the conductor in X and $\tilde{X}$ respectively. Now make an embedded resolution of $\tilde{\Sigma}$ in $\tilde{X}$ to get a diagram

$$\begin{array}{ccc} \tilde{\Delta} & \longrightarrow & \tilde{Y} \\ \downarrow & & \downarrow \\ \tilde{\Sigma} & \longrightarrow & X \end{array}$$

where $\tilde{\Delta} \longrightarrow \tilde{\Sigma}$ can be identified with the normalization of $\tilde{\Sigma}$. By the universal property of normalization, the composed map $\tilde{\Delta} \longrightarrow \Sigma$ now lifts to a map $\tilde{\Delta} \longrightarrow \Delta$, where $\Delta \longrightarrow \Sigma$ is the normalization of Σ. Now we can form a push-out diagram as in 1.4:

$$\begin{array}{ccc} \tilde{\Delta} & \longrightarrow & \tilde{Y} \\ \downarrow & & \downarrow \\ \Delta & \longrightarrow & Y \end{array}$$

Clearly, from the definition (1.7) we see that Y has only partition singularities. By the universal property of the push-out, we get an induced map $Y \longrightarrow X$. It is easily checked that this indeed is an improvement. ■

Example 2.4.

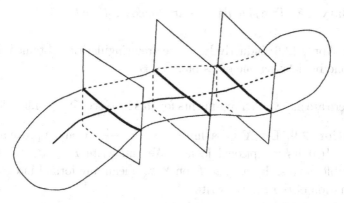

Improvement of the tree singularity of (1.10)

With this notion of improvement one can now try to build up a theory of limits along the same line as that exists for normal surface singularities. So one can define a fundamental cycle, weakly rational singularities, weakly elliptic singularities that all have properties very closely resembling those in the case of normal singularities ([58]). Let us recall here the definition of the geometric genus p_g:

Definition 2.5. Let X be a limit, and $\pi : Y \longrightarrow X$ an improvement. The geometric genus p_g of X is defined to be

$$p_g(X) = \dim \left(R^1 \pi_*(\mathcal{O}_Y) \right).$$

It is shown in [58], (2.5.28) that this p_g is semi-continuous under deformation. To compute it in examples, the following simple result is useful:

Proposition 2.6. If $\widetilde{X} \longrightarrow X$ is a finite, generically 1-1 mapping of limits, and $\widetilde{\Sigma}$, Σ as in 1.5, then one has

$$p_g(X) = p_g(\widetilde{X}) + \delta_{\widetilde{\Sigma}} - \delta_{\Sigma}.$$

(Application of this to the normalization $n : \widetilde{X} \longrightarrow X$ shows that indeed p_g is independent of the chosen improvement.)

Proof. This is straightforward, see [58], (2.3.5). ■

Definition 2.7. A limit X is called weakly rational (also called semi-rational), if and only if $p_g = 0$.

Corollary 2.8. *Tree singularities are weakly rational.*

Proof. Apply (2.6) inductively to the tree singularities obtained from the given one by deleting one of its planes. ■

There are many other arguments for this fact, see [58], (4.4.6), [25], (1.4).

Definition 2.9. Let X be a limit, $\pi : Y \longrightarrow X$ an improvement and $E = \pi^{-1}(p)$ its exceptional locus. We can write $E = \cup_{i=1}^{r} E_i$ with E_i irreducible curves. By a *cycle* F on Y we mean any formal integral linear combination of the E_i. We write

$$F = \sum_{i=1}^{r} n_i E_i$$

Such a cycle determines a unique Weil-divisor (i.e. an in general non-reduced subscheme of codimension one) of Y, that we will denote by the same symbol F. The cycle F is called a Cartier-cycle, if the corresponding divisor in fact is a Cartier divisor on Y.

Definition 2.10. Let X be a limit, and $\pi : Y \longrightarrow X$ an improvement with exceptional set E and let $F \hookrightarrow Y$ the subscheme determined by a cycle on Y. Associated to this we consider the following functors.

(1) Def (Y), the deformations of Y.

(2) Imp (Y), the deformations of Y that blow down to deformations of X. This is analogous to the functor Res of [6], or B, of [65].

(3) Def $(F \setminus Y)$, the deformations of Y for which F can be lifted as a trivial family.

This last functor is analogous to functors TR_Z considered in [65]. On Y we have a finite set P of special points, $P = \Delta \cap E$, where Δ is the singular locus of Y. We also define local functors:

(1) $\mathrm{Def}\,(Y)_P = \Pi_{p \in P}\,\mathrm{Def}\,\big((Y,p)\big)$

(2) $\mathrm{Def}\,(F \setminus Y)_P = \Pi_{p \in P}\,\mathrm{Def}\,\big((F,p) \setminus (Y,p)\big).$

All these functors are connected and semi-homogeneous. In the case that Y is smooth, these have a hull. In general, none of these will be smooth.

Proposition 2.11.

(1) *The localization maps*

$$\mathrm{Def}\,(Y) \longrightarrow \mathrm{Def}\,(Y)_P \quad and \quad \mathrm{Def}\,(F \setminus Y) \longrightarrow \mathrm{Def}\,(F \setminus Y)_P$$

are smooth.

(2) $\mathrm{Def}\,(Y)_P$ *is smooth if and only if X has finite dimensional T^2.*

(3) *There are inclusions* $\mathrm{Imp}\,(Y) \subset \mathrm{Def}\,(Y)$ *and* $\mathrm{Def}\,(F \setminus Y) \subset \mathrm{Def}\,(Y).$

If F is "big enough", one has $\mathrm{Def}\,(F \setminus Y) \subset \mathrm{Imp}\,(Y).$

Proof. Statement (1) means in particular that local deformations can be globalized, even if we want to lift the cycle F. From the local-to-global spectral sequence and using the fact that $H^2(\mathcal{F}) = 0$ for any coherent sheaf $\mathcal{F}$ on Y one gets:

$$H^0(\Theta_Y) \approx \boldsymbol{T}^0(Y)$$
$$0 \longrightarrow H^1(\Theta_Y) \longrightarrow \boldsymbol{T}^1(Y) \longrightarrow H^0(\mathcal{T}_Y^1) \longrightarrow 0$$
$$\boldsymbol{T}^2(Y) \approx H^0(\mathcal{T}_Y^2).$$

From this it follows that the map $\mathrm{Def}\,(Y) \longrightarrow \mathrm{Def}\,(Y)_P$ is surjective on the level of tangent spaces, and injective (even isomorphism) on obstruction spaces. Hence the transformation is smooth. In exactly the same way one gets for $F \setminus Y$:

$$H^0\big(\Theta_Y(-F)\big) \approx \boldsymbol{T}^0(F \setminus Y)$$
$$0 \longrightarrow H^1\big(\Theta_Y(-F)\big) \longrightarrow \boldsymbol{T}^1(F \setminus Y) \longrightarrow H^0(\mathcal{T}_{F \setminus Y}^1) \longrightarrow O$$
$$\boldsymbol{T}^2(F \setminus Y) \approx H^0(\mathcal{T}_{F \setminus Y}^2)$$

and the same conclusion.

Statement (2) follows directly from (2.1): T_X^2 finite dimensional means that X has transverse $Y(2)$ or $Y(3)$. Hence on the improvement we find the partitiion singularities $X(1,1)$, $X(2)$, $X(1,1,1)$, $X(2,1)$ or $X(3)$. The first two are hypersurfaces, the other three Cohen–Macaulay of codimension two. So in all these cases we have $T_Y^2 = 0$, hence Def (Y) is smooth. As to statement (3) we remark that the inclusion Def $(F \setminus Y) \subset$ Def (Y) is due to the fact that the embedding of F in Y is unique, even infinitesimally, as a consequence of the negativity of E. That for big F, we get in fact Def $(F \setminus Y) \subset$ Imp (Y) follows from the fact that a deformation $\mathcal{Y} \longrightarrow S$ of Y over S blows down to a deformation of X if and only if $H^1(\mathcal{O}_{Y_s})$ is constant for all $s \in S$ (see [42], [62]). We say that F is *big enough* if the canonical surjection $H^1(\mathcal{O}_Y) \longrightarrow H^1(\mathcal{O}_F)$ is an isomorphism. (As $p_g = H^1(\mathcal{O}_Y)$ is finite dimensional it follows that there are such F.) ∎

Remark 2.12. The most important case one encounters is of course the case that the limit has only transverse double points, so we have only $X(1,1) = \mathbf{A}_\infty$ and $X(2) = \mathbf{D}_\infty$ singularities on the improvement and so Def (Y) and Def $(F \setminus Y)$ are smooth. If furthermore the singularity is weakly rational, as is the case for our tree singularities, then one also has Def $(Y) = $ Imp (Y).

In [58] a series of determinal deformations of the partition singularities $X(\pi)$ was constructed. To be more precise, we have:

Proposition 2.13. *Let* $X(\pi)$, $\pi = (\pi_1, \pi_2, \ldots, \pi_r)$ *be a partition singularity. Let* $(\nu_1, \nu_2, \ldots, \nu_r)$, $\nu_i \geq 0$, *a collection of* r *numbers. Then there exists a deformation*

$$\phi : \mathcal{X}(\pi; \nu) \longrightarrow \Lambda_\pi := \mathbb{C}^r$$

such that for generic $\lambda \in \Lambda_\pi$ *the fibre* $X_\lambda(\pi, \nu) := \phi^{-1}(\lambda)$ *has the following properties:*

(1) $X_\lambda(\pi, \nu)$ *has an isolated singularity at the origin.*

(2) *The resolution graph of the minimal resolution of* $X_\lambda(\pi, \nu)$ *has the following structure:*

    (a) *All the curves are isomorphic to* $\mathbb{P}^1$.

    (b) *There is a central curve* C, *with* $(C.C) = -m$.

(c) *There are* r *chains of curves*

$$C_{i,1}, C_{1,2}, \ldots, C_{i,p_i}$$

where $i = 1, 2, \ldots, r$ and $p_i = \pi_i + \nu_i - 1$.

(d) $(C.C_{i,\pi_i}) = 1$.

(3) *The subscheme* F_n, *defined by* $x^n = 0$ *lifts trivially over* Λ_π *if*

$$n \leq \sum_{i=1}^{r} (\nu_i + 1).$$

Proof. This is essentially [58], (1.3.12). It is obtained by perturbing a matrix defining $X(\pi)$ in a very specific way. From this representation it is possible to read off all the information. ∎

Remark 2.14. In the case that one or more of the ν_i is equal to 0, the fibre $X_\lambda(\pi, \nu)$ has in general more singularities. For example, $X_\lambda(\pi, 0)$ has as singularities the $(-m)$-singularity, together with $A_{\pi_i - 1}$-singularities. This all fits with the description under (2.13) 2). Note also the special classes

$$X(1, 1; a, b) = A_{a+b+1} \quad \text{and} \quad X(2; a) = D_{a+2}.$$

So indeed these series associated to the partition singularities are a generalization of the A and D series. But from the construction as a partition singularity we unfortunately get A as a two-index series. We will strictly hold to the following equations for the A and the D series: $A_k : yz - x^{k+1}$ and $D_k : z^2 - x.(y^2 - x^{k-2})$, and so $D_3 \approx A_3$ and the deformation of A_∞ to A_{-1} represents the generator of the T^1, etc.

Definition 2.15. Let X be a limit, and $\pi : Y \longrightarrow X$ an improvement and F a sufficiently big divisor.

Roots: In [65] the notion of *root* for a normal surface singularity was introduced. It is an attempt to characterize those cycles on a resolution that can arise as specialisation of a connected smooth curve. Analogueously we call a Cartier cycle $R \subset Y$ a root iff $\chi(\mathcal{O}_R) \leq 1$. Essentially by [65], lemma (1.2), the set of roots is always finite. A root is called *indecomposable* if it is not the sum of two other roots. In particular, each exceptional curve, not passing through any of the special points $P = E \cap \Delta$, is an indecomposable root. The diagram with vertices the indecomposable roots, and with edges corresponding to intersections of roots (computed as cycles on the normalization) we call the root diagram $RD(Y)$. Note that if Y is a resolution, then $RD(Y)$ is nothing but the dual resolution graph.

Modifications: Let $s \in P$ be a special point. Then $(Y, s) \approx (X(\pi), 0)$ for some $\pi = \pi(s) = (\pi_1(s), \ldots, \pi_{r(s)}(s))$. In particular, the normalization of Y at s consists of $r(s)$ smooth planes. For each $s \in P$ and $i = 1, 2, \ldots, r(s)$ there are elementary modifications

$$Y_{\varepsilon_i(s)} \longrightarrow Y$$

by blowing up in the i-th piece of the normalization of (Y, s). Note that on $Y_{\varepsilon_i(s)}$ there is a unique point $\tilde{s}$ over s at which $(Y_{\varepsilon_i(s)}, \tilde{s}) \approx (X(\pi(s)), 0)$. In order to simplify notations we will identify the sets of special points on Y and $Y_{\varepsilon_i(s)}$. In this way we can iterate or compose these elementary transformations. The semi-group spanned by them we denote by

$$\mathcal{N}(Y) = \bigoplus_{s \in P} \bigoplus_{i=1}^{r(s)} \mathbb{N}.\varepsilon_i(s).$$

If $\nu = (\nu(s))_{s \in P} \in \mathcal{N}(Y)$, then we denote the space obtained by this composition of elementary transformations by $Y_\nu \longrightarrow Y$. Finally, for a Cartier cycle F on Y we put

$$\mathcal{N}(Y, F) = \left\{ \nu \in \mathcal{N}(Y) \; \middle| \; \sum_{i=1}^{r(s)} (\nu_i(s) + 1) \geq \text{coeff}\,(F, s) \right\}.$$

(Here coeff (F, s) is the coefficient of E_i in F for any E_i that contains s.)

A particular transformation is the blow-up $b : \tilde{Y} \longrightarrow Y$ of Y at s. One can check that on $\tilde{Y}$ there is again a partition singularity of type $\pi(s)$ at some point $\tilde{s}$ lying over s, together with $r(s)$ singularities, of type $\boldsymbol{A}_{\pi_i(s)-1}$. Let $\mu : Y_\mu \longrightarrow Y$ be the space obtained from Y by first blowing up at all the special points, and then resolve the resulting $\boldsymbol{A}$-singularities. The use of this blow-up is that on Y_μ there will be an unique indecomposable root $R(\tilde{s})$ passing through $\tilde{s}$. For details we refer to [58].

Series Deformations: For each $\nu \in \mathcal{N}(Y, F)$ we also get an element $\xi(\nu) \in \text{Def}\,(Y)_P(\Lambda)$ by putting together the local deformations of (2.13)

$$\xi_s(\nu(s)) : \mathcal{X}(\pi(s), \nu(s)) \longrightarrow \Lambda_{\pi(s)}$$

where $\nu = \big(\nu(s)\big)_{s \in P}$ and $\Lambda = \Pi_{s \in P}\Lambda_{\pi(s)}$.

All this was set up in such a way that the following theorem is true:

Theorem on Series 2.16. *Let* $Y \longrightarrow X$ *be an improvement of a limit, and* F *a sufficiently big divisor with* $\operatorname{supp} F = E$. *For each* $\nu \in \mathcal{N}(Y, F)$ *there is a deformation*

$$\mathcal{X}(\nu) \longrightarrow \Lambda$$

of X, *such that for generic* $\lambda \in \Lambda$ *the fibre* $X_\lambda(\nu)$ *has an isolated singularity with resolution graph*

$$\Gamma\big(X_\lambda(\nu)\big) = RD\big((Y_\lambda)_\nu\big).$$

We call the singularities $X_\lambda(\nu)$ *the members of the series.*

Proof. The local deformations $\xi(\nu)$ can be lifted to global deformations of Y, fixing F, by (2.11). Because F is big enough, this deformation can be blown down to give a deformation of X. At first, this is only a formal deformation, over the formal completion of Λ. But by an application of the Approximation Theorem (see appendix) we can get an honest deformation over a neighborhood of zero in Λ that approximates arbitrarily good the given formal one. Because the support of F is assumed to be the full exceptional divisor, all the irreducible components E_i lift. Furthermore, locally around each point $s \in P$ we have a standard situation, producing a resolution graph as in (2.13). It is an exercise to verify that the new roots on Y_μ are the curves of the A-singularities, together with the indecomposable root $R(s)$ mentioned in (2.15). This root lifts to the central curve of the local resolution. ∎

Remark 2.17. This is a rather weak theorem. We do not claim that any singularity can be degenerated to a limit, nor do we claim that all singularities with a given graph do occur as fibre $X_\lambda(\nu)$. Although this is rather plausible, it is much harder to prove. Our statement is really not much more than a statement about graphs, stated in a slightly fancy way. Note also that the construction depends on the improvement $Y \longrightarrow X$ in the following way: if we blow up further to Y_ν, $\nu \in \mathcal{N}(Y)$, then one gets as series members the $X_\lambda(\nu + \mu)$, $\mu \in \mathcal{N}(Y, F)$.

Examples 2.18. By far the most important cases are where we have only A_∞ and D_∞ singularities on the improvement. We illustrate the theorem with two pictures, that hopefully will clearify everything.

   A-case:

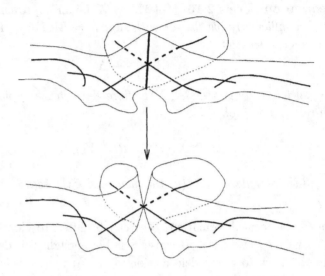

Deformation of A_∞ to A_1 on improvement

   D-case:

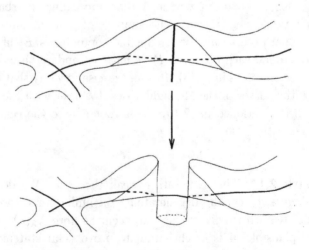

Deformation of D_∞ to D_2 on improvement

We will draw improvement graphs using an obvious extension of the usual rules for drawing a resolution graph: the presence of an A_∞-singularity is indicated by a double bar. A D_∞ is indicated by a double barred arrow. We give two examples to illustrate (2.16):

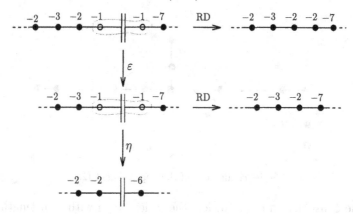

Series formation on the **A**-case

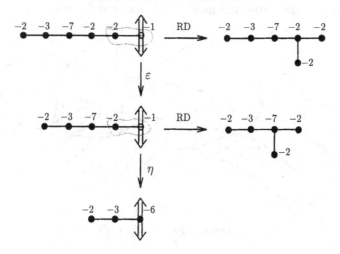

Series formation on the **D**-case

The vertical maps between the graphs are elementary modifications, obtained by blowing up the special point of the improvement. The curves enclosed by the dotted line make up the unique indecomposable root that contains the special point. The root diagrams at the right hand side are the resolution graphs of the series deformation. Blowing up further brings us to higher members of the series.

Example 2.19. The series members belonging to the improvement of (2.4) have as resolution graphs:

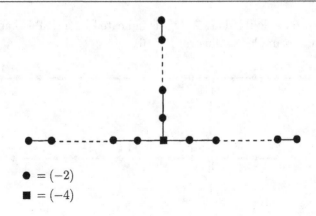

$\bullet = (-2)$

$\blacksquare = (-4)$

Series coming out of the improvement (2.4)

The construction of (2.16) lets the series begin with arm lengths equal to one, but clearly we also could let the series start one step earlier, by deforming on the improvement to A_0. It looks as follows:

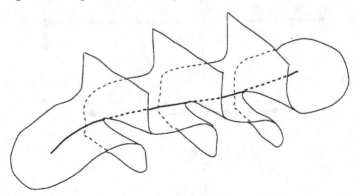

Deforming to A_0: the (-4)

We will not make fuzz about the beginning of a series. Does the A_k-series start with $k = 1$, $k = 0$, or $k = -1$?

In order to link up these series deformations with the *equations* of X we consider one more functor.

Definition 2.20. Let $\mathcal{X} \longrightarrow S$ be a deformation of X over S, with a section $\sigma : S \longrightarrow \mathcal{X}$. Let m_σ be the ideal of $\sigma(S) \subset \mathcal{X}$. The subscheme $J_\sigma^n(\mathcal{X})$ defined by the ideal m_σ^n we call the "n-jet of $\mathcal{X}$ along the section σ". We let $\mathrm{Sec}^n(X)(S) :=$ deformations of X over S with section, such that the n-jet of X along the section is deformed trivially over S, modulo isomorphism.

There is no difficulty in showing that this functor is connected and semi-homogeneous (see Appendix).

Proposition 2.21. *Let $\rho : Y \longrightarrow X$ be an improvement of a limit X. For all n there exists an F big enough such that the blow-down map $\mathrm{Def}\,(F \setminus Y) \longrightarrow \mathrm{Def}\,(X)$ factors over $\mathrm{Sec}^n(X) \longrightarrow \mathrm{Def}\,(X)$.*

Proof. We assume that the pull-back of the maximal ideal $\rho^*(\boldsymbol{m}_X)$ is invertible, say $= \mathcal{O}_Y(-Z)$. Now take $m \geq n$ such that $\boldsymbol{m}_X^m$ annihilates the sky-scraper sheaf $R^1\rho_*(\mathcal{O}_X)$, and put $F = m.Z$. This F does the job. ∎

3. STABILITY

Before formulating the stability theorem for series deformations, let us quickly discuss a situation in which the tangent-cohomological aspect of the stability phenomenon can be readily understood.

Consider a limit X, and choose a *slicing* for it. By slicing, we mean that we consider X together with a non-constant map $\rho : X \longrightarrow S$, where S is a germ of a smooth curve. So X is sliced into curves $Y_s = \rho^{-1}(s)$, if appropriate representatives for X and S are chosen. We let $Y = \rho^{-1}(0)$. Now consider a one-parameter smoothing of X:

$$\begin{array}{ccc} X & \longrightarrow & \mathcal{X} \\ \downarrow & & \pi \downarrow \\ \{0\} & \longrightarrow & T \end{array}$$

By lifting the function ρ to $\mathcal{X}$ we get a combined two-parameter deformation $\phi : \mathcal{X} \longrightarrow S \times T$ of the curve singularity Y.

Lemma 3.1. *If $\dim(T_X^2) < \infty$, then $T_{\mathcal{X}/S \times T}^k$ $(k = 2, 3)$ are artinian $\mathcal{O}_{S \times T}$-modules.*

Proof. We will use general properties of the cotangent complex, for which we refer to [20] and in particular to [11] for a nice summary of the most important facts. In general, the support of the $T_{\mathcal{X}/S \times T}^k$ as $\mathcal{O}_{S \times T}$-modules is contained in the discriminant D of the map $\mathcal{X} \longrightarrow S \times T$, which in our case consists of the axis $S \times \{0\}$, possibly together with some other curve C. For $p \in C - \{0\}$ the fibre $Y_{(s,t)}$ lies as a hypersurface in the smooth surface $X_t = \pi^{-1}(t)$, so we see that for $k \geq 2$ the module $T^k_{\mathcal{X}/S\times T}$ is supported on $S \times \{0\}$. As the limit X is assumed to have a finite T^2, it follows that X has only double or triple points transverse to Σ, and from this it follows that $T^2_{\mathcal{X}/S\times T}$ is concentrated over $\{0\}$, hence is artinian. But also $\operatorname{supp}(T^3_{\mathcal{X}/S\times T}) = \{0\}$, because transverse to the S axis in $S \times T$ we have a smoothing of the curve singularity $Y(3)$. As the support of T^3 of the family is concentrated over the S-axis, and vanish at the general point because $T^2_{Y(3)}$ vanishes: the usual argument. ∎

Now consider the maps:

$$\iota_n : S \longrightarrow S \times T; \quad s \longmapsto \left(s, \lambda.s^{n+1}\right), \quad \lambda \neq 0$$

and let the image $\operatorname{Im}(\iota_n)$ be defined by $t_n \in \mathcal{O}_{S\times T}$. We can pull-back the family $\phi : \mathcal{X} \longrightarrow S \times T$ over the maps ι_n to get sliced surfaces $X_n \longrightarrow S$. These $X_n \longrightarrow S$ can be seen as slicings of a series of isolated surface singularities. If we let run λ over a smooth curve germ Λ, we obtain a one parameter deformations $\mathcal{X}_n \longrightarrow \Lambda$ of X with fibres the X_n. We have that $\mathcal{X}_n \longrightarrow \Lambda \in \operatorname{Sec}^n(X)(\Lambda)$, because the equations of X and X_n are the same up to order n as they are obtained by pulling back via maps that are the same up to order n.

We now can see that the obstruction spaces of the X_n stabilize in the following sense:

Proposition 3.2.

$$\lim_{n\to\infty} \dim\left(T^2_{X_n}\right) = \dim\left(T^2_X\right).$$

Proof. There exists a long exact sequence

$$\cdots \longrightarrow T^1_{\mathcal{X}/S\times T} \xrightarrow{t_n\cdot} T^1_{\mathcal{X}/S\times T} \longrightarrow T^1_{X_n/S} \longrightarrow T^2_{\mathcal{X}/S\times T}$$

$$\xrightarrow{t_n\cdot} T^2_{\mathcal{X}/S\times T} \longrightarrow T^2_{X_n/S}\cdots.$$

Here $t_n \in \mathcal{O}_{S\times T}$ is as before. We have seen that $T^k_{\mathcal{X}/S\times T}$ for $k = 2,3$ are artinian $\mathcal{O}_{S\times T}$-modules. As a consequence, we see that $\operatorname{Ker}(t_n\cdot)$ and

Coker $(t_n\cdot)$, where $t_n\cdot\ :\ T^k_{\mathcal{X}/S\times T}\ \longrightarrow\ T^k_{\mathcal{X}/S\times T}$ stabilize for $n\gg 0$. By comparing with the exact sequence

$$\cdots \longrightarrow T^1_{\mathcal{X}/S\times T} \xrightarrow{t\cdot} T^1_{\mathcal{X}/S\times T} \longrightarrow T^1_{X/S} \longrightarrow T^2_{\mathcal{X}/S\times T}$$

$$\xrightarrow{t\cdot} T^2_{\mathcal{X}/S\times T} \longrightarrow T^2_{X/S}\cdots$$

we can conclude

$$\lim_{n\to\infty} \dim\left(T^2_{X_n/S}\right) = \dim\left(T^2_{X/S}\right).$$

Because S is smooth one has $T^k_{X/S} = T^k_X$ for $k \geq 2$. (c. f. [11], (1.3.1).)

(In the case that X has only transverse double points, the same argument shows that in fact all the $T^k_{X_n}$ for $k \geq 2$ will stabilize.) ∎

Remark 3.3. The above arguments show that "there are series such that high in the series T^2 (and even T^k) stabilizes, if the T^2 of the limit is finite". This is much weaker than the statement that this will happen for all deformations in $\mathrm{Sec}^n(X)$, for $n \gg 0$. Although this sounds rather probable, I have been unable to establish this. As it would be quite useful in practice, it seems worth trying to prove this in general.

As T^2_X stabilizes, is seems natural to expect that the equations of the base space also will stabilize. These stable equations then would be the equations for the base space of the limit, and these equations should not depend on the coordinates corresponding to the series deformations. So, although infinite dimensional, the base space of a limit should have some sort of finite dimensional *core*. This in fact we can prove, if we define the core in the following way:

Definition 3.4. Two semi-homogeneous functors F and G are called "the same up to a smooth factor" if there exists a semi-homogeneous functor H and smooth natural transformation $H \longrightarrow F$ and $H \longrightarrow G$. Being the same up to a smooth factor clearly is an equivalence relation.

A semi-homogeneous functor F is said to have a core iff it is the same up to a smooth factor as a (pro)-representable one. The (equivalence class) of this pro-representable functor we call the core $\mathrm{Core}\,(F)$ of F.

Theorem of the Core 3.5. *Let X be a limit with $\dim\left(T^2_X\right) < \infty$. then $\mathrm{Def}\,(X)$ has a core.*

Proof. We choose an embedding of X into some $\mathbb{C}^N$ and a linear projection $L : \mathbb{C}^N \longrightarrow \mathbb{C}^3$. Denote $L_{|X}$ by ν and put $Y = \nu(X) \subset \mathbb{C}^3$. For a generic choice of L the resulting map $X \xrightarrow{\nu} Y$ will be generically $1 - 1$. Let $\mathcal{C}$ be the conductor of the map ν, and let $\widetilde{\Sigma} \subset \widetilde{Y}$ and $\Sigma \subset Y$ be the locus of the conductor in X, respectively Y. So we have a diagram:

$$
\begin{array}{ccccc}
\widetilde{\Sigma} & \subset & X & \subset & \mathbb{C}^N \\
\downarrow & & \downarrow \nu & & \downarrow L \\
\Sigma & \subset & Y & \subset & \mathbb{C}^3
\end{array}
$$

We put the obvious structure sheaves on $\widetilde{\Sigma}$ and Σ (c.f. (1.5)): $\mathcal{O}_{\widetilde{\Sigma}} = \mathcal{O}_X/\mathcal{C}$ and $\mathcal{O}_\Sigma = \mathcal{O}_Y/\mathcal{C}$. Now, Σ will consist of two parts of different geometric origin:

(1) Σ_1: the image of the double points of the map ν. This will be an ordinary double curve on Y, and hence Σ_1 is reduced.

(2) Σ_2: the image of the curve along which X has points of multiplicity three. Transverse to such a point Y has a D_4-singularity, as it is locally the projection of the space curve $Y(3)$ to the plane. A calculation shows that the conductor structure on Σ_2 is also reduced. Note this is no longer the case if we project $Y(m)$, $m \geq 4$, so it is here that the finiteness of T_X^2 comes in. (The curve along which X has multiplicity two maps to an ordinary double curve of Y, so there is no conductor coming from this part.)

In [21] and [22] the functor of admissible deformations Def (Σ, Y) of the pair $\Sigma \hookrightarrow Y$ was studied. Loosely speaking, Def (Σ, Y) consists of deformations of Σ and Y, such that Σ stays inside the singular locus over the deformation. A fundamental result of [22] was that there is a natural equivalence of functors

$$\mathrm{Def}\left(X \longrightarrow \mathbb{C}^3\right) \approx \mathrm{Def}\,(\Sigma, Y).$$

Here Def $\left(X \longrightarrow \mathbb{C}^3\right)$ is the functor of deformations of the diagram

$$X \longrightarrow \mathbb{C}^3.$$

This functor equivalence follows essentially from

$$\mathrm{Hom}_Y(\mathcal{C}, \mathcal{C}) \approx \mathrm{Hom}_Y(\mathcal{C}, \mathcal{O}_Y) \approx \mathcal{O}_X$$

(see [22]), which means that we can recover the $\mathcal{O}_Y$-module structure and the ring structure of $\mathcal{O}_X$ from the inclusion $\mathcal{C} \hookrightarrow \mathcal{O}_Y$. Note that essentially because $\mathbb{C}^3$ is a smooth space, the forgetful transformation

$$\mathrm{Def}\left(X \longrightarrow \mathbb{C}^3\right) \longrightarrow \mathrm{Def}\left(X\right)$$

is smooth: there are no obstructions to lifting the three coordinate functions, defining the map to $\mathbb{C}^3$, along with X.

Also, in [21], the notion of I^2-equivalence on the functor $\mathrm{Def}\left(\Sigma, Y\right)$ was introduced. If Σ is defined in $\mathbb{C}^3$ by an ideal I, and Y is defined by a function f, then two deformations over S described by (I_S, f_S) and (I_S, g_S) are called I^2-equivalent if $f_S - g_S \in I_S^2$. I^2-equivalence is an admissible equivalence relation in the sense of [10], which means that the quotient map

$$\mathrm{Def}\left(\Sigma, Y\right) \longrightarrow M(\Sigma, Y)$$

is smooth, and the functor $M(\Sigma, Y)$ of I^2-equivalence classes of admissible deformations is semi-homogeneous. Combining these things, we arrive at a diagram

$$
\begin{array}{ccc}
\mathrm{Def}\left(X \longrightarrow \mathbb{C}^3\right) & \approx & \mathrm{Def}\left(\Sigma, Y\right) \\
\downarrow & & \downarrow \\
\mathrm{Def}\left(X\right) & & M(\Sigma, Y)
\end{array}
$$

The tangent space $M^1(\Sigma, Y) = M(\Sigma, Y)\left(\mathbb{C}[\varepsilon]\right)$ sits in an exact sequence (see [21])

$$0 \longrightarrow I^{(2)}/\left(I^2, \theta_I(f)\right) \longrightarrow M^1(\Sigma, Y) \longrightarrow T_\Sigma^1 \longrightarrow \cdots .$$

Here $I^{(2)}$ is the second symbolic power of I, and $\theta_I(f)$ is the ideal generated by $\theta(f)$, where $\theta \in \Theta_I := \{\theta | \theta(I) \subset I\}$. Because Σ is a reduced curve germ, we have

$$\dim\left(T_\Sigma^1\right) \leq \infty$$

and

$$\dim\left(I^{(2)}/I^2\right) \leq \infty,$$

so it follows that $\dim\left(M^1(\Sigma, Y)\right) \leq \infty$. It now follows from Schlessingers theorem that the functor $M(\Sigma, Y)$ has a hull. In other words, $\mathrm{Def}\left(X\right)$ has a core. ∎

Remark 3.6.

(1) We proved the theorem only for surfaces, but clearly something very general is going on. It is natural to expect the theorem to be true for all analytic germs (X, p) such that for some representative X of (X, p) and all $q \in X - \{p\}$ one has that $\text{Def}\big((X, q)\big)$ is smooth. It would be very interesting to prove this in general.

(2) Intuitively it is "clear" that the deformations "high in the series" should give rise to a trivial factor in the base space. One might be tempted to argue along the following lines:

By naturality of the obstruction element of $ob(\xi, \xi') \in T^2$ for $\xi, \xi' \in T^1$ we see that $ob(a.\xi, \xi') = 0$ for all $a \in \text{Ann}\,(T^2)$. So to first order, the subspace $\text{Ann}\,(T^2).T^1 \subset T^1$ is not obstructed against anything. But in general there will be higher order obstructions, or higher order Massey-products (see [29]) non-vanishing, and it is easy enough to give examples where this really happens. Theorem (3.7) states somehow that there is an end to all these Massey-products. It would be interesting to prove the theorem in such a set-up.

The next theorem tells us that this core is really the base space of any series member high in the series.

Stability Theorem 3.7. *Let X be a limit with finite dimensional T^2. Then there is a number n_0 such that for any $n \geq n_0$ and any fibre $X' = X_s$ of a deformation $\mathcal{X} \longrightarrow S \in \text{Sec}^n(S)$ one has:*

$$\text{Core}\,(X) = \text{Core}\,(X').$$

Proof. The idea of the proof is the same as that of (3.5), but for simplicity we assume that X has only transverse $\boldsymbol{A}_1$ singularities. So we again let $X \subset \mathbb{C}^N$ and let $L : \mathbb{C}^N \longrightarrow \mathbb{C}^3$ be a generic linear projection, and we get the diagram

$$
\begin{array}{ccccc}
\widetilde{\Sigma} & \subset & X & \subset & \mathbb{C}^N \\
\downarrow & & \downarrow & & \downarrow \\
\Sigma & \subset & Y & \subset & \mathbb{C}^3
\end{array}
$$

Let $\mathcal{J} \subset \mathcal{O}_{(\mathbb{C}^N, 0)} =: \mathcal{O}_N$ be the ideal of X, so the ideal of Y is

$$\mathcal{J} \cap \mathcal{O}_{(\mathbb{C}^3, 0)} = (f) \subset \mathcal{O}_3 := \mathcal{O}_{(\mathbb{C}^3, 0)}.$$

Let $l : \mathbb{C}^3 \longrightarrow \mathbb{C}$ be a generic linear function, and let P be the polar curve, that is, the critical locus of the map $(f, l) : \mathbb{C}^3 \longrightarrow \mathbb{C}^2$. So, P is defined by two generic partials of f, say $P = V(\phi); \ \phi = (\partial f/\partial x, \partial f/\partial y)$. Hence, P is an isolated complete intersection curve singularity. Let $I \subset \mathcal{O}_3$ be the ideal of Σ. Clearly we have the inclusion $\Sigma \subset P$. Let us put $m_k := m_{(\mathbb{C},0)} \subset \mathcal{O}_{\mathbb{C}^k,0})$.

Now, because Σ is reduced, there is an integer p such that

$$I^{(2)} \cap m_3^p \subset I^2.$$

Because P is an isolated complete intersection singularity, it is finitely determined, so we can find an integer q such that

$$\forall \phi' \text{ with } j^q(\phi) = j^q(\phi') \ \exists h : (\mathbb{C}^3, 0) \longrightarrow (\mathbb{C}^3, 0)$$

$$\text{such that } V(\phi \circ h) = V(\phi').$$

Because the map $X \longrightarrow Y$ is finite, it follows that for all k there is a $n = n(k) \geq k$ such that

$$\left(f, m_3^k \right) \supset \left(\mathcal{J}, m_N^n \right) \cap \mathcal{O}_3.$$

Finally, we let

$$n_0 := n(k); \quad k = \max(p, q+1).$$

Consider a fibre X' of a deformation $\xi \in \operatorname{Sec}^n(X)(S)$. By projection this family $\mathcal{X} \longrightarrow S$ we get families $Y_S \longrightarrow S$, $\Sigma_S \longrightarrow S$ and $P_S \longrightarrow S$. Now because $n \geq n_0$, we have that for each $s \in S$

$$\left(f_s, m_3^k \right) \supset \left(\mathcal{J}_s, m_N^n \right) \cap \mathcal{O}_3 = \left(\mathcal{J}, m_N^n \right) \cap \mathcal{O}_3 \supset \left(f, m_3^n \right).$$

So one has: $f - f_s \in m_3^k \subset m_3^{q+1}$. From this it follows that $j^q(P) = j^q(P_s)$, and hence, P and P_s are isomorphic. We can find a (family of) coordinate transformations, trivializing this family $P_S \longrightarrow S$. Because $\Sigma_S \longrightarrow S$ is a sub-curve (over S) of $P_S \longrightarrow S$, it follows that $\Sigma_S \longrightarrow S$ also can be assumed to be the trivial family. Let I be the ideal of Σ_S in $\mathbb{C}^3 \times S$. We then have

$$f - f_s \in m_3^p \cap I^{(2)} \subset I^2$$

because $p \leq n$. Hence, for each $s \in S$ we have that Y_s is I^2-equivalent to Y. Hence Y and Y_s have the same core. ∎

Remark 3.8. That we really need the family to connect X and X' seems to be a technicality that could be removed with some more care. Also, it is clear from the above argument that one can directly compare the base spaces of fibres X_s and $X_{s'}$ without using the core of the limit as intermediary. Remark also that the result as formulated above is not very practical, because it does not give a hint as to the value of n_0 in terms of X. It would be very useful to have a more effective version of the theorem.

4. TREE SINGULARITIES

In this section we will illustrate some of the results of chapter 2 and chapter 3 with the example of the tree singularities. We discuss the geometric content of the generators for T^1 and T^2 that were found in [25]. Furthermore, we describe the simplest class of tree singularities in more detail, to know those which have a simple star as tree. The deformation theory of these singularities leads to the study of configurations of smooth curves in the plane. It offers some insight in the complexity of deformation theory of rational surface singularities. The resulting picture method for understanding the component structure is the subject of a separate paper together with T. de Jong ([26]).

IMPROVEMENTS OF TREE SINGULARITIES

As in chapter 1, we let $v(T)$, $e(T)$, $o(T)$ and $c(T)$ be the sets of vertices, edges, oriented edges and corners of the tree T. The edges correspond to the irreducible components of the double curve of X, the oriented edges to their inverse images on the normalization. We will use

$$\Sigma_{pq} \subset X_q \quad \text{and} \quad \Sigma_{qp} \subset X_p$$

to denote these curves mapping to $\Sigma_{\{p,q\}}$, $\{p,q\} \in e(T)$ in X. Tree singularities have improvements that are easy to understand: take any embedded resolution of $\cup\Sigma_{qp} \subset X_p$, so we get a diagram

$$
\begin{array}{ccc}
\Delta_{qp} & \hookrightarrow & Y_p \\
\downarrow & & \downarrow \\
\Sigma_{qp} & \hookrightarrow & X_p
\end{array}
$$

An improvement is obtained by gluing back:

$$\coprod_{(qp)\in o(T)} \Delta_{qp} \quad \hookrightarrow \quad \coprod_{p\in v(T)} Y_p$$
$$\downarrow \qquad\qquad\qquad \downarrow$$
$$\coprod_{\{p,q\}\in e(T)} \Delta_{\{p,q\}} \quad \hookrightarrow \qquad Y$$

The corresponding improvement graph can be characterized by a certain function on the set $o(T)$ of oriented edges

$$\lambda(q,p) = \text{ length of chain from } L_p \text{ to } \Delta_{qp}.$$

We recall here that L_p is the line given by $x = 0$ in the plane X_p, and is transverse the all the other curves.

The series deformations correspond to deforming each of the double curves of the improvement, as explained in chapter 2. When we deform around $\Delta_{\{p,q\}}$, to $A_{\nu(p,q)}$ we get a chain between L_p and L_q of length equal to

$$l(p,q) = \lambda(p,q) + \nu(p,q) + \lambda(q,p).$$

There are two more or less canonical improvements to consider:

$\boldsymbol{M}$: Take for $X_p \longrightarrow Y_p$ the minimal good embedded resolution of $\cup \Sigma_{qp} \subset X_p$. We thus arrive at the minimal good improvement $Y \longrightarrow X$. In this case one has:

$$\lambda(q,p) = \max_r \left(\rho(r,q;p) \right).$$

$\boldsymbol{B}$: Blow-up points of X to arrive at the blow-up model. In this case we have:

$$\lambda(p,q) = \max_{r,s} \left(\rho(r,q;p), \rho(s,p;q) \right) = \lambda(q,p).$$

We now come to the relation between the notion of limit tree of [25] of a rational surface singularity with reduced fundamental cycle, and the series of deformations of tree singularities.

Proposition 4.1. *Let* $X = X(T)$ *be a tree singularity of multiplicity* m, *and let* $Y \longrightarrow X$ *the* $\boldsymbol{B}$*-improvement with exceptional divisor* E. *Let* $X' = X_\lambda(\nu)$, $\nu \in \mathcal{N}(Y, E)$ *a series member, as in (2.16). Then:*

(1) X' *is a rational surface singularity with reduced fundamental cycle and of multiplicity* m.

(2) *The tree T is a limit tree for X'.*

(3) *The blow-up tree $BT(3)$ of X' is equal to the blow-up tree $BT(3)$ of X.*

Proof. (1) is clear, because X' will be a singularity with hyperplane section the curve $Y(m)$. This condition is equivalent to having reduced fundamental cycle. For (2) we have to recall the definition of a limit tree from [25], Definition (1.12). There T is called a limit tree for X' if the following conditions hold.

(0) The set of vertices of $\boldsymbol{T}$ is equal to the set of $\mathcal{H}$ of irreducible components of the hyperplane section of the singularity.

(1) If $\{p, r\}$ and $\{q, r\}$ are edges of T, then

$$\rho(p, q; r) \le \rho(q, r; p)$$

$$\rho(p, q; r) \le \rho(r, p; q)$$

(2) For r and $s \in \boldsymbol{C}_{pq}$, $\{p, r\} \in e(\boldsymbol{T})$ one has:

$$\rho(p, q; r) = \rho(p, s; r).$$

(3) If p, q, r are not on a chain, and if d is the unique center of p, q, r, then one has

$$\rho(p, q; r) \ge \rho(p, q; d).$$

We recall here that the function $\rho(p, q; r)$ is the overlap function of X', that is the number of curves in the minimal resolution of X' that are common to the chains from p to r and q to r. Let us verify these conditions. It is convenient to make a picture; it is more instructive than a formal proof.

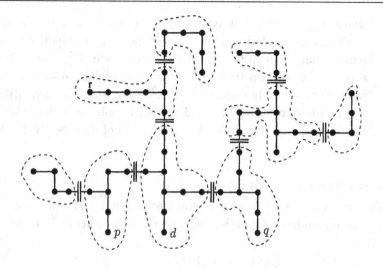

Just some improvement graph

(1) One has

$$\rho(p,q;r) \le \min\big(\lambda(p,r),\,\lambda(q,r)\big)$$

$$\rho(q,r;p) \ge \nu(p,r) + \lambda(r,p)$$

$$\rho(r,p;q) \ge \nu(q,r) + \lambda(q,r)$$

So the inequality is satisfied if $\nu(p,r)$ and $\nu(q,r) \ge \lambda(p,r) - \lambda(r,p)$. But by the symmetry of the λ-function on the $\boldsymbol{B}$-model, this is zero, so it is satisfied for all series members.

(2) This condition is fulfilled for trivial reasons: the function $\rho(p,q;r)$ is determined by the curves that lie in Y_p.

(3) Using (2), we may assume that $\{p,d\}$, $\{q,d\}$ and $\{r,d\}$ are edges of T. In this case one has (see picture):

$$\rho(p,q;r) = \lambda(d,r) + \nu(d,r) + A$$
$$\rho(p,q;d) \le \lambda(r,d) + A$$

for some A that can be positive or negative. Hence it follows that

$$\rho(p,q;r) - \rho(p,q;d) \ge \lambda(d,r) - \lambda(r,d) + \nu(d,r)$$

which is $\geq \nu(d,r) \geq 0$ by symmetry of λ. Recall that the blow-up tree of a rational singularity is the tree with nodes corresponding to singularities that occur in the resolution process, see [25]. The statement about these blow-up trees is evident: the minimal resolution of the series member is obtained by first deforming the $\boldsymbol{A}_\infty$-singularities to $\boldsymbol{A}_\nu$, and then resolving these. This clearly only changes the blow-up tree by nodes corresponding to singularities of multiplicity 2. ∎

DEFORMATIONS OF TREE SINGULARITIES

Associated to each edge $\{p,q\}$ of a limit tree of a rational surface singularity with reduced fundamental cycle, there are three elements in T^1 constructed in [25]:

$$\sigma(p,q), \quad \tau(p,q) = \tau(q,p), \quad \sigma(q,p).$$

These $3.(m-1)$ elements generate the T^1 and are subject to m relations, one for each vertex of T:

$$\sum_{p\in\nu(T)} \sigma(p,q) = 0.$$

Let us briefly indicate the method of proof, used in [25]. From the explicit equations and the choice of a limit tree T for X, one constructs elements in the normal module

$$\mathrm{Hom}\big(I/I^2, \mathcal{O}_X\big).$$

To show that these project onto generators of T^1, one uses the exact sequence relating X with its general hyperplane section, which is the curve $Y(m)$:

$$\ldots T^1_{X/S} \xrightarrow{x.} T^1_{X/S} \longrightarrow T^1_{Y(m)} \longrightarrow \ldots.$$

Here $X \longrightarrow S$ is the slicing of X defined by $x \in \mathcal{O}_X$. So we can test for independence by restriction to $x = 0$, and work inside $T^1_{Y(m)}$, which is a very simple space to understand. In this way one can show that the explicitly constructed elements in fact generate.

For tree singularities one of course has analogous elements, and their relations can be described in the same way. It turns out that there is a nice geometrical description for these T^1-generators, and in fact, we first found these geometrical elements for tree singularities, and then found the result for arbitrary rational surface singularities with reduced fundamental cycle by lifting back.

To define the τ-deformations, we take an edge $\{p,q\}$. The corresponding planes X_p and X_q intersect in a smooth curve $\Sigma_{\{p,q\}}$. So $X_p \cup X_q$ is isomorphic to an $\boldsymbol{A}_\infty$-singularity.

Lemma 4.2. *In the coordinates* (1.12)

$$z_{pq}z_{qp} = 0,$$

consider the deformation

$$z_{pq}z_{qp} = f(x)$$

of this $\boldsymbol{A}_\infty$-singularity. If

$$\operatorname{ord}_x(f) \geq s(p,q) := \max_{r,s}\big(\rho(r,p;q),\ \rho(s,q;p)\big),$$

then all the curves $\Sigma_{rp}, \Sigma_{sq}; \{r,p\}, \{s,q\} \in e(T)$ lift over this deformation.

Proof. Let $y = z_{pq}$, $z = z_{qp}$. The $\boldsymbol{A}_\infty$ singularity is described by $yz = 0$ in coordinates x, y, z. A smooth curve in the x, z plane transverse to $x = 0$ can be taken as defined by the ideal $\big(z, y - g(x)\big)$. If we deform the $\boldsymbol{A}_\infty$ and lift the curve, then after coordinate transformation we may in fact suppose that the curve is constant. Hence we must have $(y + \varepsilon.\alpha).z + \varepsilon.\beta.\big(y - g(x)\big) = yz + \varepsilon.f$, hence modulo (y,z) we have that $f \in (g)$. From this the lemma follows. ∎

One now can define elements

$$\tau(p,q) \in T_X^1$$

in the following way: deform the $\boldsymbol{A}_\infty$-singularity $X_p \cup X_q$ to $\boldsymbol{A}_{s(p,q)-1}$, where

$$s(p,q) := \max_{r,s}\big(\rho(r,p;q),\ \rho(s,q;p)\big).$$

Lemma 4.2 tells us that we can lift all the curves over this deformation. Now take such a lift, and glue back all the planes to these curves. In this way one gets a deformation of the tree singularity X. By construction, $\operatorname{supp}\big(\mathcal{O}_X.\tau(p,q)\big) = \Sigma_{\{p,q\}}$. For each element $x^n.\tau(p,q)$ one has a one-parameter deformation, that on the level of equations is characterized by:

$$z_{pq}z_{qp} = \lambda.x^{n+s(p,q)}.$$

(We note that this description fits with formula (3.10) of [25] for the τ-generators.) These τ-deformations are related to the series deformations in the following way: as mentioned before, deforming on the $(\boldsymbol{B})$-improvement from $\boldsymbol{A}_\infty$ to $\boldsymbol{A}_\nu$ produces a chain of length $\lambda(p,q) + \nu(p,q) + \lambda(p,q) = 2.s(p,q) + \nu(p,q)$. This means that the corresponding T^1-element corresponds to

$$x^{s(p,q)+\nu(p,q)+1}.\tau(p,q).$$

So one sees that the series deformations are contained in the space spanned by the τ's, and form in there a space of finite codimension. Note also that it follows from [25] that the data $(T,\phi,\boldsymbol{f})$ (see (1.13)) describe a rational surface singularity with reduced fundamental cycle exactly if $\mathrm{ord}_x(f_{pq}) \geq s(p,q)$. This gives an alternative way of thinking about the τ-deformations.

Apart from these series deformations, there is for each $(p,q) \in o(T)$ another sort of deformation, that is geometrically even easier to understand than the τ's:

$$\sigma(q,p) : \text{Move the curve } \Sigma_{qp} \text{ in the plane } X_p.$$

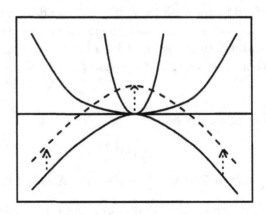

Shifting Σ_{qp} in X_p

These σ's can also be seen as deformations of the decoration:

$$\sigma(q,p) : a_{qp} \longmapsto a_{qp} + \varepsilon; \quad \phi(r,q;p) \longmapsto \phi(r,q;p) - \varepsilon.$$

From the interpretation as shiftings of the curves, it now becomes obvious that one has the relations

$$\sum_{p\in\nu(T)} \sigma(p,q) = 0.$$

These just express the fact that shifting in a given plane X_p all the curves Σ_{qp} by the same amount gives a trivial deformation of the tree singularity. It is also clear that the shift deformations are unobstructed among each other.

Obstruction Spaces For Tree Singularities

The generators for T^1 had a natural interpretation in terms of the edges of T. The generators for the obstruction space T^2 have a nice simple description in terms of the ... non-edges of T! to be more precise, there are elements

$$\Omega(p, q) \in T_X^2$$

for each ordered pair $(p, q) \notin o(T)$. It is easy to see that the number of such oriented non-edges is

$$m(m - 1) - 2(m - 1) = (m - 1)(m - 2)$$

Furthermore, for each edge $\{p, q\} \in e(T)$ we have a linear relation between the Ω's:

$$\sum_{(p,r;q)\in c(T)} \Omega(p, r) + \sum_{(s,q;p)\in c(T)} \Omega(s, q).$$

By [11], the number of generators of T^2 is $(m - 1)(m - 3)$, which is indeed the same as $(m - 1)(m - 2) - (m - 1)$.

Let us quickly describe these elements. Recall that T^2 is by definition

$$T_X^2 = \mathrm{Hom}\,(\mathcal{R}/\mathcal{R}_0, \mathcal{O}_X)/\mathrm{Hom}\,(\mathcal{F}, \mathcal{O}_X)$$

where

$$0 \longrightarrow \mathcal{R} \longrightarrow \mathcal{F} \longrightarrow \mathcal{O} \longrightarrow \mathcal{O}_X \longrightarrow 0$$

is a presentation of $\mathcal{O}_X$ as module over the ambient space. $\mathcal{F}$ is the free module on a set of generators of the ideal of X, $\mathcal{R}$ the module of relations between them, and $\mathcal{R}_0$ the sub-module of the Koszul-relations. In our case $\mathcal{O} := \mathbb{C}\{x, y_1, \ldots, y_m\}$, and equations are provided by (1.12). It is useful to use the notation of the canonical equations, as explained in (1.13). So we write $z_{rp} = z_{r'q}$ in case that $r' \in C_{rq}$, etc. the equations are then simply written as $z_{pq}z_{qp} = 0$, $p \neq q \in v(t)$, and we will use the symbol $[pq]$ to denote the corresponding elements in $\mathcal{F}$. The module of relations is generated by the symbols

$$[p, q; r] := z_{rp}[qr] - z_{rq}[pr] + \phi(p, q; r)[pq].$$

A T^2-element is represented by a homomorphism $\mathcal{R} \longrightarrow \mathcal{O}_X$. Consider the homomorphism

$$\Psi(p,q) := \sum_{a|p \in C_{qa}} z_{qa}[qa]^*.$$

Here $[qa]^* \in \mathcal{F}$ denotes the corresponding element in the dual basis.

A straightforward calculation shows the following:

$\Psi(p,q)\big([r,s;t]\big) = 0$ unless $t = q$ and the points p, q, r and s lie on a chain in T with r or s between p and q. In that case one has:
$\Psi(p,q)\big([r,s;t]\big) = \pm z_{qr}z_{qs}$, ($+$ if $s \in C_{pq}$, $-$ if $r \in C_{pq}$).

But note that if $r \in C_{pq}$, then

$$z_{qr}z_{qs} = \big(z_{sr} + \phi(q,s;r)\big).z_{qs} = \phi(q,s;r).z_{qs}$$

because of the equations. This means that the values of the homomorphism $\Psi(p,q)$ are divisible by some power of x. The power is

$$\rho(p,q) := \min_{r \in C_{pq}, r \neq p,q} \big(\rho(p,q;r)\big),$$

the minimum vanishing order of ϕ-functions of corners "belonging to the chain from p to q". Now choose for each (p,q) an r such that $\rho(p,q;r) = \rho(p,q)$, and define homomorphism

$$\Omega(p,q) := \big[(1/\phi(p,q;r))\,\Psi(p,q)\big]$$

($[-] =$ class of in the T^2). Whereas the class of the Ψ's are trivial in the T^2, this is no longer true for the Ω's; in fact they form a system of *generators* for T^2 and this leads to a very beautiful geometrical description of the structure of this module.

The fact that (p,q) is not an edge of T means that the corresponding planes X_p and X_q intersect in a fat point. By an easy explicit computation, one can check that the ideal of this intersection $\mathcal{I}_{pq} = (z_{rs}|r \in C_{sp} \cup C_{sq})$ annihilates the element $\Omega(p,q)$ and so the submodule $\mathcal{O}_X.\Omega(p,q)$ of the T^2 generated $\Omega(p,q)$ is supported on the fat point $X_p \cap X_q$. This is analoguous to the situation with the generators $\sigma(p,q), \tau(p,q)$ of the T^1 that are supported on the intersection curve $X_p \cap X_q$ for edges (p,q) of T.

For a corner $(p,q;r) \in c(T)$, there is only one ϕ on the chain between p and q, and it is easy to see the aforementioned linear relations between then, that arise from an edge.

One can show with the same hyperplane section trick that the Ω's form a generating set for the T^2. But the above explicit description of T^2 annihilating elements will give us an upper bound for the dimension T^2 of the form $\dim(T_X^2) \leq N\big((T,\rho)\big)$, where $N\big((T,\rho)\big)$ is a number that only depends on the discrete data of the limit tree.

In general it is much easier to find lower bounds for the dimension of T^2. for this one has to exhibit the fact that the X under consideration is complicated. One can do this by finding hyperplane sections with high smoothing codimension, or by finding a deformation to a singularity with many singularities. In [25] we proved (theorem 2.13)):

Theorem. *Let X be a rational surface singularity of multiplicity m with reduced fundamental cycle. Let $X_1, X_2, \ldots, X_r$ be the singular points of the first blow up $\widehat{X}$ of X. Then there exists a one-parameter deformation over the Artin-component, such that for $s \neq 0$ the fibre X_s has as singularities $X_1, \ldots, X_r$, together with one singularity, isomorphic to the cone over the rational normal curve of degree $(-m)$.*

By an application of the semi-continuity of $\dim T_X^2$ one gets

$$\dim T_X^2 \geq (m-1)(m-3) + \dim T_{\widehat{X}}^2,$$

and so

$$\dim T^2 \geq N(\Gamma),$$

where $N(\Gamma)$ is a number that is easily determined from the resolution graph of X by iteration of the inequality.

In fact, the above theorem is also true for rational surface singularities whose fundamental cycle is reduced except possibly at the (-2) curves. This follows from Laufer's theory of deformations over the Artin-component ([31]), (3.7); one takes as roots the fundamental cycle Z, together with the unions of all curves E_i such that $E_i.Z = 0$, and it is well conceivable that it is true for all rational surface singularities. But rational surfaces with reduced fundamental cycle, and also the tree singularities, have the special property that the above inequalities in fact are *equalities*. Basically this follows from

$$N(\Gamma) = N\big((T,\rho)\big),$$

a purely combinatorial fact. The proof is given in [25], (3.27). In terms of the tree the idea is simple: the $\Omega(p,q)$ live on the fat point, and after each blow-up, the length of this scheme drops by one. The T^2 element lives so long, until a further blow up will separate the planes. In this way the blow-up formula for T^2 looks very natural and obvious. Note that because the blow-up tree $BT(3)$ of X and any series member X' of the $(\boldsymbol{B})$-model are the same, one has stability of T^2.

$$\dim\left(T_X^2\right) = \dim\left(T_{X'}^2\right).$$

BASE SPACES AND CHAIN EQUATIONS

We now give a description of the "base space" of a tree singularity. These are completely analogous to the equations for the rational surface singularities with reduced fundamental cycle. So let X be a tree singularity, or a rational surface singularity, described by the data $(T, \phi, \boldsymbol{f})$ as described in (1.12), (1.13). We describe $\mathrm{Def}\,(X)(S)$, for any base S as follows.

Let for each $\{p, q\} \in e(T)$ and for each $(p, q; r) \in c(T)$ be given functions

$$F_{pq} \quad \text{and} \quad \Psi(p, q; r) \in S\{x\} := S \otimes_{\mathbb{C}} \mathbb{C}\{x\}$$

restricting to f_{pq} and $\phi(p, q; r)$ respectively. As we have seen, the F's correspond to series and the Ψ to shift deformations. The F's and Ψ's are heavily obstructed against each other. In fact, a reinterpretation of [25], (4.9) is:

Theorem 4.3. *The system* (T, Ψ, F) *describes a flat deformation over* S *if and only if for each oriented chain*

$$p_0, p_1, p_2, \ldots, p_{k-1}, p_k \; ; \{p_i, p_{i+1}\} \in e(T)$$

the following continued fraction "exists as power series in x*":*

$$\cfrac{F_1}{\Psi_1 + \cfrac{F_2}{\Psi_2 + \cfrac{F_3}{\ddots \Psi_{k-1} + \frac{F_k}{\Psi_k}}}}.$$

Here $F_i := F_{p_i, p_{i+1}}$ and $\Psi_i := \Psi_{(p_{i-1}, p_{i+1}; p_i)}$.

Remark 4.4. What happens in practice is that one has some arbitrary deformation Ψ, F given over some power series ring $R := \mathbb{C}\{a\} = \mathbb{C}\{a_1, a_2, \ldots\}$. Each chain gives rise to some ideal in $\mathbb{C}\{a\}$, as follows: every time we have to make a division A/B in the continued fraction, we consider Weierstraß division with remainder:

$$A = \mathcal{Q}.B + \mathcal{R}$$

and then equate to zero the coefficients of the x powers in $\mathcal{R}$. In this way, every chain c defines a unique ideal $J(c)$ in R, such that the continued fraction exists over $R/J(c)$. Note that $c_1 \subset c_2$ implies that $J(c_1) \subset J(c_2)$. so the ideals are build up inductively, starting from the corners $(p, q; r)$. We have seen that the obstruction space T_X^2 was generated by certain elements

$$\Omega(p, q), \quad p, q \notin e(T).$$

Of course, this is no coincidence. It was shown in [25] that the different coefficients of remainders that have to be equated to zero exactly correspond to the elements of T^2. As cyclic quotients have limit trees that are linear chains, and a component structure that is directly related to properties of continued fractions, [13], it is very tempting to try to relate these two types of fractions in some direct way.

THE CASE OF A STAR

Let us analyze further the simplest tree singularities, to know those for which the tree is a star. I.e., there is one central plane X_c, and all other planes intersect X_c in a curve. Example (1.10) is of this type. Let us denote the other planes by X_i, X_j, etc and introduce the short-hand notation $\Sigma_i := \Sigma_{i,c}$, $F_i := F_{i,c}$, $\Psi(i, j) := \Psi(i, j; c)$, etc. So the situation is that we have a bunch of curves Σ_i in the central plane, and planes X_i glued to it. Note that for a star the only non-trivial chains are the corners $(i, j; c)$. So the chain equations become simply:

$$\Psi(i, j) \text{ divides } F_i, \quad \forall i, j.$$

These conditions have a very simple geometrical interpretation in terms of the curves Σ_i. As the curve Σ_i is described by the equation $y = a_{ic}$, the x-coordinates of the intersection points of Σ_i and Σ_j are precisely the roots of the function $\Psi(i, j)$. If we consider the function F_i as a function on Σ_i, then this condition just means

$$(\Sigma_i.\Sigma_j) \subset (F_i), \quad \forall i, j.$$

Here (F_i) denotes the sub-scheme of zero's of the function F_i. Everything now can be understood in terms of these curves and points on these curves. A versal deformation can be described as follows. Let $S := S(F_1) \times \cdots (F_k) \times S(\Sigma)$, where $S(F_i) =$ unfolding space of $F_i \approx \mathbb{C}^{\mathrm{ord}(F_i)}$ and $S(\Sigma) := \delta$-constant stratum in the semi-universal deformation of $\Sigma = \cup_i (\Sigma_i)$. so this is a smooth space. Now look in S for the stratum $\Lambda \subset S$ over which the condition holds.

PICTURES AND COMPONENTS

Due to the geometrical nature of the condition one can in the simplest cases understand the component of Λ without any computations. We give some examples.

Example 4.5. We take as a first example the famous Pinkham example, the cone over the (-4), [41]. By (2.19), we can see it as the beginning of a three index series, degenerating into the limit that was described in (1.10).

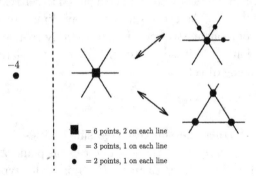

= 6 points, 2 on each line
= 3 points, 1 on each line
= 2 points, 1 on each line

Pinkham's example

Example 4.6. In a similar way we can see the (-5) as the beginning of a four index series, converging to the tree singularity corresponding to the configuration with four lines in a plane.

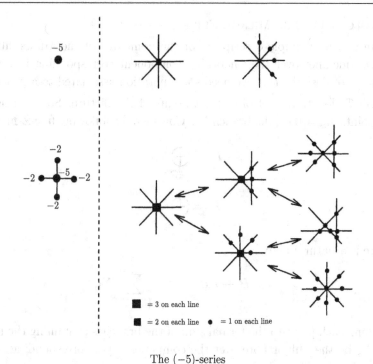

The (-5)-series

We define a picture (of $(\Sigma, \boldsymbol{f})$) to be a pair $(\Sigma_s, \boldsymbol{f}_s)$ for some $s \in \Lambda$ such that

(1) Σ_s consists of pairwise transverse intersecting curves.

(2) The zero's of each $f_{i,s}$ are all simple.

It is more or less clear that the combinatorially different pictures correspond to the components of Λ. This also makes it clear that in general there are many components, as long as we take $\mathrm{ord}_x(f_i)$ big enough, that is, high-up in the series: each stratum in the δ-constant deformation of Σ gives a new component. In fact, one can see from the description of Λ that

$$\mathrm{ord}_x(f_i) \geq \sum_{j \neq i} \rho(i, j)$$

is sufficient for base space stability. Here we clearly see that in general T^2 become stable much earlier than the base space itself. For details we refer to the paper [26].

HOMOLOGY OF THE MILNOR FIBRE

There is a nice simple description of the homology of the Milnor fibre of a series member over the smoothing component corresponding to given a picture. To describe this, we need some notation associated to a picture.

Let $\mathcal{P}$ denote the set of distinct points of the picture. So it consists of the points $f_{i,s} = 0$ on the branch Σ_i. Consider the following free $\mathbb{Z}$-modules

$$P := \bigoplus_{p \in \mathcal{P}} \mathbb{Z}.p$$

$$L := \bigoplus_i \mathbb{Z}.\Sigma_i$$

There is a natural map

$$I : P \longrightarrow L \quad p \longmapsto \sum_{\{i \mid p \in \Sigma_i\}} \Sigma_i,$$

mapping each point to the formal sum of the branches containing the point. Let X_s be the Milnor fibre over the component of Λ corresponding to the given picture. Then one has:

Theorem 4.7. *Let X' be a rational surface singularity with reduced fundamental cycle described by the data $(T, \phi, \boldsymbol{f})$, where the tree T is a simple star. Let M be its Milnor fibre over a smoothing component corresponding to a picture with incidence map $\boldsymbol{I}$. Then one has:*

$$H_1(M) = \operatorname{Coker}(\boldsymbol{I}) \quad \text{and} \quad H_2(X_s) = \operatorname{Ker}(\boldsymbol{I}).$$

Proof. (Sketch, for details see [26].) Associated to X' there is a tree singularity X with data (T, ϕ). First we have to take an appropriate small ball, and intersect X with this ball. The space we obtain is topologically a ball D in the central plane, to which we glue some other 4-disc.

We are given a picture as described above. So we can find a one-parameter deformation $X_s \longrightarrow S$ of X, with data $(\Sigma_S, \boldsymbol{f}_S)$, such that for $s \in S - \{0\}$ $(\Sigma_{i,s}, f_{i,s})$ is a picture in the above sense. We can associate to this a two-parameter deformation of X: first use the family Σ_S to shift the curves in the appropriate positions. So this is described by the data $(T, \phi_S, 0)$. Then use $t.f_{is}$ to smooth out the singularities of the spaces X_s. In other words, the two-parameter deformation of X is given by the data $(T, \phi_S, t.\boldsymbol{f}_S)$, $t \in T$, over $S \times T$.

To see clearly what happens it is convenient to blow up D in all the point $\mathcal{P}$. We call this blown-up disc $\widetilde{D}$. Its is homotopy equivalent to a bouquet of spheres. On $\widetilde{D}$ we have the strict transform $\widetilde{\Sigma}$ of the curve Σ. Because all the curves were supposed to intersect transversely, $\widetilde{\Sigma}$ consists of a collection of disjoint curves, each isomorphic to a 2-disc. Now glue the planes back to $\widetilde{D}$. Transverse to each point of $\widetilde{\Sigma}$ this space has an A_1-singularity. For the triangle picture of (4.5) it looks something like:

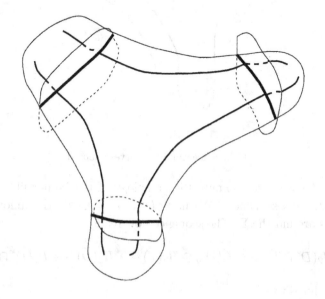

The Milnor fibre X_s is obtained from this space by smoothing out simultaneously these singularities. (This corresponds to deforming A_∞ into A_{-1}.) By contracting the discs that were glued in the direction of the central disc, we see that the Milnor fibre is noting but $\widetilde{D}^*$, the space obtained from $\widetilde{D}$ by removing s small tubular neighbourhood $\widetilde{T}$ of $\widetilde{\Sigma}$.

Smoothing out and contracting

For the above example it now looks like:

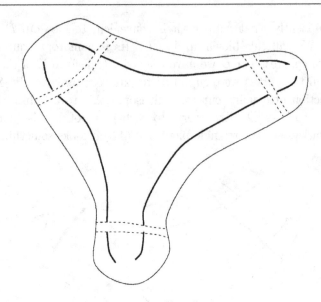

Cylinders around the curves removed

Now it is easy to compute the homology of the Milnor fibre using the Mayer–Vietoris sequence. We have $\widetilde{D}^* \cap \widetilde{T} = \cup \partial \widetilde{T}_i$, the union of small cylinders around the $\widetilde{\Sigma}$. The sequence now reads:

$$\ldots H_2(\widetilde{D}^* \cap \widetilde{T}) \longrightarrow H_2(\widetilde{D}^*) \oplus H_2(\widetilde{T}) \longrightarrow H_2(\widetilde{D}) \longrightarrow H_1(\widetilde{D}^* \cap \widetilde{T}) \ldots$$

which reduces to

$$0 \longrightarrow H_2(\widetilde{D}^*) \longrightarrow H_2(\widetilde{D}) \longrightarrow H_1(\widetilde{D}^* \cap \widetilde{T}) \longrightarrow H_1(\widetilde{D}^*) \longrightarrow 0.$$

Now, $H_2(\widetilde{D}) = P$ and $H_1(\widetilde{D}^* \cap \widetilde{T}) = L$, where this last isomorphism is set up by mapping the cycle γ_i that runs around Σ_i in the positive direction, to the generator Σ_i of the module L. From the geometrical description of the boundary map in the Mayer–Vietoris sequence we get that indeed the resulting map $P \longrightarrow L$ is given by the incidence matrix. ∎

Remark 4.8. The picture belonging to the small component of Pinkham's example consists of a triangle, see example 4.5. Hence we get as incidence matrix:

$$\begin{pmatrix} 0 & 1 & 1 \\ 1 & 0 & 1 \\ 1 & 1 & 0 \end{pmatrix}.$$

So we see coker $(I) = \mathbb{Z}/2$, ker $(I) = 0$. For the Artin-component one gets as matrix:

$$\begin{pmatrix} 1 & 1 & 0 & 0 \\ 1 & 0 & 1 & 0 \\ 1 & 0 & 0 & 1 \end{pmatrix}.$$

So here coker $(I) = 0$, ker $(I) = \mathbb{Z}$.

Remark 4.9. The onset of stability can be observed very nicely in terms of the points and curves. The set of points $\mathcal{P}$ of a picture decomposes naturally into two pieces: the *imprisoned points* $\cup \Sigma_i \cap \Sigma_j$ and the complementary set of *free points*. If on each curve there is at least one free point, then $H_1 = 0$. From this point on, the Milnor fibre in the series changes only by wedging it with some two-spheres, and so has become stable. The condition for having at least one free point on each branch is that

$$\operatorname{ord}_x(f_i) > \rho(i) := \sum_{j \neq i} \rho(i, j).$$

Note that this point is also exactly the point where the base space itself stabilizes! I do not know whether this relation between base spaces stability and Milnor fibre stability has a more general scope, but it is very well possible. On the other extreme, pictures with the same number of points as curves give rise to smoothings with $\mu = 0$, because a priori we know that the map I must be of maximal rank, as the cokernel must be torsion.

APPENDIX

We review some basics of deformation theory. We will be very sketchy and this is only meant to be a refresher. For more details we refer to the original literature, like [1], [7], [8], [10], [15], [16], [19], [20], [29], [35], [45], [46], [48]. Let X be any germ of an analytic space.

THE DEFORMATION FUNCTOR

A deformation of X over a germ $(S, 0)$ is pull-back diagram

$$\begin{array}{ccc} X & \hookrightarrow & \mathcal{X} \\ \downarrow & & \downarrow \\ \{0\} & \hookrightarrow & S \end{array}$$

where $\mathcal{X} \longrightarrow S$ is a flat map. There is an obvious notion of isomorphism between deformations over the same base S.

The deformation functor of X, Def (X), is the functor

$$\text{Def}\,(X) : \boldsymbol{C} \longrightarrow \textbf{Set}$$

$$A \longmapsto \{\,\text{Deformations of } X \text{ over Spec}\,(A)\,\} \,/\, \text{Isomorphism.}$$

Here $\boldsymbol{C}$ is the category of Artinian $\mathbb{C}$-algebras and **Set** is the category of sets. The category $\boldsymbol{C}$ sits naturally in the category $\boldsymbol{C}_{\text{an}}$ of local analytic $\mathbb{C}$-algebras, which in turn sits in $\widehat{\boldsymbol{C}}$, the formal $\mathbb{C}$-algebras. Our functor in fact is the restriction of a functor defined on these bigger categories, but we will not introduce extra notation to distinguish these functors.

Note the following easy application of the Artin approximation theorem:

Proposition 4.10 (see also [10], (3.1.3.4)). *Let X be any germ of an analytic space and let $\widehat{\xi} \in \text{Def}\,(X)\big(\mathbb{C}[[s]]\big)$ a formal deformation. Then for all $n \in \boldsymbol{N}$ there exists a deformation $\xi \in \text{Def}\,(X)(\mathbb{C}\{s\})$ such that the restrictions of $\widehat{\xi}$ and ξ over $\mathbb{C}[[s]]/(s^n) = \mathbb{C}\{s\}/(s^n)$ are the same.*

Proof. Let $X \subset \mathbb{C}^N$, and let $\mathcal{O} = \mathcal{O}_{(\mathbb{C}^N,0)} = \mathbb{C}\{x_1, x_2, \dots, x_N\}$. Let a presentation of $\mathcal{O}_X$ be given as:

$$\mathcal{O}^\beta \xrightarrow{\;\rho\;} \mathcal{O}^\alpha \xrightarrow{\;\phi\;} \mathcal{O} \longrightarrow \mathcal{O}_X \longrightarrow 0.$$

A formal deformation ξ is given by a $1 \times \alpha$ matrix f and an $\alpha \times \beta$ matrix r over the ring $\mathcal{O}[[s]] := \mathbb{C}[[s]] \otimes_{\mathbb{C}} \mathcal{O}$ that satisfy:

1) $fr = 0$ and 2) $\phi(-) = f(s = 0, -)$, $\rho(-) = r(s = 0, -)$.

Now consider the ring $\mathcal{O}\{s\}[F, R] := \mathbb{C}\{s, x_1, x_2, \dots, x_n\}[F, R]$, where $F = (F_1, \dots, F_\alpha)$, $R = (\dots, R_{i,j}, \dots)$. In here we have the ideal generated by the components of the matrix $F \cdot R$. The formal deformation ξ gives us a solution $(f, r) \in \mathcal{O}[[s]][F, R]$. By the Artin approximation theorem one now obtains for every $n \in \boldsymbol{N}$ a solution $(\widetilde{f}, \widetilde{r})$ in the ring $\mathcal{O}\{s\}[F, R]$ such that $\widetilde{f} - f = 0$ modulo $\boldsymbol{m}^n$ and $\widetilde{r} - r = 0$ modulo $\boldsymbol{m}^n$, i.e. we have a convergent deformation of X approximating the given formal one. ∎

FUNCTORS OF ARTIN RINGS

We study functors $F : \boldsymbol{C} \longrightarrow \textbf{Set}$. Any $R \in ob(\boldsymbol{C})$ gives us a functor $h_R : \boldsymbol{C} \longrightarrow \textbf{Set}$ via $A \longmapsto \text{Hom}\,(R, A)$. There is a tautological isomorphism

$$F(R) \xrightarrow{\;\approx\;} \text{Hom}\,(h_R, F).$$

In particular, any couple $\mathcal{R} = \big(R, \xi \in F(R) \big)$ gives a map

$$\phi_R : h_R \longrightarrow F.$$

If there is a couple $\mathcal{R}$ such that $\phi_{\mathcal{R}}$ is an isomorphism, then one says that F is (pro) representable. This is usually a much to strong condition on the functor F.

Definition 4.11. A functor $F : C \longrightarrow \mathbf{Set}$ is connected iff $F(\mathbb{C}) = \{.\}$. A functor is called semi-homogeneous if the following two "Schlessinger conditions" are satisfied:

If we have a diagram

$$
\begin{array}{ccc}
 & & A'' \\
 & & \downarrow \\
A' & \longrightarrow & A
\end{array}
$$

then the canonical map $F(A' \times_A A'') \longrightarrow F(A') \times_{F(A)} F(A'')$ is

H.1) surjective if $A'' \longrightarrow A$ is a small surjection;

H.2) bijective if $A = \mathbb{C}$ and $A'' = \mathbb{C}[\varepsilon]$.

Here $\mathbb{C}[\varepsilon] := \mathbb{C}[\varepsilon]/(\varepsilon^2)$ and a small surjection is a map $\alpha : A'' \longrightarrow A$ such that $\ker(\alpha).m_{A''} = 0$.

For such a functor the tangent space $T_F^1 := F\big(\mathbb{C}[\varepsilon]\big)$ acquires in a natural way the structure of a $\mathbb{C}$-vectorspace.

Definition 4.12. A transformation of functors $F \longrightarrow G$ is called smooth if for all small surjections, (hence for all surjections) $B \longrightarrow A$ the induced map

$$F(B) \longrightarrow F(A) \times_{G(A)} G(B)$$

is surjective.

If a transformation ϕ is smooth, and induces an isomorphism $T_F^1 \longrightarrow T_G^1$, then ϕ is called minimal smooth. A functor F is called smooth, if the final transformation $F \longrightarrow h_{\mathbb{C}}$ is smooth. A transformation $h_S \longrightarrow h_R$ is smooth if and only if $S = R[[x]]$, the composition of smooth transformations is again smooth, the pull-back of a smooth transformation over an arbitrary transformation is again smooth, etc. So smooth maps are surjections in a very strong sense.

Schlessinger's theorem: If F is a connected semi-homogeneous functor then there exists a minimal smooth transformation:

$$\phi_{\mathcal{R}} : h_R \longrightarrow F$$

if and only if:

H.3) T_F^1 is finite dimensional.

Under these circumstances one says that R is a *hull* for F.

Associated to a map $X \xrightarrow{f} Y$, one can consider six deformation functors: Def $(X \xrightarrow{f} Y)$, the deformations of X, Y and f simultaneously, Def (X/Y), deformations of X, f but keeping Y fixed, Def $(X \backslash Y)$, deformations of Y, f, but keeping X fixed, Def (f), deformations only of f, keeping both X and Y fixed. Apart from these one also has Def (X) and Def (Y). There are six cotangent complexes associated to these functors and their homology and cohomology groups sit in various exact sequence, described in detail in the thesis of R. Buchweitz, [10].

In case that X is not a germ, but a global space, there are global $\boldsymbol{T}_X^i$ and local $\mathcal{T}_X^i$ sheaves, related by a usual local-to-global spectral sequence:

$$E_2^{p,q} = H^p(X, \mathcal{T}^q) \Rightarrow \boldsymbol{T}_X^{p+q}.$$

THE BASE SPACE OF A LIMIT

We have seen that the base space of a semi-universal deformation of a singularity X appears formally as

$$\mathcal{B}_X = Ob^{-1}(0)$$

for the obstruction map

$$Ob : T_X^1 \longrightarrow T_X^2.$$

As for a limit we have $\dim(T_X^1) = \infty$, so the base space of the semi-universal deformation should be infinite dimensional. Working with infinite dimensional spaces causes some inconveniences. There are at least three different attitudes towards these problems possible.

(1) Try to develop a honest analytic theory in infinite dimensions.

   In principle, Hausers approach is just achieving this. He uses Banach-analytic methods to construct the base space of a semi-universal deformation for isolated singularities, and his construction works also in the case the T^1 is not finite dimensional. It seems that in the important case that T^2 is finite dimensional, one can use the essentially simpler theory of Mazet [33] of analytic sets of finite definition (i.e. finite number of equations, in infinite number of variables). This would give already quite strong structural statements about the base spaces (finite number of components, etc, see [33]).

(2) Work formally in infinite number of variables.

   This is the approach taken in the book of Laudal [29].

(3) Work only with the functor. A functor that satisfies the three Schlessinger conditions has a hull, so "behaves like a finite dimensional space". If we forget about the third Schlessinger condition, we arrive at the notion of a semi-homogeneous functor and these behave more or less as spaces infinite dimension.

We will be very lazy here, and work with the functor approach (3), although a complete development of (1) seems very desirable.

Acknowledgement: In the first place I thank T. de Jong for many fruitful discussions, during which we discovered many important ideas. Also thanks to D. Bayer and M. Stillman for their great computer program MACAULAY that we used to test our wildest conjectures. The usefulness of a tree singularities was discovered however because of the *inability* of MACAULAY to do a certain calculation. I thank G.-M. Greuel for continual encouragement to write this paper. Furthermore, I thank the DFG for a stipendium allowing me to work at the MSRI in Berkeley, a beautiful place with a pleasant atmosphere. (This paper originally was written in 1993.)

REFERENCES

[1] M. André, *Homologie des Algèbres Commutatives,* Grundlehren der math. Wissenschaften **206**, Springer, Berlin (1976).

[2] J. Arndt, *Verselle deformationen zyklischer Quotientensingularitäten,* Dissertation, Universität Hamburg, 1988.

[3] V. Arnol'd, S. Guzein-Zade, A. Varchenko, *Singularities of Differentiable Maps,* Vol. I & II, Birkhäuser, Boston (1988).

[4] M. Artin, *On isolated rational singularities of surfaces,* Am. J. of Math., **88** (1966), 129–136.

[5] M. Artin, *On the Solutions of Analytic Equations,* Inv. Math., **5** (1968), 277–291.

[6] M. Artin, *Algebraic construction of Brieskorns resolutions,* J. of Alg., **29** (1974), 330–348.

[7] M. Artin, *Deformations of Singularities,* Lecture Notes on Math., **54** Tata Institute of Fundamental Research, Bombay (1976).

[8] J. Bingener, *Modulräume in der analytischen Geometrie 1 & 2,* Aspekte der Mathematik, Vieweg, (1987).

[9] E. Brieskorn, *Rationale Singularitäten komplexer Flächen,* Inv. Math., **4** (1967), 336–358.

[10] R.-O. Buchweitz, *Contributions à la Théorie des Singularités,* Thesis, Université Paris VII, (1981).

[11] K. Behnke and J. Christophersen, *Hypersurface sections and Obstructions (Rational Surface Singularities),* Comp. Math., **77** (1991), 233–258.

[12] R. Buchweitz and G.-M. Greuel, *The Milnor number and deformations of complex curve singularities,* Inv. Math., **58** (1980), 241–281.

[13] J. Christophersen, *On the Components and Discriminant of the Versall Base Space of Cyclic Quotient Singularities,* in: "Singularity Theory and its Applications", Warwick 1989, D. Mond and J. Montaldi (eds.), SLNM **1462**, Springer, Berlin, (1991).

[14] G. Fisher, *Complex Analytic Geometry,* SLNM **538**, Springer, Berlin, (1976).

[15] H. Flenner, *Über Deformationen holomorpher Abbildungen,* Osnabrücker Schriften zur Mathematik, Reihe P Preprints, Heft, **8** (1979).

[16] H. Grauert, *Über die Deformationen isolierter Singularitäten analytischer Mengen,* Inv. Math., **15** (1972), 171–198.

[17] G.-M. Greuel, *On deformations of curves and a formula of Deligne,* in: "Algebraic Geometry", Proc., La Rabida 1981, SLNM **961**, Springer, Berlin, (1983).

[18] H. Grauert and R. Remmert, *Analytische Stellenalgebren,* Grundlehren d. math. Wissens, Bd. **176,** Springer, Berlin, (1971).

[19] H. Hauser, *La Construction de la Déformation semi-universelle d'un germe de variété analytique complexe,* Ann. Scient. Éc. Norm. sup. **4** série, t. **18** (1985), 1–56.

[20] L. Illusie, *Complex Cotangente et Déformations 1, 2*, SLNM, **239** (1971), SLNM, **289** (1972), Springer, Berlin.

[21] T. de Jong and D. van Straten, *A Deformation Theory for Non-Isolated Singularities*, Abh. Math. Sem. Univ. Hamburg, **60** (1990), 177–208.

[22] T. de Jong and D. van Straten, *Deformations of the Normalization of Hypersurfaces*, Math. Ann., **288** (1990), 527–547.

[23] T. de Jong and D. van Straten, *On the Base Space of a Semi-universal Deformation of Rational Quadruple Points*, Ann. of Math., **134** (1991), 653–678.

[24] T. de Jong and D. van Straten, *Disentanglements*, in: "Singularity Theory and its Applications", Warwick 1989, D. Mond and J. Montaldi (eds.), SLNM **1462**, Springer, Berlin, (1991).

[25] T. de Jong and D. van Straten, *On the Deformation Theory of Rational Surface Singularities with Reduced Fundamental Cycle*, J. of Alg. Geometry, **3** (1994), 117–172.

[26] T. de Jong and D. van Straten, *Deformation Theory of Sandwiched Singularities*, Duke Math. J., Vol **95**, No. 3, (1998), 451–522.

[27] B. Kaup, *Über Kokerne und Pushouts in der Kategorie der komplexanalytischen Räume*, Math. Ann., **189** (1970), 60–76.

[28] J. Kollár and N. Shepherd-Barron, *Threefolds and deformations of surface singularities*, Inv. Math., **91** (1988), 299–338.

[29] A. Laudal, *Formal Moduli of Algebraic Structures*, SLNM **754**, Springer, Berlin, (1979).

[30] H. Laufer, *On Minimally Elliptic Singularities*, Amer. J. Math., **99** (1977), 1257–1295.

[31] H. Laufer, *Ambient Deformations for Exceptional Sets in Two-manifolds*, Inv. Math., **55** (1979), 1–36.

[32] S. Lichtenbaum and M. Schlesinger, *The cotangent Complex of a Morphism*, Trans. AMS, **128** (1967), 41–70.

[33] P. Mazet, *Analytic Sets in Locally Convex Spaces*, Math. Studies **89**, North-Holland, (1984).

[34] D. Mond, *On the classification of germs of maps from $\mathbb{R}^2$ to $\mathbb{R}^3$*, Proc. London Math. Soc., **50** (1985), 333–369.

[35] V. Palomodov, *Moduli and Versal Deformations of Complex Spaces*, Soviet Math. Dokl., **17** (1976), 1251–1255.

[36] R. Pellikaan, *Hypersurface Singularities and Resolutions of Jacobi Modules*, Thesis, Rijksuniversiteit Utrecht, (1985).

[37] R. Pellikaan, *Finite Determinacy of Functions with Non-Isolated Singularities*, Proc. London Math. Soc. (3), **57** (1988), 357–382.

[38] R. Pellikaan, *Deformations of Hypersurfaces with a One-Dimensional Singular Locus*, J. Pure Appl. Algebra, **67** (1990), 49–71.

[39] R. Pellikaan, *Series of Isolated Singularities,* Contemp. Math., **90** Proc. Iowa, R. Randell (ed.), (1989).

[40] R. Pellikaan, *On Hypersurfaces that are Sterns,* Comp. Math., **71** (1989), 229–240.

[41] H. Pinkham, *Deformations of algebraic Varieties with G_m-action,* Astérisque, **20** 1974.

[42] O. Riemenschneider, *Deformations of rational singularities and their resolutions,* in: Complex Analysis, Rice University Press **59** (1), 1973, 119–130.

[43] O. Riemenschneider, *Deformationen von Quotientensingularitten (nach zyklischen Gruppen),* Math. Ann., **209** (1974), 211–248.

[44] D. S. Rim, *Formal deformation theory,* SGA 7(1), Exp. VI, Springer, Berlin, (1972).

[45] G. Ruget, *Déformations des germes d'espaces analytiques,* Sém. Douady-Verdier 1971–72, Astérisque, **16** (1974), 63–81.

[46] M. Schlessinger, *Functors of Artin Rings,* Trans. Am. Math. Soc., **130** (1968), 208–222.

[47] M. Schlessinger, *Rigidity of quotient singularities,* Inv. Math., **14** (1971), 17–26.

[48] M. Schlessinger, *On Rigid Singularities,* in: Complex Analysis, Rice University Press **59**(1), 1973.

[49] R. Schrauwen, *Series of Singularities and their Topology,* Thesis, Utrecht, (1991).

[50] N. Shepherd-Barron, *Degenerations with Numerically Effective Canonical Divisor,* in: "The Birational Geometry of Degenerations", Progress in Math. 29, Birkhäuser, Basel, (1983).

[51] D. Siersma, *Isolated Line Singularities,* in: "Singularities", Arcata 1981, P. Orlik (ed.), Proc. Sym. Pure Math. **40**(2), (1983), 405–496.

[52] D. Siersma, *Singularities with Critical Locus a One-Dimensional Complete Intersection and transverse A_1,* Topology and its Applications, **27**, 51–73 (1987).

[53] D. Siersma, *The Monodromy of a Series of Singularities,* Comm. Math. Helv., **65** (1990), 181–197.

[54] J. Stevens, *Improvements of Non-isolated Surface Singularities,* J. London Soc. (2), **39** (1989), 129–144.

[55] J. Stevens, *On the Versal Deformation of Cyclic Quotient Singularities,* in: "Singularity Theory and Applications", Warwick 1989, Vol. I, D. Mond, J. Montaldi (eds.), SLNM **1426**, Springer, Berlin, (1991).

[56] J. Stevens, *Partial Resolutions of Rational Quadruple Points,* Indian J. of Math., (1991).

[57] J. Stevens, *The Versal Deformation of Universal Curve Singularities,* ESP-preprint no. **5**.

[58] D. van Straten, *Weakly Normal Surface Singularities and their Improvements,* Thesis, Leiden, (1987).

[59] G. Tjurina, *Locally semi-universal flat deformations of isolated singularities of complex spaces,* Math. USSR Izvestia, **3**(5) (1969), 967–999.

[60] G. Tjurina, *Absolute isolatedness of rational singularities and triple rational points*, Func. Anal. Appl., **2** (1968), 324–332.

[61] G. Tjurina, *Resolutions of singularities of plane deformations of double rational points*, Func. Anal. Appl., **4** (1970), 68–73.

[62] J. Wahl, *Equisingular deformations of normal surface singularities 1*, Ann. Math., **104** (1976), 325–356.

[63] J. Wahl, *Equations defining Rational Surface Singularities*, Ann. Sci. Ec. Nor. Sup., 4e série, t. **10** (1977), 231–264.

[64] J. Wahl, *Simultaneous Resolution of Rational Singularities*, Comp. Math., **38** (1979), 43–54.

[65] J. Wahl, *Simultaneous Resolution and Discriminant Loci*, Duke Math. J., **46**(2) (1979), 341–375.

[66] J. Wahl, *Elliptic deformations of Minimally elliptic Singularities*, Math. Ann., **253** (1980), 241–262.

[67] J. Wahl, *Smoothings of Normal Surface Singularities*, Topology, **20** (1981), 219–246.

[68] I. Yomdin, *Complex Surfaces with a 1-dimensional Singular Locus*, Siberian Math. J., **15**(5) (1974), 1061–1082.

Duco van Straten

Fachbereich Physik, Mathematik und
Informatik
Johannes Gutenberg-Universität
55099 Mainz
Germany

e-mail:

straten@mathematik.uni-mainz.de

Printed in the United States
By Bookmasters